**Cliff Ecology**
Pattern and Process in Cliff Ecosystems

Cliffs are present in virtually every country on earth. The lack of scientific interest in cliffs to date is in striking contrast to how common they are around the world and to the attraction they have had for humans throughout history. Cliffs provide a unique habitat, rarely investigated from an ecological viewpoint. This book aims to destroy the impression of cliffs as geological structures devoid of life, by reviewing information about the geology, geomorphology, microclimate, flora and fauna of both sea and inland cliffs. For the first time, evidence is presented to suggest that cliffs worldwide may represent an invaluable type of ecosystem, consisting of some of the least disturbed habitats on earth and contributing more to the biodiversity of a region than their surface coverage would indicate.

The Cliff Ecology Research Group was formed in 1985 within the Department of Botany at the University of Guelph. The group is an interdisciplinary team that analyses the structure and function of cliff ecosystems.

DOUG LARSON began his career studying the ecology of coastal tundra, and then studied the ecology of lichens and mosses growing on rock outcrops in southern Ontario. He has won several teaching and research awards, and has attracted wide media coverage to the new area of cliff ecology.

UTA MATTHES worked on the ecology of coastal lichens in California and currently manages projects dealing with physiological, population, and community ecology.

PETER KELLY has previously worked on arctic soil formation processes, and now concentrates on dendroecology and demography.

# CAMBRIDGE STUDIES IN ECOLOGY

**Cambridge Studies in Ecology** presents balanced, comprehensive, up-to-date, and critical reviews of selected topics within ecology, both botanical and zoological. The series is aimed at advanced final-year undergraduates, graduate students, researchers, and university teachers, as well as ecologists in industry and government research.

It encompasses a wide range of approaches and spatial, temporal, and taxonomic scales in ecology, experimental, behavioural and evolutionary studies. The emphasis throughout is on ecology related to the real world of plants and animals in the field rather than on purely theoretical abstractions and mathematical models. Some books in the series attempt to challenge existing ecological paradigms and present new concepts, empirical or theoretical models, and testable hypotheses. Others attempt to explore new approaches and present syntheses on topics of considerable importance ecologically which cut across the conventional but artificial boundaries within the science of ecology.

# Cliff Ecology

## Pattern and Process in Cliff Ecosystems

Douglas W. Larson
Uta Matthes
Peter E. Kelly

CAMBRIDGE
UNIVERSITY PRESS

CAMBRIDGE UNIVERSITY PRESS
Cambridge, New York, Melbourne, Madrid, Cape Town, Singapore, São Paulo

Cambridge University Press
The Edinburgh Building, Cambridge CB2 2RU, UK

Published in the United States of America by Cambridge University Press, New York

www.cambridge.org
Information on this title: www.cambridge.org/9780521554893

First published 2000
This digitally printed first paperback version 2005

*A catalogue record for this publication is available from the British Library*

*Library of Congress Cataloguing in Publication data*

Larson, Douglas W. (Douglas William), 1949–
Cliff ecology : pattern and process in cliff ecosystems / Douglas
W. Larson, Uta Matthes, Peter E. Kelly.
    p.    cm. – (Cambridge studies in ecology)
Includes bibliographical references (p.    ) and index.
ISBN 0–521–55489–6
1. Cliff ecology. I. Matthes, Uta, 1955–    . II. Kelly, Peter
E., 1963–    . III. Title. IV. Series.
QH541.5.C62L27   1999   99–12175 CIP
577–dc21

ISBN-13 978-0-521-55489-3 hardback
ISBN-10 0-521-55489-6 hardback

ISBN-13 978-0-521-01921-7 paperback
ISBN-10 0-521-01921-4 paperback

# Contents

# Preface

We have given ourselves the assignment of trying to write a book about places everyone sees, but no-one knows. In completing the work, we have tried to keep a number of things in mind. First, we recognize that readers have an insatiable curiosity for the truth, and within the context of natural history and ecology this is especially true because the answers to questions about non-human taxa sometimes help us interpret the significance of *Homo* to the world. This can comfort us. Second, we acknowledge the message that 'complex questions have simple, easy to understand, wrong answers'. Thus, the kinds of simple questions we ask may not provide simple answers, and in the work that follows we will try to simplify only when such efforts can provide reasonably precise and accurate versions of the truth. Given that this is the first book on the topic of cliff ecology, it may also happen that certain topics have been so under-studied that no effective summaries or syntheses can be made. When problems like this are encountered, we will try to bring them to the reader's attention. Lastly, we will try not to misrepresent to the reader the source of the motivation for doing science in general, and cliff ecology in particular – we love cliffs. Sometimes in the writing of science these motivations become lost in the intricacies of logic. You all know the wording: 'In order to test whether species packing densities could be predicted from the equilibrium theory of island biogeography we sampled . . .' which translates into English as 'islands are fascinating.' While such theoretical arguments may attract many, we also believe that many scientists, like artists, study what they do out of sheer fascination. In our case, we have found a previously unknown presettlement forest ecosystem on cliffs of the Niagara Escarpment, in southern Ontario, Canada, within sight of Canada's largest city. We have found this discovery to be immensely exciting and we will try to present a volume that captures some of that excitement.

It has taken three years for the final text to be prepared. The task proved far more challenging than we thought at the time writing began. Still, we think that the text includes some interesting information that readers familiar with level ground will be surprised to read. There is also a modicum of speculation about the significance of cliffs in an increasingly human-dominated landscape, but we feel that a book is an acceptable place for such speculation. If nothing else, it will provide an incentive for others to prove such speculations wrong. Throughout the preparation of the book, we have been struck by the many counterintuitive aspects of the ecology of cliffs, and now we feel even more compelled than ever to have this volume read by research workers, educators, graduate and undergraduate students, naturalists and professional land managers. We have tried to consider each of these audience members in the writing of the text. Thus, in some instances in which a research scientist would want more information about experimental design or statistical interpretation, there will be disappointment. To any of such readers, we invite direct correspondence with us at the University of Guelph. In other instances, there may be too much detail about particular species or geological structures to permit the professional land manager or undergraduate to continue reading with any sense of enthusiasm. To these readers we offer our apologies in advance, and say that the details that we reviewed were far more exhausting than what we present here. So at least the book is much easier to follow than the source material. It is also all in one place! To readers in the middle of this spectrum we hope that this volume illustrates how easy it is to be blind to wonderful ecological systems that stare us in the face. We really do hope that small 'cliff ecology' or 'swamp ecology' or 'stream ecology' groups start developing over the landscape. A diversity of such multidisciplinary working groups cannot help but add perspective to the already large number of intensive discipline-based studies in the same 'places'.

*Guelph, Ontario. August 1999*                                    DWL
                                                                  UM
                                                                  PEK

# Acknowledgements

Several people have aided us enormously in the production of the text: Janet Allan, Caireen Ryan, Eden Thurston and Teresa Domingues. Many individuals provided illustrations, photographs or previously unpublished data to be included in the text, and to these people we are grateful: David Currie, R.J. Small, K. Tinkler, T. Sunamura, D. Ford, J.B. Wilson, S. Pfeiffer, J. Gerrath, C. Buddle and B. Booth. This book is the product of research carried out over 13 years by many people who have helped with the work: graduate students Steve Spring, Ruth Bartlett, Kim Taylor, Chris Briand, Brian Gildner, Pampang Parakesit, Barb Booth, Jeff Matheson, Claudia Schaefer, Janet Cox, Ken Ursic, April Haig, Michele McMillan, Jeremy Lundholm; colleagues Drs Ed Cook, Joe Gerrath, Gary Walker, Ken Carey, Steve Stewart, Richard Reader, Usher Posluszny, Larry Peterson, Jeff Robins, Alan Charlton, Stefan Porembski, Jeff Nekola, and Jean Gerrath; research assistants Sarah Owen, Merrill Jeffrey, Joyce Buck, Ailish Cullen, Andrew Millward, Chris Buddle, Janet Allan, Sandra Turner, Jennifer Doubt, Jeff Outerbridge, Al MacKenzie, Amy Bournes, Adriana Stagni, Chris Henschel, Tim Keenan, Barb Best, Harold Lee, Jill Rogers, Jennifer Lukianchuk, Ceddy Nash, Cal Clark, Phil Davis, William Sears, Christoph Neeser, Cheryl Cundell, Eugenie Fitzgerald, Stacey Buss, Margy DeGruchy, Kit Howitt, Nocha Van Thielen and Guillaume Lecanu. Special thanks go to John Gerrath who has helped with many projects over the years. Parks Canada employees Mark Wiercinski, Scott Parker and Kevin Robinson were also extremely helpful. We also thank Drs Terry Gillespie, Sandy Middleton, John Birks and Paul Adam for reading the entire text and making many helpful comments about the content and presentation. We also thank Michael Usher for suggesting the idea of the book in the first place, and Alan Crowden, Vivienne Jones, Katrina Halliday and Jane Smith for help and encouragement from Cambridge University Press at all stages of production.

The research was funded by the Natural Sciences and Engineering Research Council of Canada, Environment Canada, the Ontario Heritage Foundation, the Ontario Ministry of the Environment and Energy, the Ontario Ministry of Natural Resources, Ontario Hydro, the World Wildlife Fund Canada, the Federation of Ontario Naturalists, the Richard Ivey Foundation, the Niagara Parks Commission, Tilley Endurables, and Mountain Equipment Co-op. We also thank the various conservation authorities all along the Niagara Escarpment which assisted our group by permitting access to private land. They include the Niagara Peninsula Conservation Authority, the Hamilton Region Conservation Authority, the Halton Region Conservation Authority, the Nottawasaga Conservation Authority, and the Grey-Sauble Conservation Authority. The Ontario Ministry of Natural Resources and the Ontario Ministry of the Environment and Energy also provided extra help. Lastly, we thank our family members for putting up with us while we worked on this project.

Folgend dem Windzug kommen zum Felsen
die Wolken und weichen,
unveränderlich steht aber der Fels
in der Zeit.

[Following the wind, the clouds come
but they yield to the rock,
and the rock stands
unchanged
in time.]

Anonymous carving into limestone cliff in the Fränkische Schweiz, Germany.

thir callow young, but feathered soon and fledge
they summ'd thir penns, and soaring th' air sublime
with clang dispis'd the ground, under the cloud
In prospect; there the eagle and the stork
on Cliffs and Cedar tops their eyries build

Milton (1667) *Paradise Lost*, Book VII, 420–4.

# 1 · *Introduction*

In the six decades since Sir Arthur Tansley first coined the word *ecosystem*, an enormous amount of ecological research has been carried out in every imaginable habitat on earth. Forests, grasslands, deserts, tundra, wetlands and oceans have all been mapped in their distribution on the earth's surface and have revealed their structure and some aspects of their function to ecologists. Until recently, however, vertical cliffs have been almost completely overlooked as subjects for ecological study, even though some workers in Europe have included areas of steep rock in analyses of vegetation communities. For example, McVean and Ratcliffe (1962) described plant communities for the Scottish highlands but only a handful of stands had slopes greater than 60° and only one had a slope value of 80°. In other words, cliffs as defined in this book were not really included even if subsequent authors said that they were. McVean and Ratcliffe were also aware of the difficulty in dealing with cliff vegetation at the community scale. They stated:

> To many botanists this heterogeneous cliff vegetation is the most interesting of all but to the phytosociologist it is easily the most baffling. The larger, stable ledges usually bear tall herb communities and are amenable to the normal method of analysis but the open and patchy vegetation consisting of small herbs, sedges, grasses and bryophytes is very difficult to describe. . . . We have therefore analyzed only those cliff communities which provided stands of at least the normal minimal area of 2 × 2 m. . . . Description of the micro-communities naturally confined to open rocks is best reserved for detailed studies of individual rupestral species. (McVean & Ratcliffe, 1962, p. 88)

These words reinforce the idea that even something simple like the biological description of cliffs as habitat has been difficult for ecologists to accomplish. Cliffs simply do not fit into conceptual models that work well for other habitat types.

The available literature is scattered over a wide variety of journals, monographs and books, and much of it dates from the early part of the twentieth century. Cliffs have rarely been investigated from an ecological viewpoint and even when some individual workers have shown that many important ecological problems could be addressed by focusing on cliff vegetation (Oettli, 1904), subsequent workers for the most part have ignored their contributions. In most of the literature on geology, geomorphology or hydrology that is reviewed in the following chapters, no mention is made of the vegetation or fauna of cliffs. Equally, studies of the flora and fauna rarely try to interpret the relationship between the biota and the physical characteristics of the cliff habitat.

This lack of scientific interest is in striking contrast to how common cliffs are around the world, and to the attraction cliffs have had for humans throughout history. Cliffs are present on every continent and in virtually every country on earth. In some jurisdictions (such as Utah, USA, or Madagascar), cliffs are so common that they structure the agricultural or industrial development of the entire landscape. Even regions such as the American Midwest or Bangladesh, which are perceived as flat landscapes, have isolated cliff fragments or long sections of cliff carved by the actions of rivers or coastal tides. As an example of this, the state of Iowa (USA) is well known for its flat agricultural landscape. But the entire northeast quarter of the state is a massive palaeozoic limestone plateau through which more than a dozen large rivers have cut their channels. All of these rivers have vertical limestone cliff faces somewhere along their courses, and the total length of these faces is likely to be in the hundreds of kilometres. Paullin (1932) clearly understood this fact, and also noted that these cliff-lined river courses were locations that supported virgin forest. We will return to this point frequently in subsequent chapters.

While cliffs are common, the exact area covered by them is not known because vertical surfaces do not show up on aerial photographs and maps. In the following chapters, we will present evidence that cliffs represent some of the least disturbed habitats on earth and contribute more to the biodiversity of a region than their actual surface coverage would indicate. In Europe cliffs can represent the primary habitat for species of *Saxifraga, Draba, Sorbus, Daphne, Dianthus, Campanula, and Androsace*. In addition, Ellenberg (1988) reports that 35–40 per cent of the endemic taxa of alpine regions of Europe grow only in rocky crevices on steep slopes and cliffs, while Wardle (1991) reports that this figure is 66 per cent for New Zealand. This evidence suggests that it may be impossible to establish the full extent of biodiversity in different parts of the world without careful sampling and inclusion of these vertical habitats.

Cliffs also represent a habitat that is relatively free from human disturbance. This characteristic together with the abundance of scenic vistas common in cliff environments provide the incentive for anchoring parks or nature preserves around them. Examples include the Lake District National Park in the UK (Fig. 1.1), Yosemite National Park and the Grand Canyon in the USA (Fig. 1.2), New River Gorge National Park in the USA (Fig. 1.3), the Niagara Escarpment Biosphere Reserve in Canada (Fig. 1.4), and Las Ruinas de Tulúm National Historic Site in Mexico (Fig. 1.5). All these areas have large cliffs that form the focal point for the entire management unit.

The natural appeal of cliffs has led many companies and organizations to use them to advertise their products. The Apple Computer Corporation initially marketed its series of Powerbook laptop personal computers by showing the image of a person sitting on the edge of a cliff, working on an open laptop computer. Similarly, the Chrysler Corporation (Fig. 1.6), the Bell Telephone Company, and Cathay Pacific Airlines have all recently used cliffs as marketing images. Table 1.1 illustrates the frequent use of cliff images in magazine advertising for the years 1994 and 1995. Out of 221 advertisements surveyed in four different magazines, 31.2 per cent used cliffs as the context, 25.8 per cent with the cliffs in the foreground. Such advertising exploits the degree to which humans are attracted to the dramatic views that are derived from cliff tops, and the danger associated with being very close to cliff edges. In a gruesome but effective item of marketing, the Canadian Automobile Association recently advertised its death and dismemberment insurance programme by showing a pair of hikers standing on an overhanging rock at the edge of a 150 m cliff in northern Ontario. There is probably no habitat on earth that has received so much attention from the general public but so little attention from scientists.

This book is about the ecology of cliffs and the primary objective for writing it is to ignite an interest in all aspects of cliff research. It is interesting to speculate on the reason why ecologists have been so completely 'cliff-blind'. Vertical surfaces are difficult to sample, but so are ocean floors. Cliffs are considered hostile, lifeless and therefore not worth investigating; but why have deserts and antarctic regions, which can be equally bare, escaped the judgement of being ecologically uninteresting? The truth may be that while cliffs are in everyone's backyard, they are rarely viewed as 'places' in their own right. A good example of this is the book

*Figure 1.1*  Cliffs such as this one in the Lake District, UK, have usually been viewed as habitat barren of life, and useful only for rock climbing or as an element of attractive scenery. Photo by Alan Charlton.

*Figure 1.2* View of a series of cliff faces, terraces and talus slopes in the Grand Canyon, Arizona, USA. Differential weathering rates of strong and weak strata, combined with intense erosion over long periods of time, have produced this series of cliffs. Photo by P.E. Kelly.

*Figure 1.3* View of 30 m high limestone cliffs along the New River, Virginia, USA. These cliffs support a population of presettlement *Juniperus virginiana*. Photo by D.W. Larson.

*Figure 1.4* View of 25 m high limestone cliffs of the Niagara Escarpment near Milton, Ontario, Canada. These cliffs line the edge of the Michigan Basin that runs northward to Manitoulin Island, Canada, westward through Michigan, USA, and south along the Door Peninsula, Wisconsin, terminating near Chicago, Illinois, USA. Photo by P.E. Kelly.

*Figure 1.5* View of 10–15 m high cliffs along the Atlantic coast of Mexico, near Tulúm. Stunted *Yucca* and a variety of cryptogams and succulents occur on the cliffs. Photo by A. Haig.

*Figure 1.6* Cliffs are used abundantly in commercial advertising, as illustrated in this 1996 advertisement for a sport-utility vehicle. Note the stunted cliff-edge tree to the rear of the truck. Photo courtesy of W. McCall, Chrysler Canada.

Table 1.1. *Summary of the extent to which cliffs form the image or the background image in commercial advertising*

| | Type of natural setting | | | | | |
|---|---|---|---|---|---|---|
| | Cliff | | Forest | Lake or ocean | Field | Other |
| Magazine (no. of issues) | In foreground | In background | | | | |
| *Equinox* (12) | 22 | 4 | 3 | 18 | 3 | 29 |
| *Discover* (12) | 5 | 2 | 2 | 1 | 11 | 16 |
| *Canadian Geographic* (11) | 19 | 6 | 2 | 10 | 4 | 22 |
| *Time* (16) | 11 | 0 | 2 | 13 | 8 | 7 |
| Totals | 57 | 12 | 9 | 42 | 26 | 74 |
| | 69 | | | | | |
| Percentage of total | 25.8 | 5.4 | 4.1 | 19.0 | 11.8 | 33.6 |
| | 31.2 | | | | | |

*Note:*
All 1994 and 1995 issues of three different natural history magazines (*Equinox*, *Discover*, and *Canadian Geographic*) and one news magazine (*Time*) were surveyed. The 221 such advertisements found which presented the product in some kind of natural setting were placed into one of five categories based on the type of setting: field, lake or ocean, forest, cliff, or other (such as snow, roads, sky or lawn). The group in which the context was a cliff was subdivided further into advertisements in which the cliff was in the foreground, and those in which the cliff formed the background. The number of advertisements in each category is shown below for each magazine.

entitled *The Vegetation of Wisconsin* by Curtis (1959), in which an entire chapter is devoted to cliffs and in which it was claimed that 'a cliff is a geological feature, not a biotic community type'. Curtis then contradicted this claim by stating that the cliff environment supported a specialized group of plants that included many endemics and rare species such as *Rhododendron lapponicum* that normally occur many thousands of kilometres north along the arctic shores of the Northwest Territories in Canada.

## 1.1 What is a cliff?

The word *cliff* in the English language is of Teutonic origin and defined as 'a high, steep, or overhanging face of rock' (Onions, 1968). It is similar in meaning to the word *precipice*, which is of Old French and ultimately Latin origin and given as 'an extremely steep or overhanging mass of rock'. In some countries, the word *bluff* or *rock wall* is used in an equivalent manner to *cliff*. In contrast, a *rock outcrop* is 'a portion of bedrock protruding through the soil level'. Thus, all cliffs are rock outcrops, but rock outcrops are not necessarily cliffs. We do not think it is necessary to specify the degree of consolidation of the rock. Thus, cliffs formed of sand, gravel or loess should be included in this discussion, as well as cliffs formed of igneous rock or hard sedimentary rock. For an outcrop to be called a cliff, three essential elements must be present: a level or sloping platform, or *plateau*, at the top; a *pediment* consisting of baserock at the bottom; and a vertical or near-vertical part, called the *cliff face* or *free-face*, in between (Fig. 1.7b). In addition, the face must be both tall and steep. Figure 1.7 illustrates that tallness and steepness are relative terms and are subject to interpretation. Tallness is relative to the observer and therefore introduces a species bias. Figures 1.7a and 1.7b show two rock outcrops that are equal, except in height, but only the second would be considered a cliff by the human observer, whereas both might function as cliffs to small insects. The minimum height at which an outcrop is called a cliff is not defined, but an intuitive definition is 'high enough that falling off will kill you' and this height is probably more than 3 or 4 m. From the viewpoint of the processes that produce the rock face and the organisms colonizing the surface, however, the height may be irrelevant. In fact, we have found no literature that demonstrates that cliff height *per se* influences the organisms that live on cliffs.

The relationship between the angle of the slope and the appropriateness of the term cliff is illustrated in Fig. 1.7. Overhangs or undercuts

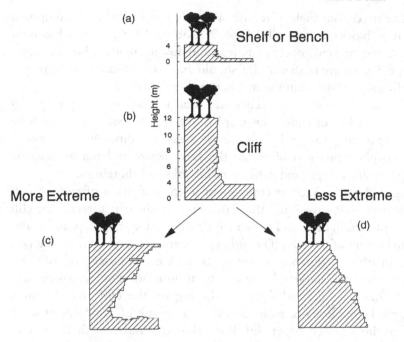

*Figure 1.7* This diagram illustrates the difficulty in defining the word 'cliff'. The strict wording of the definition includes relative words such as 'high' and 'steep' that are subject to interpretation. Comparison of the structures shown in (a) and (b) illustrates the effect of height relative to a human observer. Both structures are the same except for height, but only (b) would be described as a cliff by most people. The effect of slope angle on the concept of 'cliff' is seen by comparing the structures in (b), (c) and (d). The term 'cliff' applies to both (b) and (c); overhangs or undercuts make the cliff appear more extreme compared to a vertical free-face. Conversely, the structure shown in (d) appears less extreme and would probably be described as a steep slope.

(Fig. 1.7c) represent more extreme cliffs than vertical faces (Fig. 1.7b), while steep slopes such as that shown in Fig. 1.7d challenge the definition of a cliff. The critical distinction between a cliff and a slope may be that objects falling from cliffs usually fall through the air before they hit solid ground, whereas objects falling down slopes normally maintain at least sporadic and probably painful contact with the ground. The point here is that it is impossible to define a critical angle that separates cliffs from other structures. Furthermore, from a scientific perspective, we gain little by trying to make such strict definitions. What we can do is recognize that slope angles from 180° (the underside of an overhang) to 90° (a vertical wall) are all strictly 'cliff', whereas slope angles less than 90° are less so. The maximum angle of repose of tightly interlocked scree composed of large blocks is typically 43–45° (Ritter, 1978), and subsequent weathering or land movements result in lower angles. Thus, it seems compelling to argue that a cliff should have a minimum angle that is significantly greater than the maximum slope for scree.

The word *escarpment* is sometimes used to indicate the precipitous face along a line of cliffs. This word is defined as 'a steep slope or long cliff separating two relatively level areas of different elevations'. Geomorphologists also often use the term *terrace* or *bench* to describe steep but short slopes and cliffs on riverbanks and shorelines.

Figure 1.8 adds another component to the definition of cliffs. A cliff in the narrowest sense is just the vertical part of the structure, i.e. the cliff face. But cliffs in the broad sense include a cliff edge at the top, and a talus at the bottom of the face. The *cliff edge* is a zone extending from the face back an arbitrary distance. It consists of rock in the process of weathering, whereas the *talus* or *talus slope* is the accumulation of loose rock fragments and slabs derived from weathering on the cliff face. In some settings, large blocks of talus can create a mosaic of hundreds of small cliffs at the base of a larger cliff. Both cliff edge and talus share some of the physical characteristics of the free-face, support many of the same plants and animals, and are linked by the same ecological processes. Therefore, this book is about cliffs in the broader sense of the definition, i.e. cliff edge, free-face, and talus slope as a single unit.

Some other frequently encountered features of cliffs are also illustrated in Fig. 1.8. The *toe* is the point where the talus slope meets the pediment. Whereas the idealized cliff face is a solid vertical surface, in reality there is heterogeneity on multiple spatial scales in the form of ledges, overhangs, cracks, crevices and caves. *Ledges* are sections of the cliff face that are more or less horizontal and may be undercut, thus forming *overhangs*. Depending on the type of rock, there may be few or many fractures in a continuum of sizes from invisible microscopic *cracks* to large *crevices*. *Caves* may be formed by the expansion of these openings through the dissolution of rock. Lastly, it is entirely possible for each of these individual microsites to be immediately adjacent on a single rock face. In other words, an unfractured vertical face with no rooting space for higher plants but abundant habitat for heat-resistant and desiccation-tolerant lichens can be immediately adjacent to a small fracture plane or ledge that accumulates either liquid water or organic matter (or both) that permits the growth of grasses, sedges or small trees. The compression of all of these microhabitats in a small space leads to the peculiar (and, to some, the

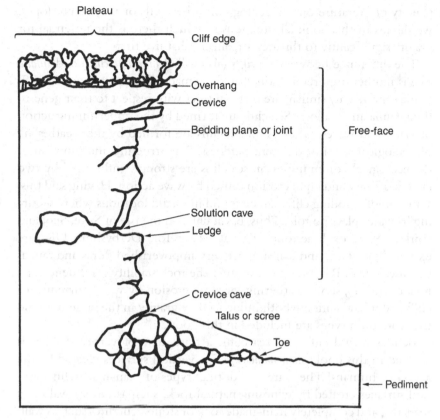

*Figure 1.8* Illustration of frequently encountered cliff features. The concept of cliff may relate to the free-face alone, or to the cliff edge, free-face and talus combined. On the face there may be cracks, crevices, ledges, overhangs and caves. The talus slope is composed of rock fragments and slabs derived from the free-face and the edge. The toe is the point where the talus slope meets the pediment, or baserock. Original artwork courtesy of C.E. Ryan.

unfathomable) situation in which wetland plants and desiccation-tolerant near-desert plants can coexist within centimetres of each other (Oettli, 1904).

One of the difficulties with applying the term 'cliff' in real-life situations is that few cliffs are homogeneous from top to bottom. Sections of rock face that are vertical or overhanging may be interspersed with sections that represent very steep slopes. Thus, cliffs in the Grand Canyon, USA include very steep slopes, ledges, true cliffs and small terraces (see Fig. 1.2). Similarly, the white cliffs of Dover in the UK and the cliffs of Gibraltar are actually a mixture of cliffs and very steep slopes. Given the paucity of literature on the ecology of either cliffs or very steep slopes, we discuss in this book all situations in which there is the potential for gravity significantly to threaten organisms that live there.

The difference between *sea cliffs* (also called *maritime cliffs*) and *inland cliffs* is another important distinction to be made. These two types of cliff, while superficially similar, are often distinct with respect to their genesis, flora, fauna and ecology. Sea cliffs are formed by the undermining action of water and waves, whereas inland cliffs are formed by the weathering of geological strata of different hardness. The structure and function of the ecological communities on sea cliffs are strongly influenced by two factors, salinity and rapid erosion caused by wave action. Having said this, some rapidly eroding cliffs can be found in inland locations where scouring by water plays no role. Thus, basalt cliffs on the Isle of Skye, Scotland (Birks, 1973) or limestone cliffs at Mam Tor, Derbyshire, UK, are extremely unstable and can support very impoverished floras and faunas because of this. Regardless of whether the rock stability is influenced by rock chemistry, slow weathering, or rapid erosion by water currents, sea cliffs and inland cliffs are both influenced by gravity in the same way, and therefore both types are included in this book.

Besides inland cliffs and sea cliffs, a third type of vertical habitat is covered in this book: *man-made cliffs*, i.e. vertical surfaces created by the action of humans. There are two distinct types of man-made cliffs: vertical surfaces created by exposing natural rock, such as quarry walls and roadcuts; and completely man-made rock or stone structures such as walls and buildings. These will be included because their flora, fauna and ecology often show striking similarities with natural cliffs and because the ecological factors controlling these communities are similar.

Shallow slopes, valley sides, mountain tops, pavements, barrens, or other forms of rock outcrop are not discussed in this book except when comparisons between them and cliffs are interesting. These habitats are covered elsewhere, for example by Larson *et al.* in a recent book by Anderson, Fralish and Baskin (1999).

## 1.2 The Niagara Escarpment and the 'ecology of place'

To a large extent, cliffs have received little direct ecological study because they are not viewed as 'places' with their own distinctive ecological structures and processes. Why do we say this? Part of the answer is that the trees of sparsely forested cliffs in different countries (for example the *Taxus*-dominated cliffs of Derbyshire and Yorkshire in the UK) are referred to by local authorities as 'scattered trees' or as 'isolated specimens'

(Rodwell, 1992) rather than as a highly recurring pattern that might suggest a whole community type. In other words, the trees that occur on cliffs are regarded as accidental occurrences outside of their normal forest communities. With this mindset, it will follow quite naturally that cliffs will be perceived as uninteresting or unimportant because they represent transition lines (having zero vertically projected area) between two places rather than 'places' in their own right.

The approach adopted in the research conducted by the Cliff Ecology Research Group begins by acknowledging that cliffs can be places (perhaps even forests) too small to be considered as 'places' by most other terrestrial ecologists. This book tries to reflect this same attitude and to survey all of the published work relevant to cliffs that we could find, and then present a coherent story about what cliffs are and how they function as 'places' within their local landscapes. As is probably unavoidable when researchers write a book about their specialty area, our own work figures prominently among the examples given throughout the text. In view of this, it is necessary to give the reader a brief introduction to our study area, the scope of our research, and our research strategy.

Over the past ten years the focus of our research has been the Niagara Escarpment (see Fig. 1.4), a band of limestone cliffs that extends over 740 km through southern Ontario, Canada. These cliffs support the oldest presettlement forest ecosystem in eastern North America (Larson & Kelly, 1991). The work we have conducted here has covered a wide range of subjects on different spatial and temporal scales, including physiological, population and community ecology, as well as microclimatology, water relations, architecture, anatomy and dendrochronology. All these disparate topics are linked by the fact that they contribute to our understanding of a specific 'place' – the Niagara Escarpment. We refer to this approach as the 'ecology of place', and we hope to show here that high levels of predictive power can be generated by adopting this approach to the study of small, relatively discrete ecosystems.

We believe that the 'ecology of place' approach could be used more widely in the study of ecosystem organization. A 1989 survey of research and researchers, sponsored by the British Ecological Society (Cherret, 1989), showed that research topics dealing with subjects on a large scale were generally perceived as more important than more specialized topics (e.g. population interactions and evolution). The lowest-ranked topics were those trying to tie narrow or specialized questions into larger-scale studies of ecosystem organization. The survey confirms the trend seen in the literature in general whereby ecological expertise grows by expanding control over subdiscipline 'home ranges' in a top-down fashion, based

Table 1.2. *Three research strategies in ecology whose advantages and disadvantages complement each other.*

| Focus of investigation | Central question | Consequence |
| --- | --- | --- |
| Processes | Role of particular processes in regulating community structure and function in different ecosystems with different taxa | Expertise develops for specific processes, but out of context with other processes |
| Taxa | Role of particular taxa or groups of similar taxa in different ecosystems responding to different processes | Expertise develops for single taxa, but out of ecological context |
| Places | Operation of all taxa and processes in context to produce a complete picture of the structure and function of a small ecosystem | Whole-system expertise develops, but only for small places |

*Note:*
The first two strategies are commonly adopted, but the third has not been popular since the expansion of specialist-based science in the 1970s.

on the tendency of science to become more and more diversified and specialized over time. In contrast, efforts to understand the way in which functional and structural hierarchies control particular pieces of the earth's surface, i.e. a bottom-up approach that avoids disciplinary specialization, are rarely seen to generate significant testable hypotheses about particular processes or taxa and thus are viewed as less capable of making significant contributions to science. Thus, researchers prefer to investigate either individual processes (such as competition or predation) in various locations and using various combinations of taxa, or individual taxa wherever they occur and with respect to all the processes affecting them. The third approach, trying to explain how a 'place' works with all processes and taxa in operation, is rarely used even though it permits a synthesis in the understanding of how processes interact with taxa (Table 1.2). In the following pages, we briefly outline the concept of 'ecology of place' and show how the study of such a 'place' can contribute to our understanding of ecosystems.

The 'ecology of place' is derived from the hierarchical concept of ecosystems (O'Neill *et al.*, 1986), which states that natural ecosystems are formed as dual hierarchies of taxa and interactive but scale-dependent processes. While this concept has had tremendous intuitive appeal in areas from biology to geomorphology (Haigh, 1987), little empirical work has been done to test it. We speculate that the reason for this delay is twofold: first, the lack of a method for conceptually and practically defining the limits of the ecosystem to be studied; and second, the resulting unavailability of a complete and real 'place' that lends itself to testing the idea.

The first of these two prerequisites has been discussed by Kolasa and Pickett (1989), who show convincingly that in order to understand the structure or function of any ecosystem (as well as predict its response to various stimuli), its operational boundaries must be initially defined. Considering the argument of Lovelock (1979), who states that the earth is comprised of obligately and strongly interacting components, some may argue that it is unrealistic to try to identify subsystems within the earth. However, the practical difficulties of trying to understand the planet as a whole have proven immense. Thus, looking for subsystems with detectable boundaries may be philosophically troubling but pragmatic. According to Kolasa and Pickett (1989), ecosystem boundaries can be found by considering the scales at which the structural and functional heterogeneity in the landscape *co-occur* (Fig. 1.9). Ecosystems do not need to be fully independent of one another, but the structural or functional components that originate outside of a particular ecosystem need to be identified.

Unfortunately, Kolasa and Pickett (1989) do not address the next step necessary for a successful hierarchical whole-system investigation, which is finding an actual ecosystem whose boundaries can be defined in this manner. Most ecosystems are too large to lend themselves to comprehensive multilevel hierarchical analysis, and few offer clear evidence of structural and functional boundaries. Lakes, for example, are structurally differentiated from the adjacent vegetated landscape but are not functionally separated from it because of their dependence on the vegetated watershed. Using a complete watershed (Bormann & Likens, 1979) avoids these problems, but introduces the new one of studying a representative subset of an ecosystem rather than a complete entity. This increases the likelihood of spatial autocorrelation and pseudoreplication problems (Hurlbert, 1984). One might argue that it is impossible to avoid such problems totally because research is always conducted on subsamples of space and time. However, some systems do exist that are more readily confined by space and time. Examples of such 'places' that are good can-

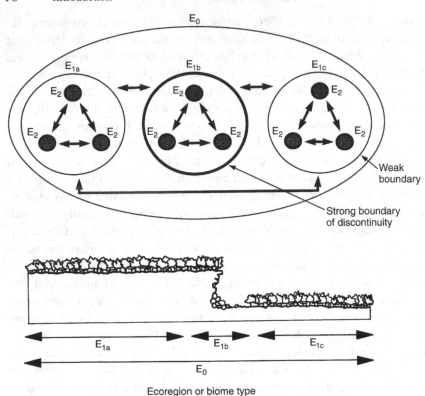

*Figure 1.9* Boundary diagram based on Kolasa and Pickett (1989) illustrating how the discreteness of entities in a hierarchical system can be used to identify potential ecosystem boundaries. The hierarchical nesting of entities is indicated by circles; integration is greater within each circle than among circles. The thickness of the circle illustrates the strength of the boundary. Horizontal lines show connections between components within the boundaries, with thicker lines representing stronger connections. In this example, there is one large-scale entity $E_0$, consisting of three subentities $E_{1a}$, $E_{1b}$ and $E_{1c}$. Each of the three subentities in turn includes three subentities $E_2$. The line surrounding $E_0$ is very thin, indicating a weak differentiation from surrounding ecosystems. $E_{1a}$ and $E_{1c}$ are similarly weakly differentiated from $E_0$, but $E_{1b}$ shows a high level of differentiation from other entities. In practical terms, $E_0$ can represent the loose assemblage of biota in the mixed-wood ecozone of Canada, with $E_{1a}$, $E_{1b}$ and $E_{1c}$ representing plateau forests bordering cliffs, cliffs themselves, and deciduous forests below cliffs, respectively. Original artwork courtesy of C.E. Ryan.

didate sites for hierarchical whole-system investigations would include alpine communities that are bounded by high peaks, lowland bogs, hedgerow communities (Küppers, 1984, 1985), mountain-bounded deserts, geothermal vent ecosystems, and, last but not least, cliffs.

One of the often-cited defences for the lack of support for whole-system study, even at this smaller scale, is the failure of the International Biological Program (IBP) to model completely the biological basis of productivity (Worthington, 1975). It must be pointed out, however, that the mandate of IBP was to 'understand the biological basis of productivity and human welfare' on earth, not to produce compartmental simulation models of major biomes or the globe. It should be acknowledged that IBP was an overwhelming success in generating large volumes of comparative data that helped in the interpretation of the earth as an integrated but complex entity with internal component entities that appear to communicate in an interdependent fashion (Lovelock, 1979). The 'ecology of place' approach is very similar to the whole-system study attempted during IBP. It differs only in having a much smaller spatial scale with more clearly defined boundaries. As a consequence, the whole place can be sampled with no need to assume that a single site is representative, and the multidisciplinary investigations can be better coordinated. The spatial and temporal dimensions of the entity, or 'place', selected for study are crucial; they must be large enough to be relevant to ecologists studying ecosystems or their component processes at other scales, but small enough to be encapsulated by a single research team.

We have chosen to follow a scaled-up version of the 'ecology of place' approach in the writing of this volume. In the initial chapters, we try to compile all published information relevant to cliffs from different geographical areas and from different specialties including geology, geomorphology, climate, flora, fauna, human impacts and conservation. By looking at cliffs at many simultaneous scales and from as many different points of view as possible, we then try to derive general patterns that apply to the structure and function of all cliffs. While we have tried to avoid interpreting individual results out of the context of 'place' as defined above, it has nevertheless become evident that cliffs worldwide appear to have structural and functional characteristics that are interesting both in terms of their striking similarities and in their differences. We hope that by the time you finish Chapter 8, you will want to pay much closer attention to the cliffs that loom over your cities, towns, campsites, cottages and natural areas. Cliffs are amazing 'places', and this is intended to be the subject matter of the rest of the book.

# 2 · Geology and geomorphology

In most discussions of terrestrial ecology, the weathering patterns and age of the bedrock are an important part of the ecological context. This is especially true on cliffs where soil formation is either absent or minimal and there is direct contact between the rock face and the biota at all times. On level ground, the underlying geology becomes less critical and the processes of soil formation more critical to plant and animal communities over time (Oettli, 1904). The persistence of the contact between the biota and the rocks on cliffs (most often without soil) makes it necessary for us to present a broad summary of the important geological and geomorphological characteristics of cliffs which may influence the abundance and distribution of the biota on the surface.

## 2.1 Bedrock composition and strength

The emergence of cliffs in the landscape is dependent on both the mechanical strength and the variability in the strength of the bedrock. These traits are built into the rock, both during the initial deposition of the materials and in the subsequent reshaping of the strata by crustal plate movement and large-scale geomorphological events such as glacial advances or the outflows from rivers. The variability built into different strata results in differences in the physical expression of weathering processes, with the removal of weaker materials and the retention of stronger ones. The pronounced series of cliffs interrupted by talus slopes lining the walls of the Grand Canyon, USA, is caused in part by this vertical geological heterogeneity in wear resistance (see Fig. 1.2).

The inherent strength of unweathered rocks derived from different materials and processes has been described by Sunamura (1992). Sunamura showed that Mesozoic–Paleozoic sedimentary rocks had compressive strength values of between 2000 and 3000 kg cm$^{-2}$, igneous

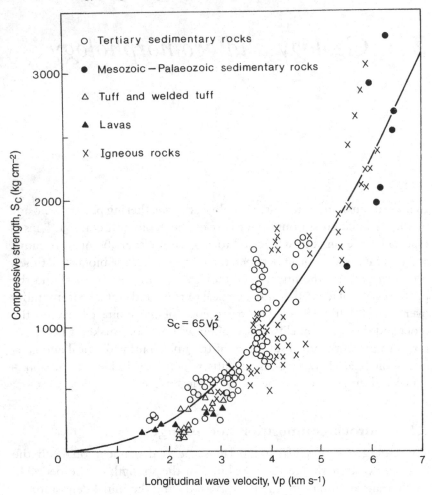

*Figure 2.1* Scatterplot of the relationship between compression strength (S$_c$; kg cm$^{-2}$) and the longitudinal wave velocity of pressure waves within rock (V$_p$; km s$^{-1}$) for rocks of different origin. Redrawn from Sunamura (1992) and used with permission of John Wiley and Sons, Publishers.

rocks had values between 500 and 3000 kg cm$^{-2}$, tertiary sedimentary rocks between 300 and 1600 kg cm$^{-2}$, tuff and welded tuff between 100 and 400 kg cm$^{-2}$, and lavas between 200 and 400 kg cm$^{-2}$ (Fig. 2.1). Continuous compression and decompression, usually from tectonic activity, can weaken rock, but once a cliff face has been exposed, weathering processes work to destabilize the rock. Cliffs in locations with a high degree of tectonic activity will be very unstable and will represent very impermanent habitat for most immobile organisms. On the other hand, such cliffs will form large talus slopes that receive continuous inputs of new material.

## 2.2 Heterogeneity

Next to the fundamental strength of the minerals in the rock, the mechanical heterogeneities caused by uneven rates of cooling, sedimentation patterns, seismic activity or crustal movement will control the rate and form of cliff development, and will also control the number and type of microhabitat features such as solution pockets, cracks, crevices and cavities. Igneous rocks formed in single massive events with few internal fracture zones or joints will be sufficiently homogeneous to weather slowly as one more-or-less rounded and consolidated unit. These rocks provide few spaces suitable for colonization by macroscopic organisms. The side walls of Inselbergs in Africa are examples of this (Porembski *et al.*, 1994). Conversely, discontinuities in igneous rocks created by separate events widely spaced in time, such as the intrusion of dykes into older rock, may provide starting points for physical and/or chemical weathering (Fig. 2.2), and the heterogeneities created may then be easily exploited as habitat by plants and animals.

In sedimentary rocks, variations between the composition and strength of individual horizontally bedded facies adds internal heterogeneity to cliff faces (Fig. 2.2b, c and d) and cliffs are often formed when a harder, more resistant, rock type overlies weaker rock that is more weathered (Fig. 2.2c and d). The form of the cliff is dependent on the dissimilarity of the various rock types. When the beds differ only slightly, the cliff face may become blurred, reduced in height and confined to the hardest rock type (Fig. 2.2d). Where differences are more pronounced, underlying rock layers may erode much faster than the surficial rock, creating a pronounced cliff or escarpment (Fig. 2.2c).

Internal variance in the strength of bedrock is also controlled by seismic activity that fractures the parent material, leading to the production of pronounced weathering planes. Karst activity that removes material and changes the internal structure of the rock by dissolution also modifies the existing variance in geological structure (Ford & Williams, 1989). Exposure of rocks to sea or lakeshore erosion (Ritter, 1978; Trenhaile, 1990; Sunamura, 1992) has similar effects, although over much shorter periods of time. A variety of shore formations including cliffs, sea-stacks, caves and talus slopes is found in locations where rock out-

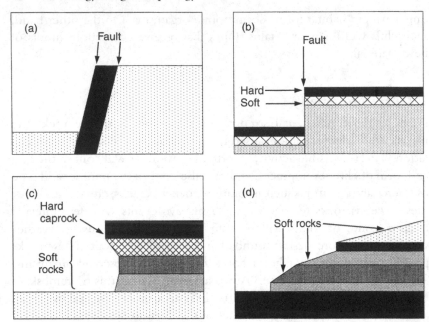

*Figure 2.2* Various configurations of bedrock strata and bedrock strength in the context of cliff formation. When non-horizontal fault lines result in upthrusts of bedrock (a), the hard rock formed at the fault commonly forms a slanted cliff surface. When faulting occurs in horizontally stratified rocks (b), the hard caprocks (darker colours) persist at the surface, and a cliff face is formed at the fault. Such cliffs can also be undercut (c) if mass wasting or erosion takes place. Slopes rather than cliffs will form if surface rocks are weaker (lighter colours) than base rocks (d). Redrawn from Trenhaile, A.S. (1990). *The Geomorphology of Canada.* Toronto: Oxford University Press.

crops face large lakes or oceans (Fig. 2.3). Figure 2.4 illustrates that in some types of rock, sea-stack formation can take place in only a six-year period. Conversely, the sea-stack formations of the Bruce Peninsula, Ontario, Canada, have formed over a period of several thousand years (Fig. 2.5). Sea-stacks and cliff faces can be formed from rock of all different chemical make-up, including igneous and sedimentary, basalts and granites to alkaline limestone and dolostone.

## 2.3 Weathering processes

Weathering processes can be divided into physical and chemical processes. Physical weathering can be imposed by the removal of loads from sudden mass wasting, dimensional changes due to water content, direct

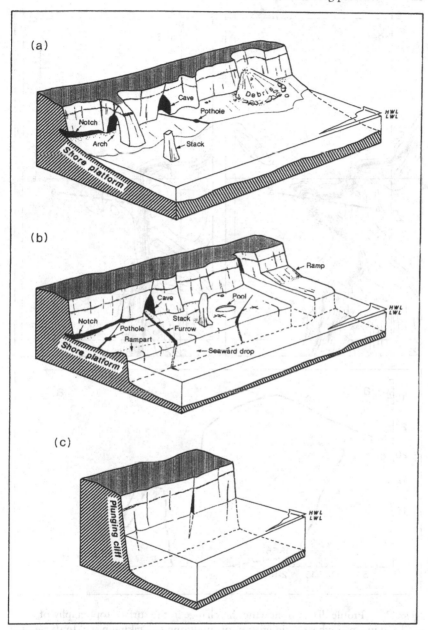

*Figure 2.3* Three configurations of rocky coasts including (a) vertical cliffs with sloping shore platforms, (b) vertical cliffs with stepped shore platforms, and (c) vertical plunging cliffs with no shore platform. Shorelines with either (a) or (b) are less stable than (c). Illustration courtesy of John Wiley and Sons, from Sunamura, T. (1992). *Geomorphology of Rocky Coasts.*

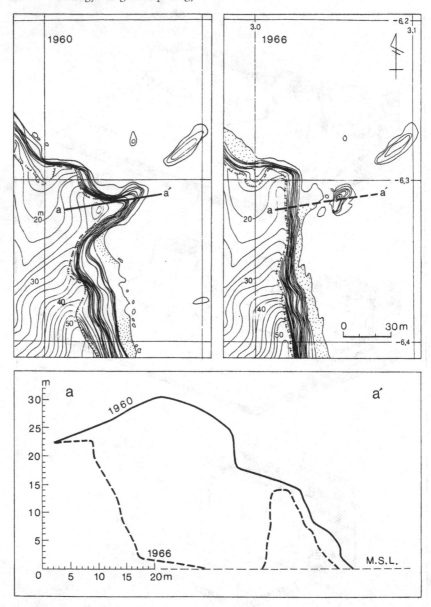

*Figure 2.4* Profile diagram showing the change in the surface topography of a headland in Japan where rapid erosion of mudstone was taking place. In six years a stepped sea cliff was transformed into a new sea cliff plus a sea-stack. Taken from Sunamura (1992) and used with permission of John Wiley and Sons, Publishers.

*Figure 2.5* A post-glacial sea-stack, Bruce Peninsula National Park, Ontario, Canada. Photo by D.W. Larson.

pressure effects from ice and mineral crystal formation, thermal stress, and frost action (Sunamura, 1992). Sudden load removal is relatively rare compared to the other more continuous physical weathering processes, and dimensional changes due to fluctuations in water content have only a small influence on the weathering of granite, gneiss, basalt and carbonate rocks. Dimensional changes may be extremely important in rock such as mudstones and shales that swell when large amounts of water are imbibed and which shrink when this water is lost (Matsukura & Yatsu, 1982). Physical weathering of rock by crystal formation and thermal stress is common, but both processes mainly influence the immediate surface of the rock and not at depth. The persistence of Egyptian carvings made in cliff faces up to 4000 years ago provides tangible evidence that thermal effects alone result in relatively minor amounts of physical weathering on exposed cliffs in arid regions (Jurmain, Nelson & Turnbaugh, 1990).

Freeze-thaw or frost action processes are the most common physical weathering mechanisms. Washburn (1973) reviews a wide variety of these processes but does not make specific reference to the formation of cliffs. There is evidence that frost action results in the plate-like fracture of a variety of types of surface rock (Ritter, 1978). On cliffs, continued freeze–thaw activity at the cliff base leads to notch formation, the creation of overhangs and the eventual collapse of the caprock into the talus (Matsukura, 1990). In some areas, such as the Niagara Escarpment, there is evidence that most of the talus slope at the foot of the limestone cliffs was formed in the immediate post-glacial environment when exposed rock was completely saturated with water, making it more vulnerable to freeze–thaw cycles (Fahey & Lefebvre, 1988). Current mass-wasting from freeze–thaw cycles appears to produce mainly abundant fine-grained material, although the occasional fall of large rock occurs. Similar results have been obtained by Luckman (1976) for a variety of rock types in the western part of Canada.

From an ecological point of view, the rate of rockfall and rock particle size have a strong influence over the organisms that occur on cliffs and on talus. Stable surfaces are retained for short periods of time on substrates that weather rapidly, therefore these cliffs and cliff edges can only support short-lived organisms. Conversely, on stable cliff faces, very few rockfall events occur per unit time and there is considerable opportunity for organisms to exploit stable habitat. For example, *Thuja occidentalis* on cliffs of the Niagara Escarpment attain maximum ages over 1800 years on stable dolomitic limestones but only maximum ages of

250 years on unstable shales (Larson & Kelly, 1991; Kelly, Cook & Larson, 1994).

Chemical weathering of rock is directly controlled by precipitation amount and chemistry, rock temperature, and geochemistry. In rock cliffs composed of silicate rocks, the rate of chemical weathering is low because silicates have a very low ion-exchange capacity. Chemical contaminants of siliceous rocks weather more rapidly ($Al^{3+}$, $Mg^{2+}$, $Fe^{2+}$, $Fe^{3+}$, $Na^+$ and $K^+$) and often leave behind a matrix of less-soluble material such as quartz. The rate of chemical weathering in rock cliffs composed of clay can vary greatly, depending on the chemical composition of the individual minerals and the existing climatic conditions.

All carbonate-based rocks are susceptible to relatively rapid chemical weathering. Water dissolves calcium carbonate ($CaCO_3$) to produce two dissociated ions ($Ca^{2+}$ and $HCO^{3-}$), which are then removed by surface flow. This solubility forms the basis of all karst geomorphology (Ford & Williams, 1989) and controls the form of limestone and dolostone cliffs. In some areas, the solubility of limestone is visible as pitted and highly weathered surfaces and in others as large cave systems formed below ground that are larger-scale, long-term products of the same process.

Figure 2.6 illustrates the influence of precipitation and temperature on the physical and chemical weathering of rock and can be extended to summarize the important aspects as they relate to the rate of cliff recession. Whereas it is true that not all rock types weather at the same rates, for any one of them the relations presented in Fig. 2.6 will still be correct. Therefore, Fig. 2.6 still applies to rock generally because most areas of the globe have mixtures of rocks of different composition and origin.

Chemical weathering increases with increases in precipitation between 750 and 2500 mm per year, and also with increasing temperature. Physical weathering such as freeze–thaw activity becomes more important at lower temperatures. Sites with the combination of high rainfall and low temperature rarely exist on the surface of the earth. Several ecologically significant predictions can be made by examining the effect of temperature and precipitation on the rates of physical and chemical weathering and the global distribution of different ecosystem types presented by Whittaker (1975).

The absence of shading on Fig. 2.6 illustrates the combinations of precipitation and temperature that produce the greatest amounts of cliff stability and, in turn, produce cliff communities with the greatest persistence and therefore the greatest tendency to hold relict communities. Such cliffs will occur in desert, dry savanna, dry grassland and relatively

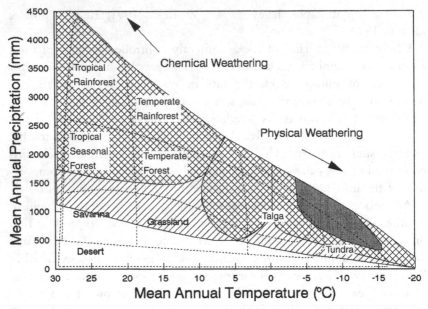

*Figure 2.6* Diagram illustrating the control of cliff stability by precipitation-dependent and temperature-dependent rock weathering rates (after Peltier, 1950; Ritter, 1978) in different biomes (after Whittaker, 1975). Chemical weathering is most rapid in places with high temperatures and high precipitation. Physical weathering is most rapid at low temperatures in places with low precipitation. By combining the rates of both processes, it can be seen that the least stable cliffs will occur in places with low temperatures and low precipitation. These conditions occur in tundra and cold taiga. More stable cliffs should occur in wet temperate and tropical forests, seasonal tropical forest, moist taiga and other tundra areas. Stable cliffs will occur in dry seasonal tropical forests and savannas, dry temperate forests and grasslands, dry taiga, and dry warm tundra. The most stable cliffs should occur in savannas and deserts, dry grasslands, and very dry taiga. Considering that productivity rises from the bottom and to the left of this figure, the most lush stable cliff communities (including macroscopic organisms) should occur most often in the area that is cross-hatched. Redrawn from Whittaker, R.N. (1975). *Communities and Ecosystems*, 2nd edn. New York: MacMillan; and Ritter, D.F. (1978). *Process Geomorphology*. Dubuque, IA: Wm C. Brown.

warm and dry tundra. The next most stable cliffs will occur in colder regions of dry savanna and warm and dry regions of temperate forest. Cliffs in tropical habitats as well as those in very cold regions having abundant precipitation should have very unstable and short-lived cliff communities. Given that habitat productivity increases with both temperature and precipitation (in an upward and left trajectory in Fig. 2.6), the results suggest that luxuriantly vegetated stable cliffs ought to be found most

often in savanna, relatively wet desert and scrubland, grassland, woodland, dry temperate forest, and dry boreal forest. Many of these predictions are easy to test at the global scale, because woody plants growing on
cliffs are available for age determinations in many countries.

## 2.4 Erosion

The erosion of cliffs is facilitated by weathering, but the forces that actually remove the materials and deposit them as talus are wind, water and
the force of gravity. In exceptionally arid regions, wind erosion can play
an important role in the formation and maintenance of cliffs. In some
instances, cliff or terrace formation can be initiated by flowing water,
then subsequently modified by wind erosion. In the Grand Canyon,
USA, a significant series of terraces and the cliffs between them were initially formed from moving water but are now sculpted by wind and ice
(Schmidt, 1987).

Sea cliffs and cliffs facing onto large lakes are eroded by wave action
(Caris, Thewessen, & Felix, 1989; Dias & Neal, 1992; Williams, Davies
& Bomboe, 1993; Komar & Shih, 1993; Jones, Cameron & Fisher, 1993).
A direct relationship exists between the average erosion rate of rocky cliffs
and the energy contained within moving wavefronts (Soons & Selby,
1982; Sunamura, 1992). Cliff recession rates for sea cliffs can be as high as
30 cm per year for mudstone cliffs in New Zealand (Healy & Kirk, 1982)
and even 300 cm per year for mudstone cliffs in Japan (Sunamura, 1992).
Only highly mobile organisms such as birds could possibly exploit such
cliffs as habitat. Ritter (1978) also reported recession rates of 25 cm per
year for chalk and 9 cm per year for sedimentary carbonate sea cliffs.
Figure 2.7 illustrates exceptionally rapid change in the profile of mudstone
cliffs in Japan (Sunamura, 1992). Such high rates only occur when there
is no protection provided by wide shore benches or platforms that absorb
wave energy. Platforms and benches that are steeply sloped (see Figs. 2.3a,
b and c) can actually accelerate erosion on sea cliff faces by breaking the
platform into rock particles which are then driven into the cliff face. Steep
cliff faces plunging great depths into the water at the shore occur on rocky
coasts where the rock surface is very hard (see Fig. 2.3c). They are especially common in locations where pre-existing vaulting, jointing or
erosion has formed the face (Trenhaile, 1987; Sunamura, 1992). Instead
of wave energy being absorbed by such surfaces, it is largely reflected.
Such cliffs can persist for many millennia without significant mass wasting.

Cliffs of the Niagara Escarpment, Ontario, Canada, include some

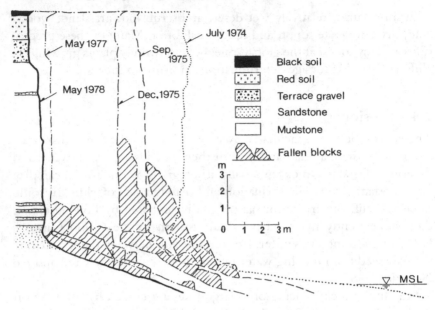

May 1977

July 1974

Sep. 1975

May 1978

Dec. 1975

| | Black soil |
| | Red soil |
| | Terrace gravel |
| | Sandstone |
| | Mudstone |
| | Fallen blocks |

MSL

*Figure 2.7* Rapid rate of cliff recession and change in cliff profile for a mudstone cliff at Kohriyama, Japan, based on studies of Aramaki (1978). Taken from Sunamura (1992), and used with the permission of the author and John Wiley and Sons, Publishers.

sections of exceptionally hard dolomitic limestone that was initially exposed by the movement of the Laurentide ice sheet, but subsequently reshaped by glacial Lake Algonquin and, later, Lake Nipissing. In some places, exposed cliffs composed of reef-forming organisms plunge to considerable depths and show no sign of current mass wasting (Tovell, 1992). Conversely, erosion of the cliff face has been more continuous in areas where sedimentary limestones and shales occur at the water's edge (Chapman & Putnam, 1973; Tovell, 1992). Discontinuities in the rock or the pattern of erosion can result in the formation of sea caves and sea-stacks (see Fig. 2.5).

Young (1972) described three types of eroded sea cliffs and proposed models to explain their formation. In the slope decline model (Fig. 2.8a), there is no inherent tendency of the free-face to maintain its edge and, consequently, it is gradually eroded to a gentle slope. The initial free-face in this case is usually the product of forces completely different from those that erode it. This model generally applies to unconsolidated rocks. In the slope replacement model (Fig. 2.8b), the angle of the talus slope is more or less maintained as the cliff face retreats, but the altitude of the contact line with the cliff face rises over time $(t_1-t_4)$ and the cliff face eventually

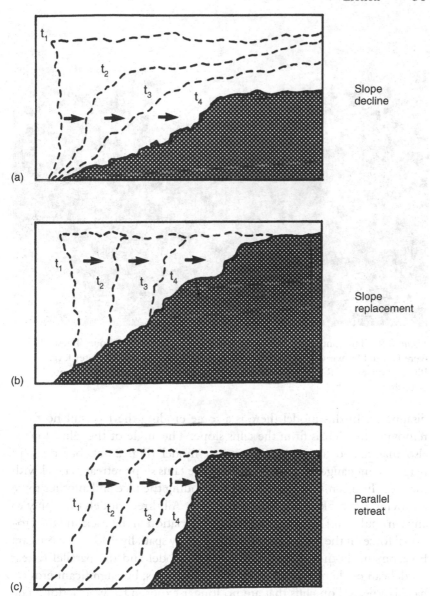

*Figure 2.8* Three hypothetical cliff recession models. In the slope decline model (a), an initial cliff is gradually eroded in both slope angle and height because the parent rock has no internal structure to resist one plane of erosion over another. In the slope replacement model (b), forces exist to cause mass wasting at the edge of the cliff only. No force exists to remove accumulated debris from the talus slope. Thus, the cliff height declines to zero over time. In the parallel retreat model (c), forces exist that cause mass wasting at the cliff edge, and mass wasting of the accumulated debris in the talus slope. Redrawn from Young (1972).

*Figure 2.9* The Jonte Gorge cut into limestone in southern France. These cliffs were formed by weathering and erosion during a time of higher water levels. Photo courtesy of R. Small, and used with permission of Longman Group, Publishers.

disappears. In this model there is a force eroding the face but no force removing the debris from the talus slope. The angle of the talus slope is also maintained in the parallel retreat model (Fig. 2.8c), but the cliff height is unchanged and the angle and the talus slope retreat parallel with the face. In this model there is a force eroding the face and another force removing the broken material in the talus. All three models can apply to any particular cliff because the structural variation in the rock and the erosional forces at the bottom of the cliff can vary spatially. Along the Niagara Escarpment, both the slope replacement model and the parallel retreat model can be demonstrated in different locations, but significant erosion has disappeared on cliffs that are no longer exposed to wave action. The rate of erosion is many orders of magnitude slower on inland cliffs than for the same or similar rocks exposed to moving water. Cliffs composed of Jurassic limestones (Figs. 2.9 and 2.10) in southern France near Millau and Meryrueis were formed by fast-flowing rivers in the early Holocene, but are no longer undergoing large-scale recession (Small, 1989).

Rivers and waterfalls are also important eroding agents that can lead to

*Figure 2.10* Close-up photograph of limestone cliffs in the Jonte Gorge, southern France, as shown in Figure 2.9. Photo courtesy of R. Small, and used with permission of Longman Group, Publishers.

cliff formation. As with moving wavefronts, rivers and waterfalls carry significant momentum and can scour rock outcrops directly, or cause the movement of debris which scours the remaining rock. One of the best-known examples of how an existing cliff face was reshaped by moving water comes from the Niagara Escarpment at Niagara Falls (Tinkler, 1986). The main cliff face of the Niagara Escarpment has retreated by no more than 10–20 m since the melting of the Laurentide ice sheet 11 000 years ago, but where the Niagara River crosses the Niagara Escarpment, the cliff face has retreated over 10 km at an annual rate of 1–2 m per year (Fig. 2.11). At present, human intervention has reduced this rate to 3 cm per year. Most waterfalls retreat in a similar fashion, such as Victoria Falls between Zambia and Zimbabwe and Angel Falls in Brazil (Small, 1989), although local variation in the strength of the bedrock can exert a strong influence on the uniformity of the process over space and time.

Regardless of the agent responsible for the momentum transfer, cliff-face erosion can result in one of three types of mass movement: a fall, a topple or a slide (Trenhaile, 1990; Sunamura, 1992; Fig. 2.12). Rockfalls occur when there is some degree of undercutting below the caprock (Fig. 2.12a). Weathering underneath the caprock weakens its internal structure, and the cliff edge collapses (Matsukura, 1990). Such caprock undercutting is extremely common on sea cliffs formed of sedimentary limestones. Falls of this type are more or less dependent on horizontal bedding planes or joints in the rock that form natural lines of weakness that are revealed when the weaker underlying rocks are eroded by smaller-scale and more frequent rockfall. Topples of rock face (Fig. 2.12b) occur when the joints in the rock are vertically inclined. Such vertical weaknesses may be present because sedimentary rocks are repositioned by faulting, or because of differential vertical weathering of horizontally bedded rock. If the cause of the topple is vertical weakness, the plates of bedrock that separate from the main face will often be thinner and more sheetlike than in cliffs where the weakness plane is caused by weathering. A rock slide from a cliff face (Fig. 2.12c and d) can be the result of a slope failure in the underlying strata, a toe failure, or a base failure. The degree of cliff sliding or slumping can be highly variable over space and time.

## 2.5 Bedrock hydrology

The fundamental principle of bedrock hydrology is that water will flow from recharge areas where it accumulates to discharge areas where it is

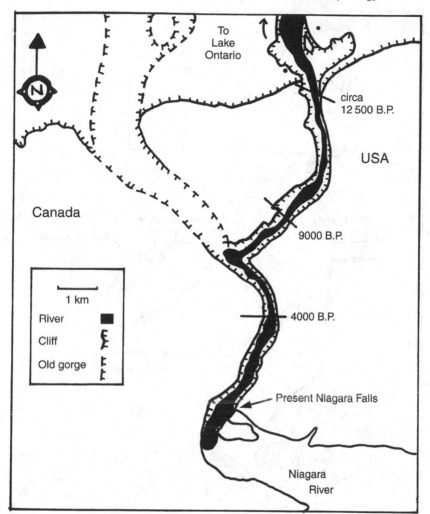

*Figure 2.11* Map of the Niagara Peninsula, Ontario, showing the recession of Niagara Falls over the last 12500 years. The cliff edge has receded over 7 km in this time. Current rates of recession are kept low by artificial flood control. Illustration modified from Tinkler (1986).

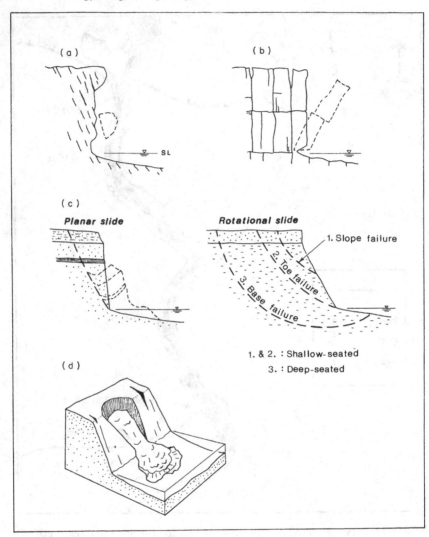

*Figure 2.12* 1. Three types of slope movement on rocky cliffs: a rock fall – often involving caprock (a); 2. a topple – involving undercut and collapse of both vertically and horizontally jointed rock (b); and 3. a slide (c) or slump (d) involving shallow or deep failures of the internal structure of the rock. Such collapses can be derived from slope, toe or base failures. Illustration courtesy of Sunamura (1992), and used with permission of John Wiley and Sons, Publishers.

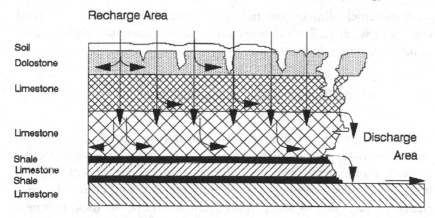

Recharge Area

Soil
Dolostone
Limestone
Limestone
Shale
Limestone
Shale
Limestone

Discharge
Area

*Figure 2.13* Bedrock hydrology of hypothetical horizontally layered bedrock with an associated cliff face. Positive hydraulic pressure redistributes percolating rainwater towards the cliff face and the discharge area towards the talus. The driest parts of the habitat are the exposed sections of rock at the cliff edge where loss of water due to evaporation is rapid. Such edges represent local deserts, while adjacent cliff faces have water continuously available. Springs of water occur more frequently above impervious strata such as shales.

released. For cliffs, the pattern of this movement has specific effects on the biota that exploit this moving water. For example, on level ground away from cliffs, the hydraulic pressures in bedrock will be more-or-less symmetrical in area (Fig 2.13). In cliff environments, hydraulic pressures will result in the movement of water towards the face, except near the cliff edge where access to percolating water is limited. Cliff faces are thus continuously supplied with water through cracks and seeps. These hydrological patterns allow us to make predictions about the distribution of plants with different patterns of dependence on water. Cliff edges should be drier than cliff faces, therefore desiccation-tolerant plants should be more frequent in these habitats. Talus slopes, in contrast, will be sites where the water is continuously released, creating small freshwater springs. Thus, all of the rocks and the air chambers among them will be permanently moist and cool deep within the cone of debris and warm and dry near the surface.

Water flow can also occur on the surface of the rock, in the immediate subsurface or soil, or it can follow existing joints in the rock. Whereas the direction and the volume of the flows can be observed and measured easily for surface and near-surface flows, water moving in joints can only be observed by using dye tracers. The inherent lag times in water movement from recharge areas to the water table to discharge areas make this work extremely difficult and time consuming (Small, 1989). The total volume of water available to move can be determined roughly by examining the water budget equation:

$$P = E + R + S$$
where $P$ = precipitation, $E$ = evaporation, $R$ = runoff, $S$ = storage.

In climate zones where evaporation is greater than precipitation, there will be less water percolating to the water table, and as a consequence the movement of minerals to plants living in or on the vertical rock will also be slower. Where evaporation is less than precipitation, cliffs will experience a greater flux of water through the matrix of the rock. It is predicted that plants rooted in cliffs should be more nutrient depleted compared to level-ground habitat where there are microbes that fix nitrogen and remobilize potassium and phosphorus. These patterns will be greatly modified in situations in which the rock outcrop receives water from a watershed in a different climatic zone. The storage of water directly in rocks is very low (see Chapter 3) and therefore most of the important factors that influence hydrological patterns are controlled by the bedding planes and jointing patterns in the rock.

Bedrock hydrology is also influenced by rock type. Whereas igneous and sedimentary rocks differ greatly in their initial geological structure, significant weathering and seismic activity create fractures and joints. In sedimentary rocks, bedding planes start off being oriented horizontally, but rock falls and topples that occur are often along vertical lines of weakness imposed by faulting and weathering. The variance in structure, strength and fracturing in all rock types appears to increase over time, so that igneous and carbonate rocks may end up having similar bedrock geology even though their initial structures were very different.

The area of greatest difference in the hydrology of rocks is in the area of karst geomorphology. This area has been covered extensively by Ford and Williams (1989) and White (1988) and is not covered here in any detail, but the important connection between cliffs and karst landscapes is that limestone cliffs are found in most locations where karst is present. Ford and Williams (1989) present a map of the location of terrestrial carbonate rock outcrops around the world (Fig. 2.14). Limestone cliffs are abundant in many of these locations.

The susceptibility of limestone rocks to dissolution makes them vulnerable to surface erosion that produces cliffs. In the late stages of karst development, the limestone plane is often dissolved except for remnant

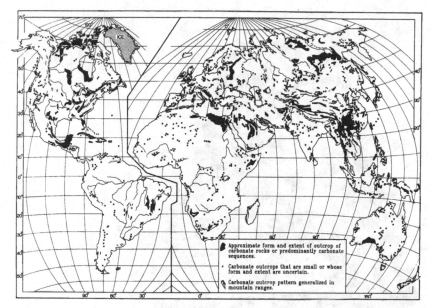

*Figure 2.14* The location of carbonate rocks around the world. Limestone cliffs are often a component of the landscape in these areas. Illustration courtesy of D.C. Ford, taken from Ford and Williams (1989) and used with permission of Unwin Hyman, Publishers.

limestone pillars or towers. These towers are particularly common in south east Asia and China (Fig. 2.15). Karst towers usually have caves which form at the cliff base. Episodes of rapid erosion produce a succession of caves up the tower as the surrounding landscape is removed. Ford and Williams (1989) called these towers time-transgressive landforms and surmised that increasingly younger flora and fauna are found on the cliff towards the cliff base. This prediction, however, has never been tested. Coastal limestone cliffs are also prevalent karst features in these areas (Fig. 2.16).

## 2.6 Summary

Cliffs assume a wide variety of forms, influenced by factors such as rock type and strength, climate and the processes of physical and chemical weathering. Freeze–thaw activity is an important physical weathering process in cliff environments. Chemical weathering is important in carbonate-based rocks where cliffs are often formed through the dissolution of limestone and dolostone. The inherent physical characteristics of the

*Figure 2.15* Inland tower karst development in China. Such outcrops are the result of dissolution of the entire limestone plane over 100–300 million years. Photo courtesy of D. Ford, and used with permission of Chapman and Hall.

rock determine its susceptibility to weathering processes, and climate controls the rate at which these processes act. Wind, water and gravity are the principal agents of erosion. The geological structure of the bedrock also controls the underlying hydrology, which in turn controls the availability of moisture in the cliff environment.

Cliff faces result in less hydraulic pressure retaining water within rock, and as a result they tend to be places where liquid water is more consistently found than in the rest of the surrounding habitat types. In contrast, the water movement through the rock will result in cliff edges being very

*Figure 2.16* Coastal tower karst development in Viet Nam. Lushly vegetated cliffs occur on these limestone towers, but areas exposed to tidal activity are devoid of most macroscopic plants. Photo courtesy of A. Matheson.

dry. Talus slopes will accumulate the water that is expressed from cliff faces and, because of the large accumulations of debris, will keep this water protected from sudden changes in temperature. They will therefore have humid and cool air chambers within them where the size of the living space for organisms is controlled by the particle size of the talus-forming materials.

# 3 · *Physical environment*

The vertical orientation of cliffs represents the primary (if obvious) difference from other landscape types, but verticality affects the environmental conditions on cliffs in a number of important ways that are not always obvious to ecologists who study the microclimate of level ground or slopes. The function of this chapter is to point out the various ways in which the physical environment of cliffs is distinct from that of horizontal surfaces. Its purpose, therefore, is not to give a complete account of all components of the physical environment. The reader is referred to standard texts, such as Monteith and Unsworth (1990), Arya (1988), Oke (1987) and Gates and Schmerl (1975), for basic information on microclimate and energy balance.

The first subsection briefly outlines the various components of the physical environment that are affected by vertical orientation of the substrate, and shows how these factors are interconnected in a complex way to make the cliff environment drastically different from surrounding level ground. More detail on each of these factors is then provided in the subsections that follow.

## 3.1 The effects of vertical orientation

Vertical orientation affects the total amount of direct radiation a surface receives and the way radiation input varies diurnally, seasonally and latitudinally. It also affects wind speeds on the surface and the amount of direct precipitation received. Radiation, wind and moisture together control the temperature of the rock. Absorption of radiant energy increases rock temperature, while wind speed controls the amount of energy that is dissipated by the heating of air and the evaporation of moisture, thus cooling the cliff surface.

The orientation of the force of gravity relative to the habitat surface is the basis for a second complex of interrelated factors that make the cliff environment distinct. Under normal circumstances, the force of gravity is perpendicular to the surface, acting to stabilize the habitat. On the surface of cliffs and steep slopes, gravity works to destabilize the habitat and remove all loosely attached material including rock, litter, soil, water and nutrients. The result is a habitat that can be expected to be well drained (although not necessarily dry, as shown below), nutrient poor, unstable, and incapable of significant soil development.

The absence of soil has further consequences beyond the inability of the habitat to accumulate water and nutrients. An exposed rock surface of any given slope or aspect can represent a more extreme thermal environment than a soil-covered surface with the same exposure. These extremes are due to the fact that the energy required to raise the temperature of a substrate is highly dependent on its moisture content, and rocks generally contain much less water than soil does. The absence of significant vegetation cover on cliffs further contributes to the tendency of cliff surfaces to show large fluctuations in environmental conditions with an amplitude, frequency and duration determined by the surrounding air mass. Finally, all the above factors are further modified by heterogeneities in the cliff surface, and by the degree to which solid rock has disintegrated.

## 3.2 Incident radiation and its controls

The total amount of radiant energy incident on any surface, whether horizontal or vertical, is directly dependent on the angle of incidence of the sun's rays on that surface. The nearer this angle is to the perpendicular, the greater will be the amount of energy received (Fig. 3.1). Of the two components that make up total irradiance, diffuse radiation is normally much smaller in magnitude and much less influenced by the angle of incidence than direct radiation. For this reason, diffuse radiation is not as important for differentiating between the radiation environment of cliffs and horizontal surfaces. The greater the proportion of total flux that is direct beam radiation, the more the angle of incidence will matter and the more the amount of radiation received by cliffs will differ from that received by horizontal surfaces. Thus, the radiation environment of cliffs will differ most from that of flat land in geographical areas where cloud cover is rare.

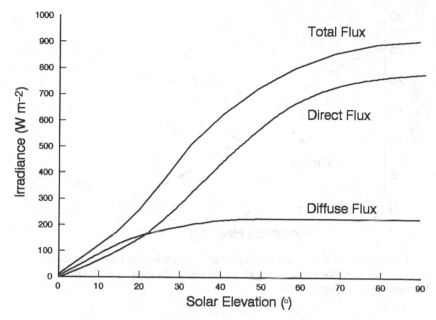

*Figure 3.1* The dependence of irradiance on the angle of solar elevation for the total flux, the direct flux and the diffuse flux. Values are for a cloudless day at 53°N latitude. Redrawn from Monteith and Unsworth (1990).

The angle of incidence of the sun's rays on a surface is controlled by three factors: the position of the sun in the sky, the aspect of the surface, and the slope of the surface. The position of the sun in the sky depends on the latitude, the date, and the time of day. Knowing these three parameters will allow one to predict the radiant flux density of a horizontal surface; this is how solar radiation is conventionally measured. These relationships are well known and can be found, for example, in Monteith and Unsworth (1990).

For surfaces that are not horizontal, the relationships are much more complicated because of two more factors that affect the angle of incidence: slope and aspect (Garnier & Ohmura, 1968). All these factors – slope, aspect, latitude, date, and time of day – interact to control the angle of incidence of the sun's rays, and thus the amount of radiation received by a cliff. It is almost impossible to present such a multivariate problem in a way that is not confusing, but we will try to illustrate the relationships from three different points of view, using three different graphs. These examples will be for the northern hemisphere; the relationships

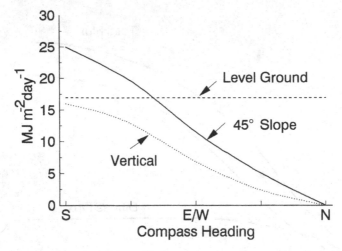

*Figure 3.2* The effects of different slopes and aspects on the total daily radiant flux received at 45°N latitude on a cloudless day at the equinoxes. Redrawn from Monteith and Unsworth (1990).

would be the same for the southern hemisphere, except that northern and southern aspects would be reversed.

Figure 3.2 shows the daily integral of direct irradiance for a specific latitude (45°N) and date (the equinoxes) as a function of slope angle and aspect. Under these circumstances, a south-facing vertical cliff receives 16 MJ m$^{-2}$ day$^{-1}$ of direct radiation, which is slightly less than the amount received by level ground. An east-facing or west-facing cliff receives only about half as much as level ground, while a north-facing cliff receives no direct radiation at all. In contrast, a south-facing 45° slope will receive about one-third more radiation than level ground, and an east-facing or west-facing slope will receive less than level ground but more than a cliff of the same aspect. Thus, all mid-latitude cliffs, regardless of their aspect, receive less radiation at the equinoxes than slopes or level ground.

Moving closer to the equator, the greater sun angle at the equinoxes would lead to level ground receiving more, and south-facing vertical cliffs receiving even less direct radiation. East-facing and west-facing cliffs near the equator would still receive less radiation than level ground, but more than south-facing cliffs because they face the rising or setting sun at an angle close to the perpendicular. The only circumstance under which vertical cliffs would receive more direct radiation than level ground would be near the poles. Here, the sun angle would be greater for south–

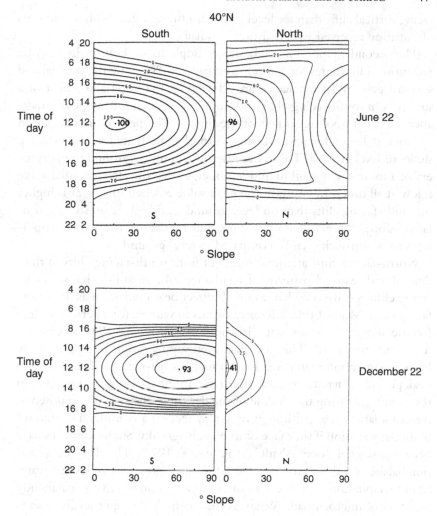

*Figure 3.3* Isopleths of hourly direct solar radiation at a constant latitude (40 °N) in mid-summer (June 22) and mid-winter (December 22) for slope angles 0° to 90° (i.e. level ground, slopes and cliffs) facing north and south. Isopleths are given as cal cm$^{-2}$ h$^{-1}$; 1 cal cm$^{-2}$ h$^{-1}$ = 0.0419 MJ m$^{-2}$ h$^{-1}$. Redrawn from Buffo, Fritschen and Murphy (1972).

facing vertical cliffs than for level ground, although the absolute amounts of radiation received would always be small.

The second graph (Fig. 3.3) shows isopleths of hourly direct solar radiation. It illustrates for a specific latitude (40°N) how the diurnal and seasonal patterns of irradiance are affected by the slope and aspect of a surface. On south-facing cliffs and slopes of all angles, daily peak irradiance occurs at noon regardless of the season. The peak value reached in summer is lower for south-facing vertical cliffs than for south-facing slopes or level ground. This is because the sun moves around the periphery of a south-facing cliff in mid-summer, so that the effective solar angle is low at all times. In contrast, the peak value reached in winter is higher on south-facing cliffs than on level ground, and highest on steep south-facing slopes. Direct-beam radiation thus moderates the winter conditions on south-facing cliffs, compared to level ground.

North-facing cliffs are quite different from south-facing cliffs in their diurnal and seasonal patterns of irradiance. On an annual basis, north-facing cliffs receive only half as much direct beam radiation as do south-facing cliffs. Most of this difference occurs in winter: for a period of time (whose length increases with latitude), north-facing cliffs receive no direct radiation at all. This poor radiation supply results in a particularly hostile winter microclimate. In summer, north-facing cliffs have two peak periods of irradiance, one occurring in the morning and another in the evening. During the middle of the day these cliffs are not exposed to direct radiation at all, although they may receive a considerable amount of diffuse radiation if they face onto open bright sky. Such exposures have been described as 'open-shade' (Stoutjesdijk, 1974). The daily net radiation balance on these cliffs may be sufficiently close to zero so that water from precipitation or dew does not evaporate, resulting in a permanently moist, cool microclimate. Whereas the open-shade superficially resembles the shade below leaf canopies, it is much more enriched in blue light and has also been called 'Blauschatten' ('blue shade') for this reason. Because photosynthetically active wavelengths are present in much higher proportions in the open-shade compared to the shade under a canopy, the open-shade is a more favourable environment for plant growth.

Figure 3.4 illustrates how latitude will modify the diurnal and seasonal patterns of irradiance on south-facing and north-facing vertical cliffs. The amount of radiation received in the summer will increase as latitude increases, both for south-facing and north-facing cliffs. North-facing cliffs will receive their twice-daily peaks of irradiance earlier and later in

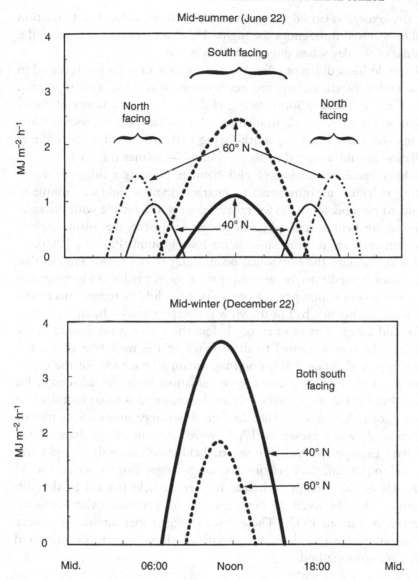

*Figure 3.4* Daily course of direct solar radiation flux in mid-summer (June 22; top panel) and mid-winter (December 22; bottom panel) at two latitudes (40°N, solid lines; 60°N, broken lines) for north-facing vertical cliffs (narrow lines) and south-facing vertical cliffs (wide lines). Data from Buffo *et al.* (1972).

the day, except at latitudes above the arctic circle where direct radiation will be received throughout the night. This leaves a longer period in the middle of the day when direct radiation is zero.

Latitude has a different effect on the amount of radiation received in mid-winter. North-facing cliffs never receive direct radiation in winter, regardless of latitude. South-facing cliffs receive fewer hours of direct radiation with lower peak irradiance values as latitude increases. At low to mid-latitudes, however, south-facing cliffs may receive more direct radiation in mid-winter than they do in mid-summer (Fig. 3.4).

The vertical orientation of cliff habitats presents a dilemma when taking radiation measurements to characterize the light environment. Standard methodology calls for measurements to be made with the radiation sensor held horizontally, but the environments that plants experience on cliffs are not the same as on level ground (Proctor, 1980). In non-cliff habitats, the horizontal positioning of the sensor ensures that the sensor is parallel to the ground surface so that radiation is integrated over the whole hemisphere of open sky. On cliffs, however, this methodology fails because half of the view is always occluded by the cliff (Fig. 3.5a) and a large part of open sky below the sensor is not included. By holding the sensor parallel to the surface of the rock (the plant's eye view, if you like: Fig. 3.5b), one may obtain a better idea of the exposure to clear sky and backscattered radiation from the adjacent talus slope, but the radiation values will not be the same as those found when the sensor is held horizontally. In fact, when measurements of photosynthetically active radiation (PAR) were taken in 60 quadrats on the Niagara Escarpment with the sensor held simultaneously upright and parallel to the cliff, the horizontal readings ranged from 5 per cent to 33 per cent of full sunlight, whereas the sensor held parallel to the cliff showed values between 28 per cent and 66 per cent (Matthes-Sears, Gerrath & Larson, 1997). These results suggest that significantly more radiation reaches the plants that occur on cliffs than can be measured using standard methods.

### 3.2.1 Radiation within rock

Recent studies have shown that a rich community of organisms, belonging to many phyla, may be present below the surface of seemingly bare rock faces. These endolithic communities occur in many different types of rock, including sandstone, quartz, flint, limestone and granite. They include many photoautotrophs, whose distribution within the rock matrix is strongly controlled by the available light. For this reason, a brief consideration of the radiation penetrating cliffs is needed here.

*Figure 3.5* Hemispherical (or fish-eye) photographs demonstrating the effect of sensor orientation on the amount of radiation measured on a cliff face. These photographs show the whole hemisphere over which radiation is integrated when a sensor is held in the same orientation. The top photograph (a) was taken with the lens oriented horizontally; the bottom photograph (b) with the lens oriented vertically and parallel to the cliff face. Photos by B. Gildner.

The form of the attenuation of light that passes through any medium is expressed in the Beer–Lambert equation:

(1) $I_e = I_o \cdot e^{-Kl}$

where $I_e$ = penetrating radiation, $I_o$ = incident radiation, $K$ = substrate dependent extinction coefficient ($mm^{-1}$), $l$ = thickness of the medium (mm), and e = base of the natural logarithm.

Photosynthetic organisms can exist only at depths ($l$) below the surface where the amount of radiation available to them ($I_e$) is large enough so that the carbon gained through photosynthesis exceeds the carbon lost through respiration. According to the above equation, this depth depends on only two variables. The first is the amount of incident radiation, $I_o$, whose controls have been discussed in detail in the previous section. The second is the extinction coefficient ($K$) of the rock.

Methods used to determine $K$ for different types of rock have included the use of rock chips or pebbles (Broady, 1981; Bell, 1993); thin serial sections (Berner & Evenari, 1978); or drilling a hole just large enough to hold a sensor from the bottom of large rock pieces, leaving progressively thinner overlying layers of rock (Matthes-Sears et al., unpublished). These measurements have shown that light attenuation through rock can be quite variable. Quartz pebbles ranging from 13 to 80 mm in thickness transmitted between 0.6 and 2.7 per cent of incident sunlight; however, some of this light may have penetrated through the sides of the stones (Broady, 1981). For light opaque and dark transparent flint, Berner and Evenari (1978) found that about 30 and 60 per cent (averaged for two different wavelengths) of incident light, respectively, penetrated to 1 mm below the surface. At 5 mm depth, 0.5 per cent of incident light was still present in light opaque flint and 14 per cent in dark transparent flint (Fig. 3.6). Sandstone and dolostone seem to be much harder for light to penetrate: 1–3 per cent of incident radiation was present at 0.2–1.2 mm depth in white sandstone, and less than 1 per cent in brown sandstone (Bell, 1993). Dolostone from the Niagara Escarpment showed penetration of only 0.5 per cent of incident light to 1 mm depth, and radiation was rarely detectable at depths greater than 4–5 mm (Matthes-Sears et al., unpublished).

The value of $K$ is dependent on the wavelength of the light, with greater transmission of the longer wavelengths compared to shorter ones (Berner & Evenari, 1978; Fig. 3.6). Properties of the rock itself that determine light penetration include rock colour, with lighter rocks trans-

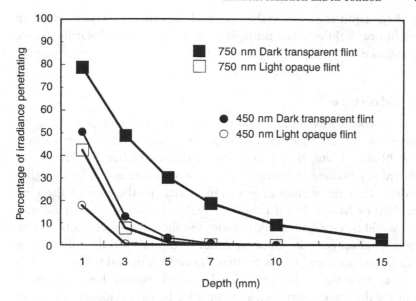

*Figure 3.6* Percentage transmission of two different wavelengths of incident radiation to various depths below the surface of two types of rock, dark transparent flint (solid symbols) and light opaque flint (open symbols). Squares are for 750 nm, circles are for 450 nm. Data from Berner and Evenari (1978).

mitting more light than dark ones (Bell, 1993). Transparency also plays an important role, which explains why dark–transparent flint let through more light than light–opaque flint (Berner & Evenari, 1978; Fig. 3.6). In addition, crystal size and porosity differ between different types of rock and this may affect their light-penetration characteristics (Bell, 1993). Finally, internal fracturing and small heterogeneities within the rock can also affect the light transmission of a rock sample. This last factor may explain why there is considerable variation among different samples of the same material (Matthes-Sears *et al.*, unpublished).

Endolithic organisms can live either completely within the solid rock matrix (cryptoendoliths), or in cracks and fissures below the surface where connections to the surface exist (chasmoendoliths). From the evidence presented above, the maximum depth at which cryptoendolithic photoautotrophs can exist is rarely more than 5 mm and will be controlled by the combination of surface irradiance and rock characteristics. The depth is expected to be largest where surface irradiance is high (e.g. south-facing cliffs in the northern hemisphere) and where the rocks are light, transparent and porous. Data from Niagara Escarpment dolostone suggest that chasmoendolithic organisms extend deeper into the rock but that their depth is more variable, probably because of variable amounts of additional light received through pores to the surface (Matthes-Sears, unpublished).

## 3.3 Moisture

Cliffs are commonly thought of as dry, desert-like habitats. Although a number of authors have stated this (Curtis, 1959; Maycock & Fahselt, 1992; Nuzzo, 1996), they have rarely clarified whether they mean that cliffs are dry because the surrounding macroclimate is dry, or that cliffs are drier than the surrounding macroclimate. In the case of the cliffs described by Maycock and Fahselt (1992), it is easy to understand why they would be classified as dry because the site experiences only 64 mm of precipitation per year. In many other cases, the cliffs described are in more humid areas and it may be that the authors intend to say that cliffs are drier than the surrounding level ground. Rarely have statements about the dryness of cliffs been accompanied by hard evidence to support this idea. We suspect that in many cases the claims are simply based on the observation that cliffs have no soil and their surface appears bare. Oettli (1904), in a comprehensive investigation of the ecology of cliffs, showed clearly, by analysing the details of the microsites where cliff plants grew, that water was rarely limiting the growth of higher plants. In this section we review the literature relevant to the question of whether cliffs are dry in relation to the surrounding landscape.

### 3.3.1 Direct precipitation

The three major factors controlling the moisture relations of cliffs are the total amount of precipitation in the geographical area, the seasonal distribution of this precipitation, and the moisture-retaining properties of the substrate. In areas where precipitation is distributed over the whole year, wet/dry cycles are short. In areas with seasonal precipitation such as savanna, cliffs are permanently wet for the rainy season, then completely dry for many months except for periods of time with dew and high atmospheric humidity. Dew may be an important source of moisture for rock surfaces, especially in the tropics where it is more frequent and occurs in greater quantity than in temperate zones (Zehnder, 1953). Whereas the macroclimatic influences are the same for vertical cliffs and for level ground, there are some important ways in which cliffs are different and these will be discussed in more detail.

Vertical cliffs always receive less direct precipitation per surface area than the level ground surrounding them (Lundqvist, 1968). In the absence of wind, the amount of rain or snow received per square metre of cliff or slope surface depends on the slope angle. The amount of precipitation received by a sloped surface is obtained by multiplying the amount received by level ground, $P_{potential}$, by the cosine of the slope angle, $\Theta$:

$$(2) \qquad P_{actual} = P_{potential} \cos \Theta$$

The steeper the angle, the more $P_{actual}$ will be reduced from $P_{potential}$, and for cliffs approaching the vertical, the amount of direct precipitation received will approach zero. For cliffs with overhangs, a rain shadow develops between the base of the cliff and the vertical projection of the edge of the free-face (Fig. 3.7). In the presence of wind, the impact of precipitation on the cliff surface will be dependent on wind speed and direction (Fig. 3.7). Wind that is directed away from the cliff face will increase the extent of the rain shadow and decrease the slope angle above which the surface receives no precipitation. The amount of this change is proportional to the wind speed. Wind directed at the cliff face will have the opposite effect.

Any precipitation intercepted by the cliff face that is not immediately absorbed by the rock will run off due to the effects of gravity. This is different from level ground, where precipitation that cannot penetrate the ground will pool and infiltrate slowly, providing moisture over a longer period of time. Similar considerations apply for snow and ice. Cliff faces and edges of the Niagara Escarpment accumulate no snow in winter (Bartlett, Matthes-Sears & Larson, 1990; 1991a; Fig. 3.8a). Sometimes, however, small seeps or aquifers can result in water being permanently expressed to frozen cliff faces during the winter, resulting in substantial accumulations of ice on these cliffs (Fig. 3.8b). Equally, cliffs adjacent to waterfalls can become coated with massive amounts of ice. The weight of this ice can result in the failure of the rock or the removal of vast amounts of accumulated vegetation when the ice sheets fall during the spring.

The ability to absorb, hold and transfer water varies tremendously among different rock types, and few cliffs consist of homogeneous material. Igneous cliffs tend to be more homogeneous than sedimentary or metamorphic cliffs, but even they invariably contain heterogeneities that greatly complicate any consideration of water relations on cliffs. For cliffs formed of sedimentary rocks, impervious layers of shale or clay often result in water seeping from cliff faces as small springs and seeps.

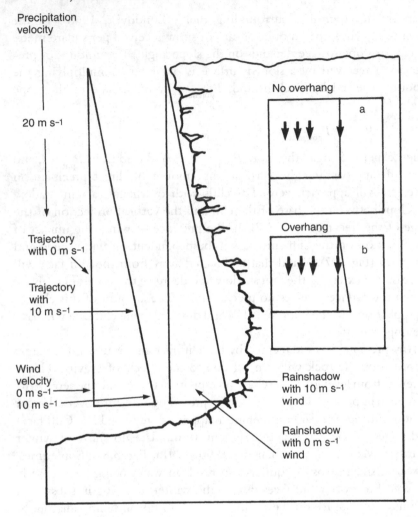

*Figure 3.7* Diagram illustrating the combined effects of cliff angle, wind speed, and wind direction in determining precipitation interception by cliff surfaces. When cliffs are completely vertical and have no overhang (inset a), direct precipitation will not contact the cliff face, but will contact all of the talus. When the cliffs include overhangs (inset b), the cliff again receives no direct precipitation but there is a rainshadow effect that also keeps the talus dry. Only when there is sufficient wind to carry the water to the base of the cliff will the talus or the cliff face receive direct precipitation. Original artwork by C.E. Ryan.

(a)                                             (b)

*Figure 3.8* (a) Cliffs of the Niagara Escarpment near Milton, Ontario, in winter. Unlike the surrounding level ground, cliff faces are snow free. (b) Ice accumulation on a granite cliff in northern Ontario. Photos by D.W. Larson.

The availability of liquid water can thus be highly variable at extremely small spatial scales across the cliff face in ways that are completely independent of incoming precipitation or evaporation.

### 3.3.2 Water availability from rock
The maximum volume of water held per volume of rock, or water storage capacity, is generally one to two orders of magnitude less for rock than for soil. Whereas sandy soils hold around 25 per cent of water by volume when fully saturated, and clay soils as much as 60 per cent, solid rock can hold only of the order of a few per cent (Lewis & Burgy, 1964; Jones & Graham, 1993). The maximum water content for different types of rock ranges from 0.05 per cent of dry weight for chert to 0.09–0.28 per cent for different limestones, 0.18 per cent for basalt, 0.13–1.2 per cent for different granites, and 1.59 per cent for schist (Rejmánek, 1971; Jones & Graham, 1993). Different types of rock have different water-storage capacities because of variation in porosity and weathering characteristics affecting drainage (Rejmánek, 1971). For example, vertical

jointing of basalt is one of the reasons for its dryness, but granite can be more wet when large fractures are absent (Rejmánek, 1971). Because weathering has a large effect on drainage and porosity, water-storage capacity varies more with weathering state than with any other factor. Solid granitic rock holds 3.1 per cent of water by volume when weakly weathered but may hold up to 27 per cent when highly weathered (Jones & Graham, 1993). The effects of rock type and weathering are further illustrated by Pentecost (1980), who compared two volcanic rock types: a hard, impervious ordovician rhyolite tuff, and a softer pumice tuff inter-mixed with calcareous sedimentary rock. Unweathered pumice tuff held twice as much water as unweathered rhyolite tuff and also absorbed and lost water faster. However, weathering more than doubled the water-holding capacity of pumice tuff, and quintupled that of rhyolite. Weathering is probably also responsible for the relatively high water-holding capacities of rock fragments within soils. For example, maximum water contents of 10.6 per cent have been reported for siltstone fragments (Montagne, Ruddell, & Ferguson, 1992), 5.9 per cent for sandstone/silt-stone fragments, and 17.3 per cent for shale fragments (Hanson & Blevins, 1979).

For organisms (such as plants) to obtain water from a substrate, it is not the water-storage capacity alone that matters (Brady, 1974). The tension with which the water is held, or matric potential, is also important because it determines how easily this water can be extracted. For example, although clay soils can hold large amounts of water, a significant part of this water is not accessible to plants (Fig. 3.9). Plant-accessible water is generally defined as water held at water potentials between −0.01 MPa (the maximum amount of water that can be held against gravity, or field capacity), and −1.5 MPa (the permanent wilting point for many mesic plants). The amount of water per unit volume or dry weight of soil that is held within this water potential range is called available water content. Whereas rocks have a small water-storage capacity compared to soils, they hold water at a low tension so that part of it is plant accessible. As a consequence, rock as a substrate is not as unfavourable for plants as its low water content would suggest. If hydrological conditions are such that the cliff surface remains hydrated over long periods of time, a small but steady amount of water remains available for plants.

The amount of water required to hydrate a given volume of rock fully is much smaller than that required to saturate the same volume of soil to the point at which plants can withdraw water from it. Under dry conditions, rocks can therefore be a more favourable substrate than soil.

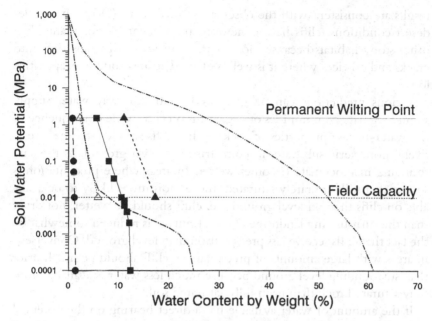

*Figure 3.9* Water release curves, showing matric potential as a function of gravimetric water content, for different types of soil and rock. Values for sand (dashed-dot line) and clay (dashed-two dot line) soils are taken from Brady (1974); for solid, very weakly weathered granitic rock (large dots) from Jones and Graham (1993); for siltstone fragments (solid squares) of size 45 × 45 × 15mm from Montagne *et al.* (1992); and for sandstone/siltstone fragments (open triangles) and shale fragments (solid triangles) of diameter 20–35 mm from Hanson and Blevins (1979). Water retained in the substrate between field capacity (−0.01 to −0.02 MPa) and the permanent wilting coefficient (−1.5 MPa) is said to be usable by plants and is called plant-available water. Matric potential is plotted on a logarithmic scale.

Although the authors know of no studies of the water potential of cliffs, there is evidence from work on shallow or stony soils to support these ideas. Fractured bedrock beneath shallow soil in xeric ecosystems may hold more plant-accessible water than the overlying soil (Jones & Graham, 1993), and many chaparral and conifer species are known to exploit such bedrock for substantial portions of their water supply (Hellmers *et al.*, 1955; Stone & Kalisz, 1991). Rock fragments or large stones can improve the water relations of soils and play a significant role in the water budgets of plants in semi-arid regions. Water is held not just between these fragments but also within them (Hanson & Blevins, 1979; Ashby *et al.*, 1984; Flint & Childs, 1984; Montagne *et al.*, 1992). These results are consistent with the observation of Zohary (1973) that under desert conditions, cliffs have conditions more favorable for plants than other stony habitats because most of the rain water that falls runs into cracks and crevices where it is well protected against sudden evaporative loss.

Various predictions can be generated about the way water supply should regulate communities of organisms on cliffs. One consequence of the water-release properties of rocks is that cliffs should become a relatively more xeric substrate, in comparison to level ground, as the surrounding macroclimate becomes wetter. In areas where precipitation is so low that soils are rarely saturated, more moisture is likely to be available on cliffs than on level ground, i.e. cliffs should be wetter, not dryer than the surrounding landscape. This advantage is reduced somewhat by the fact that cliffs receive less precipitation than level ground. Conversely, in areas with large amounts of precipitation, cliffs should be much drier than surrounding level ground because much less water is available from fully saturated rock than from fully saturated soil.

If the amount of water available has a direct bearing on the potential for high levels of above-ground productivity, then we can use the above information to arrive at the following predictions. Cliffs should never be able to support a large biomass even in humid areas, but the biomass supported by cliffs should be relatively constant from region to region in comparison to the biomass of surrounding level-ground communities. Cliffs in arid or semi-arid areas should have similar biomass to cliffs in humid areas, even though the biomass in the surrounding vegetation changes drastically. Although this makes intuitive sense, it has never been formally tested.

Related to this, cliffs should not be good habitats for plants that require large amounts of water. This may account for the facts that there are relatively few trees on cliffs, that the trees that tend to grow there are small and slow growing, and that many of them are sclerophyllous shrubs, ericads or conifers, all of which have low transpiration or stem conductance rates. Ferns are also extremely common on cliffs in all countries, and it is interesting to note that these plants require the presence of liquid water for fertilization, but they do not require abundant water for growth. In fact, Woodhouse and Nobel (1982) have shown that for the many species of ferns, the conductance values for whole plants or for stipes alone were up to six orders of magnitude lower than for conifers, and two to three orders of magnitude lower than for dicotyledonous plants. They also noted that small tracheid numbers and narrow tracheid diameters are the factors that limit stem hydraulic conductance, plant size and growth rate. Cliffs, therefore, provide an environment with exactly the right conditions to make liquid water freely available over short periods of time, while also providing cool, low-radiation, humid conditions (Bunce, 1968: Fig. 3.10) that prevent hydraulic pathways from cavitating. Because the water on cliffs is relatively easily available, plants on cliffs (with the exception of sea cliffs) do not require the ability to develop large suction pressures and thus no special root adaptations should be necessary.

### 3.3.3 Moisture relations of sea cliffs

Sea cliffs differ from inland cliffs in one important aspect of their water relations. For inland cliffs, matric suction alone (as influenced by texture, structure and porosity) determines the quantity of water the substrate can supply to plants. For sea cliffs, osmotic effects add to these matric effects, resulting in an increase in the amount of suction required to withdraw water and effectively reducing the range of available moisture. The osmotic effect of sea water (with 3 per cent NaCl) amounts to $-2.7$ MPa and thus imposes significant levels of drought (Fitter & Hay, 1987). The osmotic factor is due to the presence of salts that are deposited on sea cliffs by waves or spray. The extreme topography generates high wind speeds that act to increase the amounts of salt deposited on cliffs and the distance that salt is carried inland on sea breezes. On the other hand, rainfall will dilute deposited salt and therefore osmotic effects on sea cliffs will be least in areas with large amounts of precipitation. The timing of events that deposit salt, such as gales, in relation to the seasonal maxima of precipitation is also important in determining the maximum salt concentrations to which plants on sea cliffs are exposed. From a study on sea cliffs in Great Britain, Malloch and Okusanya (1979) concluded that the water available to most species on these cliffs was equivalent to 5–30 per cent sea water. Some species utilized the equivalent of 100 per cent sea water. Special adaptations are required for plants to survive these conditions, and the lower portions of most sea cliffs are exclusively colonized by halophytes.

Because a higher suction potential is required for plants to extract water from sea cliffs, fewer and different species should be able to grow there compared to inland cliffs in the same climate. In situations in which there is less available water, it is further predicted that there should be less

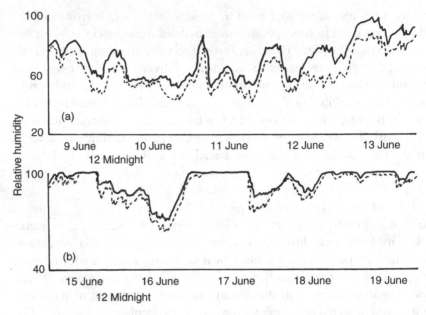

*Figure 3.10* Relative humidity on a cliff (solid lines) and in level-ground grassland (dotted lines) in Snowdonia, UK. The two panels above show variations in relative humidity during the day for two periods of five days: (a) during a dry sunny period and (b) during a moist cloudy period.

plant biomass on sea cliffs. Also, the osmotic factor should become more dominant as precipitation decreases, because high rainfall will wash salinity away. It follows that differences in the species composition and biomass between sea cliffs and inland cliffs should be greatest in dry areas and least in humid areas.

### 3.3.4 Moisture relations of man-made cliffs

Quarries, road cuts, and European cart-track canyons (Hohlwege) are man-made cliffs that resemble natural cliffs in their water relations. This is because they remain hydrologically connected to the water supply of a larger area in the same way as natural cliffs. Walls, on the other hand, are either free standing or – in the case of retaining walls – connected to an area whose hydrology has been drastically altered by human interference. As a result, walls should be much drier and less favourable for plant growth than natural cliffs. Another difference between natural and man-made cliffs that may affect their water relations is the degree to which the

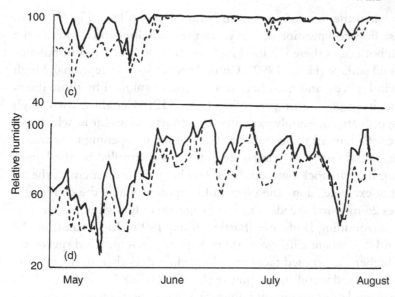

*Figure 3.10 (cont.)*
Panel (c) shows daily maximum and panel (d) shows minimum relative humidity
for the period between May and August, 1963. The results show higher humidity
throughout the year on the cliff stations. Taken from Bunce (1968), and used with
permission of Blackwell Science, Publishers.

surface is weathered. Recently abandoned quarries consist of relatively
unweathered rock that has a low water-holding capacity. Walls, on the
other hand, may have numerous cracks and crevices that can hold moisture or debris (Lisci, 1997).

Depending on the type of building material, the inclination of the
surface, and the height from the ground, walls may present a variety of
different microhabitats. Water availability is lowest on vertical surfaces
that consist of homogeneous material, and greatest on horizontal surfaces
and near ground level (Lisci & Pacini, 1993a).

## 3.4 Wind

Organisms on cliffs can be directly influenced by the momentum present
in wind and thus it is important to know the pattern of air movement
around cliffs. Fortunately, the effects of cliffs, hills, ridges and other
terrain obstructions on the distribution pattern of wind and wind speed
are well known (Grace, 1977; Oke, 1987; Arya, 1988). Flow separation

occurs when the flow of air is from behind the cliff face (Fig. 3.11). In contrast, flow compression occurs when the wind direction is towards the cliff. In both cases there is a local increase in wind speed and turbulence at the cliff surface (Hétu, 1992). Cliff edges can have exceptionally high local wind speeds, and this effect is sometimes exploited by those interested in hang-gliding or parasailing (Fig. 3.12). Certainly, these high wind speeds are commonly exploited by raptors and sea birds, which can achieve sufficient lift to become airborne simply by opening their wings. The magnitude of these wind-speed effects is controlled by the height and length of the rock outcrop that is blocking the movement of the air mass. For example, along the Niagara Escarpment where the cliff height averages 25 m, wind speeds at the cliff edge are only slightly higher than in the surrounding landscape (Bartlett *et al.*, 1990). In contrast, in the western USA where cliffs over 500 m high are present, wind speeds are much higher on exposed faces and along cliff edges than in neighbouring level-ground woodland or prairie (Maser, Rodiek & Thomas, 1979).

The second most important influence of wind speed on organisms colonizing cliffs is the effect it has on other components of the microclimate. Wind speed controls the size of the boundary layer, the thin layer of air that is directly adherent to the rock surface. All non-radiative transfers across this layer occur by molecular diffusion and are therefore very slow. The processes that are inhibited by the presence of a boundary layer include the transfer of heat between the surface and the atmosphere by convection, and the evaporation of water from the surface. Therefore, high wind speed will have a drying and cooling effect on cliffs that are damp and exposed to intense radiation fluxes, but will also have the effect of condensing water on (and therefore warming) cliffs whose surface temperatures are below the dewpoint and bathed in a warm humid airmass. This latter effect could be largely responsible for the common observation that the surface rocks of north-facing (or south-facing) cliffs in the northern (southern) hemispheres are usually moist even in the absence of direct precipitation or groundwater seepage.

## 3.5   Temperature

The temperature of a cliff surface is the direct result of its energy balance and as such is dependent on the balance between the amount of radiation energy absorbed, the amount dissipated, and the amount utilized by photosynthetic organisms within the rock. The last component is so small

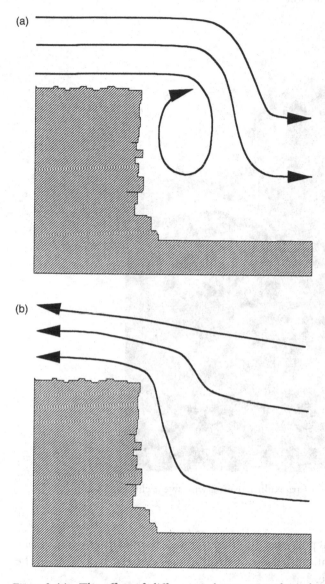

*Figure 3.11* The effect of cliffs on wind patterns and wind speeds. (a) Turbulence occurs in the lee of the cliff edge when the wind moves away from the cliff face. Conversely, in (b) higher wind speeds occur at the cliff face and on the cliff edge when the wind moves towards it. Both scenarios result in greater wind speeds and greater turbulence on cliff faces compared to surrounding habitat.

*Figure 3.12* Examples of parasailing from a windswept cliff edge south of Grenoble, France.

that, for all practical purposes, its effect on rock temperature is negligible. Any absorbed radiation that is not dissipated or utilized will increase the surface temperature.

### 3.5.1 *Factors controlling the amount of solar energy absorbed*

The amount of energy absorbed is the portion of incident radiation that is not reflected by the rock. Solar radiation incident on a rock surface will affect the temperature of that surface more than any other factor. Insolation, therefore, is a good measure of microclimate in rocky terrain and can be used to predict patterns of moisture and temperature, and the distribution of plant communities (Michalik, 1991). Factors such as aspect, slope and latitude that control incident radiation (as discussed in a previous section) are therefore important determinants of cliff temperature as well. This is illustrated by Zehnder (1953), who reported temperature measurements on concrete walls of different aspects and slopes in the tropics. For a day with peak air temperature of 31 °C, daily temperature maxima were 44 °C for a horizontal surface, 41 °C for a 90° south-exposed wall, but 48 °C for a 45° south-exposed slope. Zehnder (1953) notes that the maximum surface temperatures in the tropics are less extreme than the ones found under similar circumstances in temperate regions. Exposed cliffs at low elevations in central Europe have been reported to experience peak summer temperatures of 65 °C (Riebe, 1996) or even 70 °C (Lüth, 1993), but exact locations and methods of measurement were not given. More reasonable differences between air and surface temperatures were reported by Lundqvist (1968), who found south-facing cliff surface temperatures of 20 °C during a day in spring with air temperatures of only +0.4 °C. Such high temperatures only occur in microsites with direct exposure to the sun. In fact, Ellenberg (1988) states that south-facing vertical cliffs in central Europe experience peak summer temperatures of only 35 °C, and such moderate temperatures would be expected for steep slopes at a time of year when the solar angle is very steep throughout the day on the surface of the rock. These results are completely consistent with the radiation exposure of these surfaces, as illustrated in Figs. 3.2 and 3.4.

Besides incoming radiation, cliff surface temperatures are influenced by a number of other factors. One of these factors is the proportion of incident short-wave solar radiation that is reflected by the rock, also called the albedo. For most rocks, albedo values range between 0.10 and 0.40 and thus are similar to the values for soil-covered or vegetation-covered surfaces (Rejmánek, 1971). However, some white limestones may have an albedo as high as 0.70, meaning that only 30 per cent of incoming light is absorbed, whereas some dark granites have an albedo as low as 0.05, meaning that 95 per cent is absorbed. All other factors being equal, their low albedo would cause dark rocks to develop higher surface temperatures than light rocks. In the absence of heat dissipation by convection or evaporation, dark volcanic rock exposed to radiant flux densities equal to solar radiation should be able to acquire surface temperatures 25 °C higher than ambient temperature, based on its physical properties (Rejmánek 1971). For limestones and dolostones, the potential difference is only 12–17 °C, and for calc-silicate hornfels (erlan) it is

11.2°C. Field data of air and rock temperatures support some of these predictions (Zehnder, 1953; Rejmánek, 1971; Broady, 1981).

### 3.5.2 Energy dissipation from cliffs

The processes that dissipate energy and result in a lowering of surface temperatures are discussed next. These include reradiation, evaporation, convection and conduction. Reradiation is long-wave radiation emitted by objects as a function of their temperature and thus the temperature difference among different parts of a cliff, or between a cliff and its surrounding environment, will control the intensity of this radiation. The magnitude of the reradiation term in the energy balance equation can be large, but there is no evidence that the values for cliffs are different from those for adjacent level ground. Under certain circumstances, one could expect large increases in this term: for example, a south-facing (or north-facing) cliff in the northern (southern) hemisphere in mid-winter will be warmed significantly above the temperatures on adjacent level ground and these increased temperatures will significantly increase the loss of energy by reradiation. Conversely, the same cliffs in mid-summer will experience much lower surface temperatures than the surrounding landscape, and will therefore disperse far less energy by reradiation. In any event, the ecological significance of the reradiation term is small because it almost always reflects differences in surface temperature that have large and direct effects on organisms, as discussed below.

A second mechanism for energy dissipation is evaporation and the release of latent heat. This process is dependent on the amount of water available for evaporation, on the dryness of the atmosphere, and on wind speed. Evaporation plays a large role in controlling the surface temperatures of soils (Brady, 1974), but should be relatively less important for the surface temperatures of rock surfaces. This is because rocks do not have a large quantity of water available for evaporation and therefore the amount of energy that can be dissipated as latent heat of evaporation is much smaller. Thus, rocks (unlike moist soils) are not buffered from rapid temperature increase during high radiation loads. A further consequence of the low water content of rocks is that the amount of energy required to raise subsurface temperatures is much lower. The specific heat, or number of joules required to raise the temperature of 1 g of material by 1°K, is typically $0.8 \, J \, g^{-1} \, K^{-1}$ for dry soil, but only 0.44–0.68 for dry rocks (Rejmánek, 1971). In comparison, wet sandy soils have values of 1.48 and saturated peaty soils $3.65 \, J \, g^{-1} \, K^{-1}$. The specific heat increases with increasing water content, but because rocks hold very little water, it is always much lower for rock than for soil. As a consequence, even wet rocks are much more quickly warmed and cooled than soils and therefore undergo more rapid temperature fluctuations per gram of mass than an equal mass of soil. The general impression that people have, that rocky cliffs are huge and slowly changing reservoirs of heat, is due entirely to the sheer bulk of the exposed rock, not to the thermal properties of the material itself. Thus, when Ellenberg (1988) claims that cliffs have high surface temperatures during the night because of a high heat capacity of the rocks, the observation itself is correct but the explanation for it is not.

While the above is true for flat rock surfaces and vertical cliffs that are hydrologically isolated (such as rock islands or ledges), tall vertical cliffs may behave very differently because of their hydrology (see Chapter 2). Due to the pressures forcing water towards the face of a cliff, such cliffs may have a virtually unlimited water supply for evaporation. As a consequence, we predict that such cliffs will dissipate significant amounts of energy as latent heat long after horizontal rock surfaces would be dry. Such cliffs will tend to be moist and cool at all times relative to level-ground rocky terrain. The cliffs of the Niagara Escarpment appear to belong in this category, since atmospheric humidity is higher and air temperature lower at the cliff edge in comparison to level-ground forest only 5 m away (Bartlett et al., 1990). Even on these cliffs, however, the liquid water that is available for evaporation at the surface of the cliff is concentrated along fracture lines, crevices and other places where its flow is easier.

A third process that removes thermal energy from the cliff surface is conduction. Material that has a high thermal conductivity will quickly transfer to depths any heat generated by irradiating the surface. Compared to soil, rock of all kinds has a higher thermal conductivity (by up to an order of magnitude), regardless of water content (Fig. 3.13). This becomes important when evaporation is absent, i.e. when the rock is dry, and will counteract the effects of low water content on surface temperatures by moderating temperature extremes. A dry irradiated rock surface will be cooler than an equivalent dry soil surface because it acts as a heat sink, conducting thermal energy away from the surface. After sundown, the process will be reversed: the cliff will function as a heat source, slowly losing the absorbed heat to the cooler night air. This will keep the surface warmer than air or surrounding soil at night but cooler in the day. This heat conduction effect may also be partly responsible for the popular impression that rocks and cliffs are enormous reservoirs of heat when compared to soil. Thus, especially in dry climates, rock

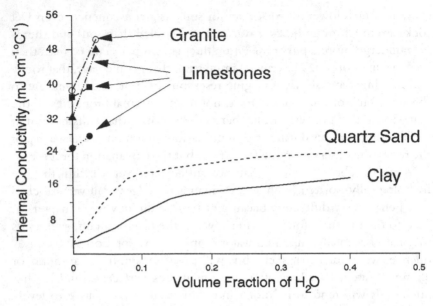

*Figure 3.13* The thermal conductivity of different types of soil and rock as a function of the volumetric water content of the material. Data for clay soil and quartz sand are taken from Sellers (1965); all other data are from Rejmánek (1971).

surfaces often have a more moderate temperature regime than adjacent soil surfaces (Williams, 1984) and we predict that their floras and faunas should reflect this. Wunder and Möseler (1996) have shown that rock surfaces in talus slopes are also prone to much smaller changes in temperature than are found in the air or in the moss colonies on the rocks. In wet locations most of the energy absorbed will be dissipated as evaporation, thus the role of conduction should be less, and the biota on cliffs should not be distinctive from the surrounding biota in terms of thermal requirements. As with evaporation, the moderating effect of conduction will be absent for small rock islands or fragments that are thermally isolated from surrounding bedrock. Such fragments should experience much more extreme temperatures than large rock outcrops and should therefore have much more distinctive biotas than neighbouring rock that is physically connected to the pediment. Smaller heterogeneities, such as cracks and crevices, should also decrease the conduction of heat, but no data have ever been collected on this.

Convection is the form of energy dissipation that involves the transfer of heat by air movement over a surface. Its magnitude is dependent on the temperature difference between the air and the rock, and on the wind speed. The higher wind speeds near cliffs should increase the amount of convective energy loss or gain, thus tending to make cliffs cool (or warm) quickly depending on the direction of the temperature gradient.

### 3.5.3 Temperature regime of cliffs

It is clear from discussing the individual components of the energy balance of cliffs that the actual temperature of a cliff of a certain slope and aspect, and its temperature relative to surrounding level ground, is a complex result of multiple interacting processes that are dependent on local conditions such as degree of thermal and hydrologic isolation, rock type, and local climate. For example, the European plant sociological literature makes frequent reference to 'warm' substrates, mostly limestones and basalts, as opposed to 'cool' substrates such as granite, based on the types of plant communities found there. Rejmánek (1971) examined the thermal properties of a range of rock types to find the underlying causes of this, and found that physical properties such as water-holding capacity, conductivity and specific heat would indeed lead to higher potential surface temperatures in certain rocks, such as basalt, than in others, such as granite. However, he also found that there was huge variation within rocks of the same type, for example limestones, and that other factors such as moisture availability were probably more important. He concluded that it was difficult to predict ecological conditions from physical thermal properties of rock.

Despite this, some general trends common to most cliffs emerge. With a few exceptions (such as cliffs facing towards the equator in winter or at high N or S latitudes in the summer), vertical cliffs tend to be cooler than surrounding level ground or slopes. This is because they receive less direct radiation, have high rates of cooling by conduction, convection, and possibly evaporation, and have no insulating snow cover in winter. The lowest temperatures that occur on north-facing cliffs in the winter are lower than the lowest temperatures of a neighbouring level soil surface, but the highest temperatures reached in the summer on south-facing or west-facing cliffs are also lower than those on neighbouring level ground. Field data collected by Michalik (1991) in Poland on a steep rocky ridge with a 60° aspect show that cliffs, especially north-facing ones, have temperatures that are more constant than for other nearby sites. While the range of temperatures is small, the rate at which temperature changes occur in cliffs should be faster compared to level ground because rock lacks the buffering effect of the high water content in soil. Thus, cliff surface temperatures should lag less behind air temperatures in their daily and seasonal course than do soil temperatures. While there are plenty of cliff temperature measurements in the literature, most studies simply report rock and air temperatures and few have specifically compared cliffs to neighbouring level-ground soil in a way that could be used to test these theoretical predictions. Taken together, the consideration of the thermal relations of cliffs leads to the prediction that cliffs will generally select for more cold-tolerant organisms when comparisons are made among local biotas.

### 3.5.4 Subsurface temperatures

While the above discussion has focused on cliff surface temperatures, the fact that cliffs contain endolithic organisms makes it necessary briefly to consider subsurface conditions as well. As long as all energy is absorbed and converted into energy components at the surface, subsurface temperatures can, under steady state, never be higher than surface temperatures. In practice, a small part of the incident radiation will be transmitted a small distance into the rock and converted to heat below the surface as it is attenuated within the rock. The physics of radiation attenuation dictates that attenuation (and thus the amount of heat released) will decrease with increasing depth into the rock. Therefore, rock temperature must also always decrease (Fig. 3.14). The results of Bell (1993) that rock subsurface temperatures were between 7°C and 12°C higher than those at rock surfaces are therefore physically impossible, unless the system is not in steady state or the rock is not homogeneous. The only way that subsurface temperatures could be higher than surface temperatures at steady state is if there was no attenuation over the first several millimetres of rock, i.e. the rock surface was completely transparent and all energy absorption took place in a darker layer underneath (Fig. 3.14).

The actual form of temperature gradients within rock has not been measured, but the gradients ought to have a form that reflects high thermal conductivity, high thermal diffusivity, and high thermal admittance as well (Oke, 1987). In a comparison with an array of materials used for the construction of buildings, Oke shows that stone has a high value of all three of these parameters, second only to steel. In fact stone has four times the thermal conductivity of glass, ten times its diffusivity and twice its admittance. Diffusivity (with units $m^2 \ s^{-1}$) is the property that defines the time required for temperature changes to travel in wave-like fashion through the material. High diffusivity is obtained when there is low water content and high degree of direct contact of substrate materials allowing for conduction (Oke, 1987). Both of these characteristics

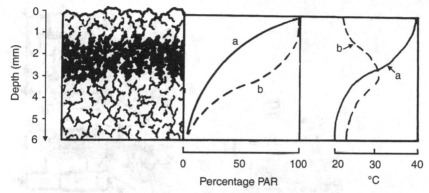

*Figure 3.14* Two theoretical radiation and temperature profiles within rock that is colonized by cryptoendolithic organisms (left panel). The middle panel shows the percentage of incident light transmitted, and the right panel shows the temperature as a function of depth below the surface. The dashed lines represent the profiles suggested by the internal heating model of Bell (1993). Such greater temperatures at depth than at the surface could only occur if there were consistent structural heterogeneities that made the surface completely transparent and the subsurface darker by comparison, leading to extinction curve (b) shown in the middle panel. With a normal extinction curve as shown by (a), the temperature must be highest at the surface. PAR, photosynthetically active radiation. Original artwork by C.E. Ryan.

apply to igneous, metamorphic and sedimentary rocks. High thermal conductivity and diffusivity together mean that rocks will absorb large volumes of heat (or cold) and transmit these pulses of energy deep within the rock. These pulses of heat or cold should be measurable on a daily basis as short-term pulses of heat penetrating to depths of 10–20 cm followed by longer term seasonal trends of heat pulses penetrating tens of metres into the rock. We believe that the formation of ice caves and cold-air (algific) slopes in temperate latitudes may reflect these long-term pulses of energy storage in rock (Nekola, Smith & Frest, 1996).

## 3.6 Effects of topographic heterogeneity on microclimate

Real cliff faces are never perfectly homogeneous flat walls, but heterogeneous on many scales from microscopic cracks to bedding planes, crevices and caves. There are also complex mixtures of table rocks, undercuts, talus slopes and pinnacle rocks that stand adjacent to the intact cliff faces. This heterogeneity produces an enormous variation in the microclimate and other physical conditions of the rock faces that are

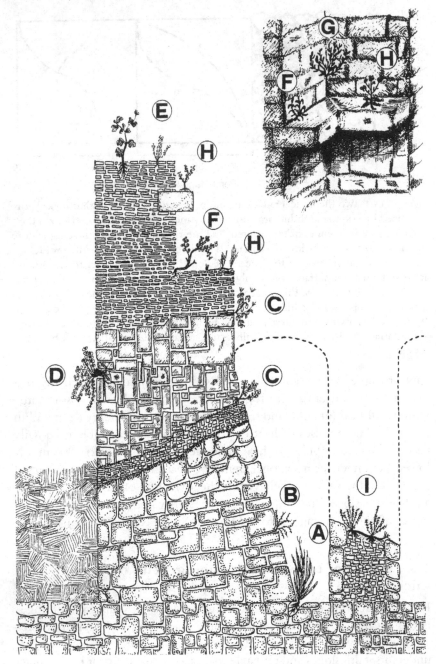

*Figure 3.15* Illustration of the variety of possible plant microhabitats present on walls. 'A', cavities on level ground; 'B', cavities at inclined angles; 'C', cavities at

present on multiple and overlapping spatial scales (Jung, 1961; Nuzzo, 1996). Oettli (1904) describes cliffs as habitats where extreme amounts of environmental variability occur at very small spatial scales leading to an intense 'compression' of a variety of habitat types into a small area. Cooper (1997) has come to the same conclusion regarding the factors that control cliff vegetation in Northern Ireland. Each small microsite will produce a unique microclimate that has properties derived in part from the characteristics of the main face, and in part from the unique characters of the microsite. As an example, a slab of collapsed caprock next to a south-facing cliff face will have one side that faces south, and the opposite side facing north. A similar slab of rock next to a north-facing cliff will also have a north and a south face, but they are not equivalent to the north and south faces of the first slab of rock (or to the north and south faces of the main cliff). The complex interactions of the factors such as moisture, wind and temperature with topographic heterogeneity make the microclimate of any cliff face difficult to predict at a small scale.

Surface heterogeneity can also affect components of the microclimate, specifically the temperature and moisture content of rock surfaces, through its effects on convection and evaporation. Rough surfaces and depressions will have a large boundary layer, while projecting edges will have a small boundary layer. One of the extremes of cliff microclimate is found in isolated rock islands, exposed ledges, and the upper vertical areas of man-made walls (Lisci & Pacini, 1993a). Such sites tend to be dry and prone to extreme and fluctuating temperatures (Fig. 3.15) because they lack the buffering provided by contiguous masses of bedrock. Heavy fracturing of walls also promotes microhabitat differentiation. Heat/cold and desiccation resistant organisms as well as ruderal species are common in these places (Lisci, 1997). At the opposite extreme is the microclimate of caves, large crevices, and spaces between boulders. These places are dark, cool, and have high atmospheric humidity. Generally these are extremely stable over time (Leclerc, Couté & Dupuy, 1983; Wunder & Möseler, 1996). Due to large boundary layers developing in recesses that never get exposed to turbulent air flow or direct solar radiation, ice and snow can

Caption for *Figure 3.15 (cont.)*
the interface between two types of building materials where the chemistry or texture may change; 'D', cavities in a vertical homogeneous face; 'E', cavities in horizontal surfaces; 'F', cavities at the intersection of vertical and horizontal surfaces; 'G', cavities where two vertical surfaces meet; 'H', substrates of porous material; 'I', substrate formed of broken stone. Taken from Lisci and Pacini (1993a), and used with the permission of the publisher.

persist in the deep recesses of some of these caves until midsummer (Jung, 1961). Organisms in these microhabitats should show the opposite bioenergetic characteristics of those found at cliff edges or on walls.

The most extreme small-scale heterogeneity in microclimate is found in the talus, where topographic heterogeneity is at a maximum (Jahns & Fritzler, 1982; Cox & Larson, 1993a, 1993b). Conditions in the talus range from fully exposed boulder surfaces to always shady, cool and moist crevices and hollows. In basalt scree in Germany, temperature differences of over 30 °C were measured at two sites only 1.8 m apart (Lange, 1953). Cold air streams rising from the base of the talus and emerging at the cliff base can increase local temperature extremes (Wunder & Möseler, 1996). The moisture available in talus is often extremely favourable, compared even to level ground: Pérez (1991) found soil moisture content and available water content greater for soil under stones and boulders than for soil in the open due to condensation of water under stones after sunset. Soil temperatures also fluctuated less under stones than in the open.

Even for superficially smooth vertical surfaces, there are differences in rock microtopography that are important for small organisms colonizing a cliff face. Pentecost (1980) noted that smooth and polished rock surfaces were rarely colonized by bryophytes and lichens and attributed this to several factors. Clefts, for example, will harbor moisture and provide shelter from radiation, wind and herbivores, while minute projections could act as nuclei for condensation droplets, which could provide additional moisture. Finally, dry sharp projections have high surface-charge density and therefore aid cohesion of diaspores.

## 3.7 Summary

Cliffs experience a range of microclimatic conditions that are very different from both level-ground rocky terrain and horizontal landscapes with soil. Some of these differences are consistent with the idea that cliffs are hostile environments (e.g. the lack of protective snow cover, the lack of a thermal or evaporative buffer in soil, and full exposure to wind relative to level ground). But at the same time, other aspects of the microclimate suggest that cliffs are much more moderate environments than people think. For example, during the growing season in temperate or subarctic latitudes, cliffs are exposed to a much more moderate radiation regime than surrounding habitat types. In fact, south-facing cliffs in the northern latitudes receive their peak amounts of radiation exposure in mid-winter. Cliffs are places that have more water availability than surrounding habitat types. This is because hydraulic pressures that keep the percolating groundwater inside the rock are absent. The water that is present in cliffs is easily available because the rock does not exert a high matrix potential of its own. But this is not to say that there is a large volume of water present. Therefore, most organisms that exploit cliffs must be able to tolerate a water supply that is easily available for only a short period of time. Once this water is exploited, percolation of water through the rock keeps the water contents of the rock just above the levels that impose drought on the organisms present. Lastly, cliffs are places where the temperature regime is much more moderate than people think. The only conditions in which cliffs appear to take on extreme temperatures are in the mid-winter conditions when direct-beam radiation is at a seasonal high and cliff-face surface temperatures are correspondingly high. All of this information suggests that cliffs cannot be simply characterized as 'dry', 'hot' or 'exposed' the way many investigators, including ourselves, have done over the years. Another important difference between cliffs and other habitat types is related to the extremely high level of microsite variation on a very small spatial scale. Complex environments with variable amounts of radiation, water and temperature also occur within the rocks of open cliff faces and influence the growth conditions of endolithic organisms. Cliffs that are made as a result of human activity share most if not all of the microclimatic properties of natural cliffs. Only when the cliff takes on the form of a very narrow wall will the environmental conditions of the man-made cliff become more extreme.

# 4 · *Flora*

Cliffs are largely inaccessible to people and their livestock and are therefore generally free from disturbances such as grazing and fire. As we have already shown, cliffs cannot support organisms with high productivity and therefore most vegetation on cliffs is small and unassuming. We believe that these features are the reasons why cliffs have attracted far less attention from biologists than other more accessible habitats with large numbers of productive macroscopic organisms. Maycock and Fahselt (1992) studied the vegetation of high arctic cliff faces and scree slopes in Canada that had previously been described as 'unvegetated'. On these surfaces they found 156 plant species, of which half were lichens, one-quarter were macroscopic higher plants and one-quarter were bryophytes. The authors offer no satisfactory explanation as to why others might have so grossly misrepresented the diversity of species in these habitats, but they hint that the appearance of low productivity has discouraged close scrutiny in the past. The same suggestion was also offered by Larson (1990) to explain the lack of prior discovery of an ancient forest of stunted *Thuja occidentalis* on the apparently bare cliffs of the bare-looking Niagara Escarpment in southern Ontario, Canada. The small size of many cliffs often results in them being viewed as 'break-points' or transition-points in landscapes, rather than as separate landscape elements. This view leads to the characterization of cliffs as the 'edges' of other places, rather than places in their own right.

We feel that all of these factors help to explain the small amount of scientific literature dealing with the vegetation of cliffs compared to the vast amount of literature on level-ground forests, grasslands, wetlands, deserts, and tundras of the world. Some have expressed to us the view that cliffs have been carefully investigated in many parts of the world. We must hasten to point out that most of this literature (as described below) deals not with cliffs proper, but rather with smaller rock outcrops or shelves that are accessible without the use of technical climbing equipment (Fig. 4.1). We acknowledge that such sites may be more 'cliff-like' than meadow-like in their ecological character (and we include a great deal of this literature here because of this), but we also claim that the flora and vegetation of large open cliff faces have been poorly characterized. Only rarely have we encountered papers or monographs where cliffs were actually climbed to facilitate qualitative or quantitative vegetation sampling. Instead of direct sampling using climbing equipment, most workers have made observations on cliffs from distant locations (Oosting & Anderson, 1937, 1939; Cooper, 1984; Debrot & de Freitas, 1993). Thus, using the definition of *cliff* from Chapter 1, we conclude that most people have not sampled cliffs in their work.

This chapter summarizes the published information on the species and communities of higher and lower plants and micro-organisms that have been reported from inland, maritime and man-made cliffs around the world. Within each type of cliff, tropical and subtropical locations are considered first, followed by temperate and subarctic regions. Alpine cliffs are not included for the most part, mainly because once the tree-line has been passed, environmental conditions are difficult everywhere and the steepness of the terrain has much less influence on the organization of the vegetation (Ellenberg, 1988). This chapter represents the first time global patterns of lower-elevation cliff vegetation have been reviewed in one place. Our intent is not to provide the reader with in-depth floristic accounts and complete species lists; these can be obtained from the original papers cited. Rather, it is to summarize and condense these findings by concentrating on the patterns shown by interesting or important species (as determined by the authors of the cited papers). Above all, the aim of this review is to highlight similarities and differences in the flora of cliffs around the world. We will delay until Chapter 6 our discussion of the experimental work that has been done on a small subset of species to explain the causes of plant distribution patterns on cliffs.

## 4.1 Inland cliffs

### 4.1.1 *Tropical and subtropical locations*

Investigations of cliff flora and vegetation are heavily biased towards the temperate zone. Despite the presence of many large and spectacular cliffs in tropical and subtropical regions, only a few of them have been studied in any detail. Among these are large outcrops of granite, sandstone and other rocks (inselbergs) in equatorial Africa (Bonardi, 1966; Porembski

*Figure 4.1* The quantitative sampling of cliffs requires the use of technical climbing equipment. Descent is best controlled by a two-rope system in which the field workers lower themselves by rappelling while a second person holds them on belay from the top. This system allows the cliff worker to be safely supported even if one of the people involved becomes unconscious. Photo by D.W. Larson.

*et al.*, 1994; Porembski, Brown & Barthlott, 1996). Some of these cliff faces are heavily colonized by a desiccation-avoidant bromeliad, *Pitcairnia feliciana*, which is the only species in this family naturally occurring in the Old World. Exposed rock is covered with cyanobacteria in the genera *Stigonema* and *Scytonema* and a wide variety of desiccation-tolerant lichens and mosses. The sedge *Bulbostylis coleotricha* and the fern *Asplenium pubescens* colonize small shaded fissures. *Pellaea doniana* occurs on overhanging rocks and boulders formed from both granite and sandstone. The rare sedge *Afrotrilepis pilosa* is also reported from such sites; this species is poikilohydric and individuals on rock outcrops can exceed 200 years of age (Hambler, 1961, 1964; Bonardi, 1966). These findings are interesting for they suggest that desiccation tolerance is a necessary condition for life in certain microhabitats on the cliff, even though neighbouring microhabitats support plants that are desiccation-intolerant.

Rock outcrops and steep cliff-like slopes of sandstone and dolerite in South Africa also support unusual assemblages of fire-intolerant and disturbance-intolerant higher plants (Rutherford, 1972; Fuls, Bredenkamp & van Rooyen, 1992, 1993). *Boscia albitrum, Ficus cordata, Ficus guerichiana* and *Leucosidea sericea* are the common woody species on these cliffs, but *Heteromorpha trifoliata, Croton gratissimus, Euclea undulata, Hibiscus engeri, Sutera acutiloba* and *Pellaea calomelanus* occur as well. Of these, only the genus *Pellaea* stands out as part of a recurring group of genera found on cliffs around the world. The vegetation of undisturbed cliffs is clearly distinct from that of disturbed sites, as shown by a detrended correspondence analysis (Fuls *et al.*, 1992, 1993). Deacon, Jury and Ellis (1992) and Marloth (1913) also describe the restriction of many moss, liverwort and fern species to steep rock cliffs and other rocky outcrops in South Africa. Among the species that co-occur with these are three species of *Widdringtonia* (Cupressaceae), including the rare fire-sensitive endemic tree *Widdringtonia cedarbergensis* (clanwilliam cedar) that now grows only on cliffs and steep rock outcrops (Manders, 1986). As other members of this family, the tree shows stem-stripped axial morphology when growing slowly on rock outcrops.

An abundance of rare and endemic plants is characteristic for the cliff flora of many other tropical or subtropical locations. For example, a newly discovered species of perennial shrub, *Schiedea attenuata*, grows only on the basalt cliff on the island of Kaua'i, Hawaii (Wagner, Weller & Sakai, 1994). In many cases these plants are slow growing and the absence of human disturbance has been important for their preservation. Cliff edges and ledges of Tian Mu Shan in Zhejiang Province, China, are one of the last natural habitats for slow-growing, deformed trees of *Ginkgo biloba* (Del Tredici, Ling & Yang, 1992; Fig. 4.2). Also, Cox (1945) reported that limestone cliffs were some of the most important habitats from which various species of *Rhododendron* were collected for commercial propagation in the nineteenth and early twentieth centuries. The protection provided by the cliff produces ecological opportunities for the flora to undergo natural development, unlike most other terrain units in Asia that have been grossly modified by human activity. Several detailed floristic accounts also exist of the massive limestone hills and karst towers bordered by vertical cliffs on the Malay Peninsula (Henderson, 1939; Chin, 1977; Whitmore, 1984). The flora appears to include relatively few lichens or mosses. Among the plants that are mainly or completely restricted to these cliffs are rare species of *Boea*, *Chirita*, *Monophyllaea*, *Vitex* and *Paraboea*. The cliffs on the Malay Peninsula are largely undisturbed by humans, and the occurrence of these rare species on the cliffs reflects both their tolerance of harsh environmental conditions and their intolerance of human disturbance. All of the authors report scattered misshapen trees on these cliffs but do not consider the possibility that such deformed trees are very old. Even if they were, the absence of a seasonal pulse of growth might make it impossible to detect annual rings.

Formal comparisons of cliff floras from different geographical regions are very rare. Zehnder (1953) concludes that the algal vegetation of rocks in the tropics is not much different from that on temperate cliffs, even though the macrovegetation differs drastically. A recent study by Alves and Kolbek (1993) examines shaded cliff communities at the bottom of cliffs in Brazil, and makes qualitative comparisons with similar data collected in southern France (Rioux & Quézel, 1949). While the individual species that make up the plant communities are very different between the locations, the life-form distributions and physiological characteristics of the species are similar (Fig. 4.3). Woody trees and shrubs (such as *Anemia oblongifolia* in Brazil and *Juniperus phoenicea* in France) tend to occur on open exposed rock where soil cover is absent. On cliff faces with shaded fissures, ferns, lichens and mosses are more common, with species such as *Trichomanes pilosum* in Brazil and *Asplenium trichomanes* in France. The authors speculate that species associated with cliffs have many functional characteristics in common. Most interesting in the comparison of these two cliff floras is that the Brazilian cliffs are formed of quartzite with some hematite and schist, whereas the French cliffs are formed of limestone. Alves and Kolbek (1993) remark that the floristic

*Figure 4.2* A human-modified cliff at Tian Mu Shaw, China, supporting natural populations of *Ginkgo biloba*. These trees are slow growing and grossly deformed. Photo courtesy of P. Del Tredici.

## Penumbral Vegetation in Brazil and France

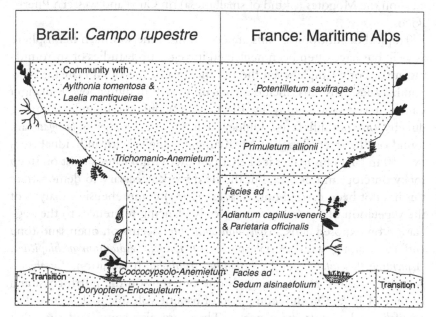

| Brazil: *Campo rupestre* | France: Maritime Alps |
|---|---|
| Community with Aylthonia tomentosa & Laelia mantiqueirae | Potentilletum saxifragae |
| Trichomanio-Anemietum | Primuletum allionii |
| | Facies ad Adiantum capillus-veneris & Parietaria officinalis |
| Transition Coccocypsolo-Anemietum Doryoptero-Eriocauletum | Facies ad Sedum alsinaefolium Transition |

*Figure 4.3* Comparison of the plant communities occurring on shaded cliffs in Brazil and southern France. While the actual species were different, similar life-forms were found in corresponding habitats on both continents. Modified from Alves and Kolbek (1993).

composition seems to vary little among sites with different mineral composition, with the possible exception of hematite.

Debrot and de Freitas (1993) examined the vascular plant cover of five large rocks (20–47 m diameter) that had fallen away from a steep, 20–50 m high, limestone escarpment in Curacao, southern Caribbean. The desiccation-avoidant plant *Tillandsia flexuosa* and the desiccation-tolerant fern *Polypodium aureum* are both common in areas inaccessible to grazing mammals, but absent where grazing occurs. This pattern suggests again that cliff floras are not tolerant of disturbance, even if they can tolerate harsh physical environments. Brewer-Carias (1986) presents a brief account of the vegetation of cliff tops and edges of tepuis (sandstone table mountains) in southern Venezuela. Dozens of highly localized and endemic species occur on the tops of these sandstone outcrops, including species of *Drosera, Brocchinia* and *Ptersozonium*, but no detailed accounts of the species restricted to the cliff faces have been published.

Graham (1973) reports that the palm genus *Gaussia* is restricted to cliff faces on the Mogotes (a kind of small mesa) on Cuba and western Puerto Rico.

Tropical or subtropical cliffs are common in the southern hemisphere as well, but have often been neglected in vegetation studies of surrounding areas. For example, even though the area of the Boyd Plateau, New South Wales, Australia, includes cliffs up to 900 m high above Kanangra Creek, a vegetation study by Black (1982) did not include sampling directly on the cliff faces. Unlogged forests of *Eucalyptus fastigata* are found on the steep slopes of Mt Krungle Bungle, with individual trees over 30 m tall. This makes it likely that other virgin forests exist on steep rocky outcrops in this area, but the field work necessary to demonstrate this has not been carried out. Beadle's (1981) comprehensive analysis of the vegetation of Australia includes only glancing references to the vegetation on exposed cliffs. *Asplenium obtusatum* occurs on open sandstone cliff faces along with several species of *Drosera*, *Adiantum aethiopicum*, *Asplenium flabellifolium* and *Cheilanthes tenuifolia*. In addition, *Eucalyptus phoenicea* is found mainly on sandstone outcrops on the north coast at a latitude of 16 °S. On cliffs composed of basaltic volcanic rock, *Cheilanthes tenuifolia* and *Culcita dubia* occur. There are also some cliff sites that support small woody plants of *Leptospermum brachyandrum*. Beadle also discusses the patterns of primary succession on Triassic sandstones in Australia. The first colonizers of bare rock are apparently members of the cyanobacterial genera *Stigonema* and *Gloeocapsa*. These organisms give the rock a distinctive bright red colour. Lichens and mosses are next in the invasion sequence, followed by the vascular plants *Lepyrodia scariosa*, *Ptilanthelium deustum* and *Leptosperma laterale*. Shrubs can become established in these microcommunities, including *Epactris pulchella*, *E. microphylla* and *Leptospermum squarrosum*. On wet rocks, *Pyrrosia rupestris* can invade. While no evidence is given as to the similarity of the invasion sequences on rocks of different types, Beadle observes that rocks of differing chemical and physical structure support different mature communities. Thus, granite cliffs support species in the genera *Andersonia*, *Anthocercis*, *Boronia*, *Darwinia* and *Leptospermum*, whereas quartzite rocks support *Regelia velutina*, *Calothamnus validus* and *Banksia quercifolia*. *Leucopogon unilateralis* is an additional species that occurs on metamorphosed sedimentary rock.

Coates and Kirkpatrick (1992) examined the ecological responses of higher plants to habitat conditions on inland sandstone cliffs in Tasmania. They found that those species of fern and higher plant that are restricted to cliffs respond little to gravity, are slow growing, and cannot be stimulated to grow faster when fertilized. The species most restricted to cliffs is *Blechnum vulcanicum*, but a variety of other species occurs on the cliffs, including *Polystichum proliferum*, *Leptospermum lanigerum* and *Asplenium flabellifolium*.

Jones, Hill and Allen (1995) have recently reported Wollemi pine (*Wollemia nobilis*, Araucariaceae) from a single narrow sandstone gorge in the Blue Mountains, New South Wales, Australia. This species represents one of the rarest plants in the world. Whereas most of the trees are found on broken talus slopes, a portion of the total population of 40 trees occurs as stunted and deformed individuals on the sandstone cliffs above the talus (D. Noble, personal communication). Unfortunately, no studies have yet been carried out on the vascular or non-vascular flora of these particular cliffs. Elsewhere in New South Wales, cliffs adjacent to waterfalls are the only sites known to support another extremely rare species. Dwarf mountain pine (*Microstrobos fitzgeraldii*) only occurs on very wet ledges and cliff faces in the upper Blue Mountains (Jones, 1994). This species shows some of the characteristics described for other rare cliff trees (Rioux & Quézel, 1949; Del Tredici 1989a, 1989b), namely very slow growth and sporadic seedling recruitment (Jones, 1994). Wade and McVean (1969) studied mountainous areas of Papua New Guinea but did not report any details of the vegetation of open cliffs, except to argue that only cliffs and swamps are habitats uninfluenced by human disturbance.

### 4.1.2 Temperate and subarctic locations

The bulk of the literature describing temperate-zone cliff vegetation is derived from Europe and eastern North America. Most of this literature comes from floristic surveys, although a few studies, such as Oettli (1904) and Jaag (1945), analyse many different aspects of the vegetation of cliffs, including patterns of community assembly and causes of mortality. Many of the genera characteristic for tropical cliffs can also be found on cliffs in the temperate zone, for example *Campanula*, *Asplenium*, *Sedum*, *Pellaea* and *Polypodium* (Jung, 1961; Wilmanns & Rupp, 1966; Lüth, 1993; Herter, 1993; Ficht et al., 1995; Stärr et al., 1995). Many rare species appear to be using the cliffs as local refuges from the forest development that took place on most level ground in Europe following deglaciation, and this is discussed further in Chapter 6. The characteristic flora is beginning to disappear from places where rock climbing and hiking have recently become common, as is shown in more detail in Chapter 7.

In eastern Europe, Michalik (1991) and Pawlowski, Medwecka-Kornás and Kornás (1966) report the occurrence of a rich community dominated by *Festuca pallens* on limestone or granite cliffs in Poland. Adamović (1909) investigated exposed limestone cliffs in Serbia south of Belgrade. In this area rich in karst activity, *Juniperus communis, Cotoneaster vulgaris, Cystopteris fragilis, Asplenium ruta-muraria, A. viride, A. trichomanes, Polypodium vulgare* and *Geranium robertianum* all occur. Pax (1908) published an early report of the vegetation of cliffs in the Carpathian Mountains. *Dianthus nitidus* was once common in high rocky areas and cliffs in these mountains, but Pax indicates significant declines in the early part of the twentieth century. *Pinus pumilo, Taxus baccata* and *Juniperus communis* were also once common in these areas but are becoming less so. Jurko and Peciar (1963) observe that limestone and dolomitic cliffs in the same region support rich communities of higher plants, lichens and mosses. In particular, cliffs support *Ctenidium molluscum, Tortella tortuosa, Rhytidiadelphus triquetrus, Anomodon viticulosis, Polypodium vulgare, Asplenium trichomanes, A. viride, A. ruta-muraria, Cystopteris fragilis, Geranium robertianum, Ribes grossularia, Campanula carpatica, C. rotundifolia, Peltigera horizontalis* and *Dermatocarpon miniatum*. Shrubs and trees include *Sorbus aucuparia, Euonymus verrucosa* and *Picea excelsa*. In south-eastern Europe, Davis (1965) makes reference to a high level of endemism in the rocky mountainous areas of Turkey. *Cupressus sempervirens, Juniperus drupaceae* and *J. excelsa* occur at and below tree-line in these areas, and *Cheilanthes fragrans* is often an associate on dry, open, rocky cliffs. *Adiantum capillaris-veneris, Asplenium trichomanes, A. viride, A. ruta-muraria, Cystopteris fragilis, Polystichum lonchitis, Polypodium vulgare* and large *Taxus baccata* built up of coalescing suckers occur on steep slopes in areas free from significant human disturbance.

Cliffs in central Europe have been the subject of a number of vegetation studies. Igneous rock faces in eastern Germany are characterized by *Dianthus gratianopolitanus, Cotoneaster integerrimus, Hieracium schmidtii, Asplenium septentrionale* and *Campanula persicifolia* (Türk, 1994). Sandstone cliffs in eastern Germany support a unique plant community dominated by *Pinus silvestris* and *Calluna vulgaris* (Jung, 1961; Riebe, 1996). For the same cliffs, Schade (1923) also describes a rich flora of bryophytes, lichens and terrestrial algae, including cryptoendolithic communities dominated by *Protococcus*. Such microscopic communities are just beginning to be appreciated for their role in ecosystem function, even though their existence has been known for a long time. Stöcker (1965) also reports species of *Calluna*, as well as *Empetrum* and *Vaccinium* on granite cliffs and open exposed rocks in central Germany. Herter (1996) distinguishes a number of phytosociological units among the vegetation of limestone cliffs in south-eastern Germany. A community character- ized by *Asplenium viride* and *Cystopteris fragilis* occupies rock crevices on moist shady rock faces. Dry and sunny faces support a *Draba aizoides– Hieracium humilis* community, while shaded rock ledges support a *Valeriana tripteris–Sesleria varia* community. Level cliff or chimney tops are charac- terized by an *Alyssum alyssoides–Sedum album* community, vertical faces immediately below the cliff edge by a *Dianthus gratianopolitanus–Festuca pallens* community. A *Sisymbrium austriacum–Asperugo procumbens* commu- nity occupies hollows at the base of cliffs and cave entrances. Five addi- tional communities are described for the talus. Each of the above communities can exist in several varieties, depending on the accompany- ing species. Herter (1996), Stärr *et al.* (1995) and Ficht *et al.* (1995) all show evidence that large numbers of rare species occur on cliffs in this area, among them *Hieracium franconicum*, a species known from fewer than ten sites worldwide. Ellenberg (1988) observes that it is true in general that cliffs tend to support populations of exceptionally rare plants, and that many of these species (such as species of *Dianthus*) are in turn restricted exclusively to cliffs. Cliffs also often support both boreal and mediterranean species that are relicts from the Holocene (Türk, 1994; Stärr *et al.*, 1995; Riebe, 1996). Ficht *et al.* (1995), Lüth (1993) and Türk (1994) all agree that apart from bogs, cliffs are the only 'primary' habitats in Germany, i.e. habitats that have escaped significant amounts of human disturbance and support remnants of the original vegetation. In light of this, Lüth (1993) wonders why these environments have attracted so little scientific interest. An interesting hypothesis proposed by Lüth (1993) is that many plants (as well as animals) that are now common in open land- scapes were originally much more narrowly restricted to cliffs, with *Juniperus communis* as an example.

In the Jura of south-eastern France, Richard (1972) reports a variety of rocky habitat types including open limestone cliffs. *Asplenium viride* and *A. ruta-muraria* occur on wet, exposed rock faces along with various species of *Hieracium*, including *H. tomentosum*. Also reported are *Pinus syl- vestris, Sorbus aria, Arctostaphylos uva-ursi,* and *Quercus petraea. Geranium robertianum* and *Phyllitis scolopendrium* are also found on rocky slopes at the bases of many of the cliffs, along with *Epilobium montanum, Oxalis aceto- sella* and *Asplenium trichomanes.* The relict alpine plant *Campanula cochlea- riifolia* is also reported. Sutter (1969) reports a high degree of endemism among the plants that occur in rocky mountainous regions of the south-eastern Alps. This idea is also well presented by Ellenberg (1988), who concludes that rocks and cliffs, especially in the southern Alps, are especially rich in endemic rock-crevice plants. Ellenberg claims that 35–40 per cent of the endemic plants in central Europe occur in such habitats. He also claims that some well-known woody plants, such as *Taxus baccata*, occur only (or mainly) on steep rocky slopes that escaped disturbance, fire and logging from people. An early study by Vogler (1904) supports this idea. Vogler found that *T. baccata* was quite common on steep slopes and cliffs throughout Switzerland, despite the prevailing impression that the species was threatened by local extinction. Several other conifers, most notably *Juniperus phoenicea, J. oxycedrus* and *J. communis,* occur throughout central and southern Europe as slow-growing plants in ancient woodlands on cliffs to elevations of 350 m (Larson *et al.*, 1999).

In western Europe and the British Isles, studies of cliff and rock outcrop vegetation have been numerous and show trends that are similar to one another in many ways. Rodwell (1992) reports a small number of grass-dominated and herb-dominated communities that only occur on steep rock. The *Saxifraga aizoides–Alchemilla glabra* banks community and the *Luzula sylvatica–Vaccinium myrtillus* tall herb community both occur on exposed cool and damp rock outcrops at low to middle elevation (to maximum 800 m). *Cystopteris montana* and *C. fragilis* also occur in this community. These communities all occur in places that are free from human disturbance, fire and grazing, and where there is continuous contact between open rock, water and the vegetation. Up to 700 mg calcium per gram of soil has been reported from some of these ledges, but pH values are still near neutral (6.8–7.5). When such sites become overtly vertical, *Asplenium ruta-muraria, Cystopteris fragilis, Polystichum lonchitis* and *Pinguicula vulgaris* become common. Rodwell also reports a different vegetation community type on rock ledges that includes *Angelica sylvestris, Empetrum nigrum, Juniperus communis, Cystopteris fragilis* and *Asplenium viride.* McVean and Ratcliffe (1962), Ratcliffe (1960) and Birks (1973) also present descriptions of vegetation communities in different parts of the British Isles. The tall herb community is recognized by all of these authors as a community type that appears on steep rocks in places that deter sheep grazing. Whereas soil development does not take place in such settings, thick accumulations of organic matter are usually found, with the higher plants rooted directly in the humus. Both Birks (1973) and Ratcliffe (1960) report dwarf forms of *Juniperus communis* growing on cliffs and broken ground from sea level to elevations of 330 m. In these communities near the sea, *Grimmia maritima* and *Asplenium marinum* occur as well as *Potentilla caulescens*. Other, more sheltered settings include *Asplenium trichomanes, Fissidens cristatus, Cystopteris fragilis* and *Phyllitis scolopendrium*. Calcareous ledges and cliffs support *Saxifraga aizoides, Pinguicula vulgaris, Asplenium viride* and *Polystichum lonchitis* (Huntley, 1979). Constant, though not abundant, species in the more inland woodland cliff outcrops include *Betula pubescens, Corylus avellana, Sorbus aucuparia, Ilex aquifolium, Prunus padus, Quercus petraea* and *Juniperus communis*. Birks (1973, 1988) agrees with others that these stands are restricted to areas that are inaccessible to animal grazing. He also reports that interspecific plant competition is low in these areas and that the vegetation of rock outcrops and cliffs tends to be characteristic of sites both much further north and further south of the British Isles, suggesting that cliffs support truly relict vegetation from all episodes of climatic change.

For the most part, cliffs in the British Isles are viewed as habitats that never supported a significant woody component (Pigott & Walters, 1954; Rodwell, 1992). Despite this, there have been a number of studies of the few trees or shrubs that occur on cliffs or rock outcrops and in many cases these individual trees are quite rare or important components of the landscape that escaped significant habitat destruction by people (Pigott & Huntley, 1980). Pigott (1969) and Pigott and Huntley (1978, 1980) have reported on the habitat characteristics and longevity (Pigott, 1989) of *Tilia cordata* and *T. platyphyllos* on large outcrops of dolomitic limestone rock near Matlock Dale and elsewhere in central England. These trees grow slowly, for exceptionally long periods of time, and with highly deformed axes as a result of their growth conditions. Pigott and Walters (1954) report that *Helianthemum apenninum, Buxus sempervirens, Sorbus aria* and *Taxus baccata* occur on many of the same outcrops. Similar descriptions are given by Jackson and Sheldon (1949). Pigott and Pigott (1993) have also reported on the habitat characteristics of *Tilia cordata* in the southern portion of its range in southern France. Here, as in the extreme northern part of its range, its distribution is restricted to vertical north-facing cliffs whose water supply is nearly continuous over time.

At Markland Grips, Sheffield, UK, species that are abundant on limestone cliffs include *Hedera helix, Rubus fruticosus* and *Solanum dulcamara*, as well as the shrubs *Sambucus nigra* and *Cornus sanguinea* (Jackson & Sheldon, 1949; Table 4.1). Saplings and seedlings of *Betula* spp. and *Taxus baccata* are also common, especially on the cliff top and face. The authors report finding old individuals of *T. baccata*, but no increment coring was performed to precisely measure their age. We returned to this site in the summer of 1997 and found that living *Taxus baccata* with estimated ages

Table 4.1. *Species composition of cliff top, cliff face and scree slopes of Markland Grips*

| Composition of vegetation of cliff top | | | |
|---|---|---|---|
| **Trees** | | | |
| Betula spp. | l.d. | Taxus baccata | l.d. |
| Quercus robur | f. | Ulmus glabra | f. |
| Fraxinus excelsior | f. | Tilia platyphyllos | o. |
| Sorbus aucuparia | o. | | |
| **Shrubs** | | | |
| Corylus avellana | d. | Ligustrum vulgare | l.a |
| Cornus sanguinea | l.a. | Crataegus monogyna | f. |
| Viburnum opulus | f. | Sambucus nigra | o.–l.f. |
| Rhamnus frangula | r. | | |
| **Field layer under shrubs** | | | |
| Mercurialis perennis | l.d. | Hedera helix | l.d. |
| Sanicula europaea | l.f. | Brachypodium sylvaticum | l.f. |
| Rubus fruticosus | o. | Viola riviniana | o. |
| **Field layer in openings** | | | |
| Brachypodium sylvaticum | f. | Carex flacca | f. |
| Scabiosa columbaria | o. | Centaurea nigra | o. |
| Poterium sanguisorba | o. | Serratula tinctoria | l. |
| Carex montana | l. | Avena pubescens | l. |

| Composition of vegetation of cliff face | | | |
|---|---|---|---|
| **Trees** | | | |
| Taxus baccata | f. | Ulmus glabra | o.–f. |
| Betula spp. | o. | Fraxinus excelsior | o. |
| Quercus robur | o. | Tilia platyphyllos | o. |
| Sorbus aucuparia | o. | | |
| **Shrubs** | | | |
| Sambucus nigra | o. | Crataegus monogyna | o. |
| Ilex aquifolium | o. | Viburnum opulus | o. |
| Corylus avellana | o. | Cornus sanguinea | o. |
| Ligustrum vulgare | o. | Rosa spp. | o. |
| **Other angiosperms** | | | |
| Hedera helix | l.v.a. | Rubus fruticosus | l.a. |
| Lactuca muralis | a. | Sanicula europaea | f. |
| Mercurialis perennis | f. | Brachypodium sylvaticum | f. |
| Melica uniflora | f. | Geranium robertianum | f. |
| Solanum dulcamara | f. | Viola riviniana | f. |
| Urtica dioica | f. | Stachys sylvatica | f. |
| Taraxacum officinale | o. | Epilobium montanum | o. |

Table 4.1 (*cont.*)

| Composition of vegetation of cliff top | | | |
|---|---|---|---|
| Dactylis glomerata | o. | Solidago virgaurea | o. |
| Fragaria vesca | o. | Senecio jacobaea | o. |
| Hieracium spp. | o. | Campanula rotundifolia | o. |
| Lonicera periclymenum | o. | Deschampsia caespitosa | o. |
| Arctium vulgare | r. | Cirsium vulgare | r. |
| Bromus ramosus | r. | Serratula tinctoria | r. |
| Stellaria media | r. | Primula veris | r. |

Ferns
| | | |
|---|---|
| Asplenium ruta-muraria | Cystopteris fragilis |
| Scolopendrium vulgare | Dryopteris filix-mas |
| Polystichum aculeatum | Dryopteris dilatata |

| Composition of vegetation of scree slope | | | |
|---|---|---|---|

Trees
| | | | |
|---|---|---|---|
| Ulmus glabra | l.d. | Fraxinus excelsior | f. |
| Betula spp. | f. | Tilia platyphyllos | o. |
| Acer campestre | o. | Quercus robur | o. |

Shrubs
| | | | |
|---|---|---|---|
| Sambucus nigra | a. | Corylus avellana | a. |
| Cornus sanguinea | f. | Crataegus monogyna | f. |
| Prunus spinosa | o. | Ligustrum vulgare | o. |

Ground layer
| | | | |
|---|---|---|---|
| Mercurialis perennis | d. | Scilla non-scripta | l.d. |
| Sanicula europaea | l.d. | Asperula odorata | l.d. |
| Hedera helix | f. | Rubus fruticosus | f. |
| Geranium robertianum | f. | Melandrium dioicum | f. |
| Viola riviniana | f. | Viola reichenbachiana | f. |
| Lamium galeobdolon | f. | Arum maculatum | f. |
| Geum urbanum | | Oxalis acetosella | |
| Potentilla fragariastrum | | Fragaria vesca | |
| Veronica chamaedrys | | Veronica montana | |
| Glechoma hederacea | | Anemone nemorosa | |
| Orchis fuchsii | | Arctium vulgare | |
| Carex sylvatica | | Dactylis glomerata | |
| Brachypodium sylvaticum | | Poa nemoralis | |

*Notes:*
Taken from Jackson and Sheldon (1949). d. = dominant; a. = abundant;
f. = frequent; o. = occasional; l. = local; r. = rare; v. = very.

to 1052 years still occur on these cliffs (Larson *et al.*, 1999). The larger trees and shrubs become rooted in cracks and crevices on the cliff and eventually cause rockfall to occur, exposing their root systems and gradually resulting in the recession of the cliff margin. Jackson and Sheldon (1949) note that cliff-edge and cliff-face vegetation appeared undisturbed in comparison to the highly disturbed vegetation of the plateau on top of the cliff.

Bunce (1968) investigated the restriction of certain species to cliffs of dolerite, rhyolite and rhyolite tuff on Mount Snowdon, UK, and the factors controlling species distribution on these cliffs. He indicates that the cliff community is essentially closed and that there is no floristic continuity between cliff and surrounding habitats. The cliff flora includes a rare arctic-alpine component (*Silene acaulis, Saxifraga oppositifolia*), as well as lowland species and the maritime plant *Armeria maritima*. *Vaccinium myrtillus* occurs in oligotrophic stands along with *Vaccinium vitis-ideaea* and *Dryopteris filix-mas*. Common desiccation-tolerant mosses include *Rhytidiadelphus loreus* and *Dicranum scoparium*, and ledges are colonized by *Campanula rotundifolia* and the succulent *Sedum rosea*. No evidence was found in this study that localized subcommunities exist on the cliffs, even though some species clearly differ in their habitat preferences. For example, *Saxifraga oppositifolia* tends to be more common in wet eutrophic places near the bottom of tall cliffs, while *Campanula rotundifolia* is more widespread on the cliff face.

An explanation for the arctic-alpine component of many cliff floras is that climate fluctuations during glaciation and deglaciation resulted in the restriction of previously widespread species to mountains and cliffs. They are therefore relic species which have survived on cliffs as isolated populations (Crawford, 1989). There is palynological evidence to support this notion (Godwin, 1956; Birks, 1988). The presence of many relic species further suggests that cliffs are protected from rapid change and that the vegetation is relatively constant. Bunce (1968) is one of many authors to conclude that cliffs are refuges also in the sense that they provide protection from human influences, grazing animals and competing species. He notes that many of the arctic-alpine species are frequent in areas of low competition, such as crevices where limited space prevents the establishment of competing plants. The restriction to cliffs of other particularly rare British plants, including *Adiantum capillus-veneris, Dianthus gratianopolitanus, Sorbus domestica* and *Pyrus cordata,* may also reflect their intolerance of competition and/or herbivory.

A quantitative analysis of the flora on basic igneous rocks in Scotland suggested that there are three community types (Jarvis, 1974). They are partly distinguished on the basis of aspect and soil conditions, but show considerable overlap in species composition. The first community type, which is found on rocks facing north and west, includes the species *Solidago virgaurea*, *Dicranum scoparium* and *Viola rivinana*. South-facing or west-facing cliffs are characterized by the winter annual *Aira praecox* and the desiccation-tolerant moss *Polytrichum piliferum*. In the third group, also on south-facing rock faces, are *Lychnis viscaria* and *Umbilicus rupestris*. *Campanula rotundifolia*, on the other hand, is present across all community types. *Umbilicus rupestris* is also mentioned by Crawford (1989) as a rare species restricted to damp cliffs on the west and south sides of the British Isles.

The idea that distinct communities of lower plants exist on exposed rock surfaces was explored by Allen (1971), who conducted a quantitative analysis of epilithic algal communities on natural limestone cliff faces and walls of an abandoned quarry in north Wales. Allen found that the species that occurred on the rocks displayed continuous variation at a variety of spatial scales, and that the concept of 'community-type' was only useful as an item of convenience to people trying to study these systems. He also concluded that discrete ecosystem boundaries may be evident at one spatial scale but may disappear when another is examined. This idea was important because most of the vegetation sampling protocols used in phytosociology require that sample plots of a certain uniform shape and size be used, and cliffs rarely display communities that are easily sampled using these methods. McVean and Ratcliffe (1962) as well as Stöcker (1965) were also aware of this difficulty, as discussed in Chapter 1.

Cliff vegetation studies have also been conducted in northern Europe. Communities of endolithic algae dominated by *Gloeocapsa* are described for cliffs in Fennoscandia (Häeyren, 1940), continuing a trend that is evident worldwide. Holmen (1965) reports that exposed calcareous cliffs in the middle of spruce-fir forest in Sweden support a herb community dominated by *Cystopteris montana*, *Sedum annuum* and *Polystichum lonchitis*. Coastal granite cliffs investigated by Hallberg and Ivarsson (1965) are dominated by the lichens *Rhizocarpon constrictum*, *Ramalina siliquosa*, *Xanthoria parietina*, and several species of *Umbilicaria*, in addition to *Sedum acre*. Faegri (1960) reports various species of *Asplenium* on both acid and alkaline rocks in Norway, with most of these species showing no clear substrate specialization. *Geranium lucidum* and *G. sanguineum* are always associated with warm, south-facing cliffs. When growing on south-facing cliffs, *Hedera helix* was found much larger than when growing on trees or on the ground. The possibility that cliff-dwelling *H. helix* may also be much older was not considered by Faegri. The author also noted that *Phyllitis scolopendrium* and *Taxus baccata* in Norway are almost completely restricted to rock outcrops and cliffs in shaded forest settings. Nordhagen (1943) presents another comprehensive account of cliff vegetation in central Norway. As reported above, these areas support a variety of herbs, including *Saxifraga cotyledon*, *S. oppositifolia* and *S. aizoides*, as well as *Epilobium collinum*, *Silene rupestris* and *Campanula rotundifolia*. The ferns *Polypodium vulgare*, *Woodsia ilvensis*, *Asplenium septentrionale*, *A. trichomanes*, *A. ruta-muraria*, *A. viride* and *Dryopteris phegopteris* also occur. In locations where the rock walls collapse, a sparse woody vegetation including *Juniperus communis* var. *nana* establishes and is accompanied by *Carex rupestris*, *Sedum rosea* and *Dicranum fuscescens*.

Lundqvist (1968) presents a comprehensive account of the vegetation and ecological characteristics of steep hillsides and cliffs at latitude 66–68°N in Sweden. Karlsson (1973) provides a similar account for Swedish Lappland in Sarek National Park and includes individual site descriptions and pH measurements for stands supporting particular species. *Juniperus communis*, *Betula pubescens*, *Festuca ovina*, *Sorbus aucuparia*, *Grimmia hartmanii* var. *anomala*, *Saxifraga aizoides*, *Prunus padus*, *Arctostaphylos uva-ursi*, *Tortula ruralis*, *Campanula rotundifolia*, *Cystopteris fragilis*, *Polypodium vulgare*, *Asplenium viride* and *Dryas octopetala* all grow on exposed cliff faces of Mt Aistjajakk (Lundqvist, 1968). Despite the predominantly acid rock substrate, most of the species investigated by Karlsson grew in soil with near-neutral pH values that usually ranged from 5.4–7.0. *Rosa majalis* is very rare in northern Sweden, but Karlsson discovered it on a high, south-facing cliff on a ledge shared with a stunted juniper. *Woodsia alpina*, *W. glabella*, *Asplenium viride*, *Polypodium vulgare* and *Campanula rotundifolia* are other species that occur on these ledges. While this author acknowledges that most large cliffs could not be sampled without rock-climbing equipment, he nonetheless provides interesting accounts of many rare and important species that are more or less confined to steep rock outcrops and cliffs in this area. Among the 400 taxa included in this volume, there are several references to ancient (>300 year) *Juniperus communis* that occur on exposed cliffs, cliff ledges and the upper parts of talus slopes. Reference is also repeatedly made to the inability of cliff vegetation to compete with the more productive communities of surrounding level ground, but no quantitative data are provided to support this claim.

Kallio, Laine and Mäkinen (1971) report that in northern Finland, *Juniperus communis* can obtain ages in excess of 1000 years in locations where growth is constrained and where cambial mortality occurs around the axis of the plant, leading to stunted and grossly deformed plants. Kallio, Laine and Mäkinen (1969) also state that cliffs near the Kevo Subarctic Research Station support stunted communities that include *Juniperus communis, Woodsia ilvensis, Campanula rotundifolia, Cerastium alpinum, Poa glauca, P. nemoralis, Polypodium vulgare, Asplenium viride, Pinguicula alpina, P. vulgaris, Saxifraga oppositifolia, S. aizoides* and *Cystopteris fragilis*. This assemblage of species bears a striking ecological similarity to groups described previously for other geographical regions.

In southern Europe, exposed calcareous cliffs can be found in Spain and their vegetation has been recently investigated using multivariate statistical methods (Escudero & Pajarón, 1994, 1996; Escudero, 1996). Nine different community types can be distinguished and their distribution is mainly controlled by aspect and slope. The most common species are *Lonicera pyrenaica, Koeleria vallesiana, Silene saxifraga, Rhamnus alpina* and *Globularia repens*. Non-vascular plants were not included in this study. The flora of sandstone and conglomerate outcrops is differentiated from the calcareous flora in this area (Escudero, Gavilán & Pajarón, 1994).

On calcareous cliffs and steep slopes around the perimeter of the Mediterranean Sea, a number of woody plants grow on bare rock which some of them actively penetrate with their roots (Oppenheimer, 1956, 1957). These species include *Pistacia lentiscus, Ceratonia siliqua* and *Rhamnus palaestina*. Lichens and mosses are also common. A more comprehensive study of the cliffs in this region was conducted by Davis (1951), who reports a very rich flora of obligate cliff species, the majority of which are chamaephytes and very few of which are therophytes. No annuals grow on vertical exposures and only a few are found in the crevices of sloping rock, including several species of *Campanula*. *Asplenium adiantum-nigrum* var. *virgili*, the desiccation-tolerant fern *Polypodium vulgare*, and the succulent *Sedum laconium* occur on both sheltered and exposed faces. Another group of species is found only on vertical and overhanging faces, among them *Dianthus pendulus, Centauria speciosa* and *Podonosma syriacum*. The last-mentioned species is conspicuously absent from basalt rocks in the area, which instead are dominated by *Rosularia lineata*. The only tree species on the exposed cliffs is *Cupressus*

*Figure 4.4* *Cupressus sempervirens* L. var. *horizontalis* on a limestone cliff in northern Cyprus. Photo taken from Davis (1951), and used with permission of Blackwell Science Ltd, Publishers.

*sempervirens* var. *horizontalis* (Fig. 4.4). It grows slowly from large cracks and crevices and becomes deformed from rockfall and gravity working directly on the main axis. Step-crevices in rock support remnant populations of trees such as *Quercus coccifera*. Davis (1951, p. 77) states that:

> Natural selection in a cliff habitat presumably acts in favour of lignification and longevity. In a cliff community most of the holes suitable for ecesis are already occupied, so that it is extremely difficult for the individual to produce progeny. Seed is the only method of reproduction; stolons and runners are not adapted to the habitat conditions, and plants possessing them are not found in Mediterranean crevice communities. The scarcity of seedlings shows that several years may go by, especially on vertical and overhanging rock, without an individual establishing progeny. This fact (which is true of other biologically closed communities) must be unfavourable for the persistence of annuals in these exacting habitats, but will certainly favour the selection of long-lived perennials.

Thus, he argues that cliffs are biologically closed communities. He concludes his analysis by stating that cliffs in the Mediterranean serve as refugia from unfavourable climatic change, from competition by more aggressive level-ground vegetation, and from grazing. The high rate of endemism is due to periodic climate change causing the local extinction of some cliff species but not others, followed by a very slow, perhaps unmeasurable, rate of reinvasion. Lavranos (1995) has come to similar conclusions to explain the occurrence of rare species of *Aloe* on cliffs in Oman. Zohary (1973) also reports that under desert conditions in north Africa and the Middle East, cliffs and rock outcrops produce conditions more favourable to plant life than other stony habitats. He argues that rock crevices accumulate both soil and water, and that their steep slopes lessen the radiation input that produces evaporation. Among the plants that exploit such cliffs are large deformed *Juniperus phoenicea* as well as *Oreganum dayi, Heliotropium rotundifolium, Helianthemum ventosum* and *Centaurea eryngioides*.

In North America, studies of cliff vegetation are almost exclusively restricted to the eastern part of the continent. Despite the abundance of spectacular cliffs in the western USA and western Canada, investigations of their flora are almost completely lacking. One of the few is by Everett and Robson (1991), who reported the occurrence of rare cliff-dwelling species such as *Petrophyton cinerascens* and *Erigeron basalticus* in Washington State, USA. A second one is by Wentworth (1981), who compared the vegetation on granite and limestone in rocky terrain in the Mule Mountains of Arizona. While this study did show some degree of species separation as a function of rock type, the particular property of the rock that resulted in this separation was not found. Camp and Knight (1997) reported extremely high levels of plant and avian biodiversity on cliffs in Joshua Tree National Monument, USA. *Pinus monophylla* and *Quercus cornelius-mulleri* were found exclusively on cliffs, along with an array of perennial vascular plants. The authors noted that cliffs have rarely been included in land management decisions in the western USA because managers have viewed these sites as zones that separate two landscape features, rather than habitats in their own right.

While studies of cliffs in eastern North America are relatively abundant, most of the work dates from the middle of the twentieth century and much of it is qualitative or focused on single plant growth forms. For example, investigations of the lichen vegetation on limestone cliffs in Ontario (Yarranton & Green, 1966) and in Wisconsin (Foote, 1966) make no reference to other plant groups. One of the better known North American cliff floras is that of granite outcrops in the south-eastern and south-central USA. Many of the plant species that are characteristic for such outcrops belong to genera that are found on cliffs and rock outcrops in other parts of the world. Such species include *Asplenium platyneuron*, *Juniperus virginiana*, *Polypodium polypodioides* and *Bulbostylis capillaris* in Georgia (Bostick, 1971); and *Juniperus ashei*, *Selaginella riddellii*, *Cheilanthes lindheimeri*, *Pellaea ternifolia* and *Campanula reverchonii* in Texas (Walters & Wyatt, 1982). The latter study also found a large number of endemic species.

In a majority of the older studies on granite outcrops, the main focus is the process of succession and the question of how quickly it progresses at such sites (e.g. Oosting & Anderson, 1937, 1939; Keever, Oosting & Anderson, 1951). Some interest in succession on granite outcrops continues (Uno & Collins, 1987; Houle & Delwaide, 1991). Cryptogams of various species are described, with members of the genus *Cladonia* being interpreted as invaders that are followed by *Dicranum scoparium* and other mosses. Grasses, sedges and forbs follow later in the invasion sequence. Woody species such as *Chionanthus virginica* and slow-growing *Juniperus virginiana* then establish in cracks and crevices with small amounts of soil that accumulates from the death of mosses and lichens. Increment cores taken from small but maturing trees on these outcrops show that growth is very steady but exceptionally slow (Oosting & Anderson, 1937). The vegetation surrounding open rock outcrops and cliff faces is described as being completely different in growth rate, character, composition and origin. However, recent models of successional development make it clear that Oosting misinterpreted the form and rate of successional development on rock by assuming that the distinct patches of vegetation in space represented patches in time (Drury & Nisbet, 1973; Grime, 1979). Most evidence now suggests that differences in life-forms and biomass of vegetation on these outcrops are a result of spatially variable habitat quality that is present from the beginning, and that cliff and rock outcrop communities change very slowly over time (Smiley & George, 1974). This makes it unnecessary to invoke successional theory to explain repeating patterns in the community structure of granite outcrops (Winterringer & Vestal, 1956; Burbanck & Platt, 1964; Shure & Ragsdale, 1977; Phillips, 1981). Evidence supporting the conclusion that rock outcrops support stable slow-growing climax communities is provided by Bostick (1971), who showed that the vegetation of a granite outcrop changed only imperceptibly over a 30-year period.

Limestone cliffs are common throughout the south-eastern and mid-western part of the USA and into Ontario and Quebec, Canada, and their vegetation has been investigated in a number of places. One of the more comprehensive accounts is that by Cowles (1901) on the vegetation near Chicago. It makes detailed reference to the vascular flora of cliffs near Starved Rock on the Illinois River and reports that the vegetation of steep slopes, rocky gorges and cliffs is remarkably alike despite differences in aspect and location. Based on the growth form of the plants, Cowles interpreted both exposed and shaded cliffs to be dry. Species that occur on such sites include *Selaginella rupestris*, *Campanula rotundifolia* and *Pellaea atropurpurea*. Deformed and slow-growing arborvitae (*Thuja occidentalis*) also grow on these rock faces. A similar community is found on the limestone cliffs that line the Mississippi River in north-western Illinois, except that *T. occidentalis* is replaced by *Juniperus virginiana* (Nuzzo, 1996; Fig. 4.5). Trees shown on the cliffs in this photograph were sampled by us in the summer of 1997, and proved to have ages extending to 380 years (Larson *et al.*, 1999).

Cliffs of dolomitic limestone also occur in south-eastern Iowa, USA (Fig. 4.6) and support *Juniperus virginiana*, *Cystopteris bulbifera* and *Adiantum pedatum* on ledges and in shady crevices (Lammers, 1980). Dry exposed surfaces are dominated by *Pellaea glabella*, *Aralia racemosa*, *Aquilegia canadensis* and various rock cresses (*Arabis* spp.). Hotchkiss *et al.* (1986) and Walker (1987) report a very distinctive community of rare or disjunct plants on exposed limestone cliffs in Kentucky, Tennessee, North Carolina and Ohio. This community is characterized by *Pellaea atropurpurea*, *Carex eburnea*, *Sedum ternatum*, *Asplenium platyneuron*, *A. ruta-muraria*, *A. ebenoides*, *Cystopteris bulbifera* and *Aquilegia canadensis*. Most

*Figure 4.5* A limestone cliff in Mississippi Palisades Park, Illinois, USA. These cliffs were the location of several studies on the impact of human disturbance (Nuzzo, 1995, 1996). Photo courtesy of J.A. Gerrath.

*Figure 4.6* A limestone cliff at Starr's Cave, south-eastern Iowa, USA. Photo taken from Lammers (1980), and used with permission of the Iowa Academy of Science.

cliffs with aspect <350° and >10° also support *Juniperus virginiana*. Twisted, inverted and upward-turned trees of *Thuja occidentalis* occur only on high limestone or sandstone cliff faces of direct northern aspect. This restricted distribution is a sign of the more northern, colder and wetter affinity of this species, which exists there in the southern disjunct portion of its range. A detailed genetic analysis of these cedar populations showed them to be relics that once were in the centre of the species range (Young, 1996).

On high elevation cliffs and rock outcrops in the southern Appalachian Mountains (western North Carolina and eastern Tennessee), Wiser (1994) found a large number of species that are restricted to open rocky slopes. Among them are *Selaginella tortipila*, *Scirpus caespitosus*, *Campanula rotundifolia*, as well as three species of *Asplenium*. One of the few woody plants restricted to rock outcrops is *Juniperus communis* var. *depressa*. These outcrops are composed of a variety of metamorphic and igneous rocks.

An unusual escarpment flora on a carbonate rock outcrop in western Quebec, Canada, is described by Brunton and Lafontaine (1974). *Xanthoria elegans* brightly colours the rock, making the escarpment visible from a considerable distance. The trees *Thuja occidentalis* and *Juniperus virginiana* are rooted on the cliff faces and appear to be very slow growing and deformed. The ferns *Cystopteris bulbifera*, *C. fragilis*, *Cryptogramma stelleri*, *Asplenium trichomanes*, *A. platyneuron* and *Pellaea atropurpurea* are found in crevices and on ledges, as is the rock-cress *Arabis holboelli* and the arctic species *Draba cana*. The desiccation-tolerant fern *Polypodium virginianum* occurs on large rocks at the base of the cliffs. The authors conclude that 'We believe that the southern species survive today on the escarpment because (1) they became established on the escarpment during the warm period when conditions became more suitable, (2) the calcareous nature of the sites described is particularly favourable for these species, (3) they are in refugia where competition from contemporary species is minimized and (4) their locations permit maximum advantage from the south orientation of the cliffs.' Table 4.2 illustrates some of the relationships between habitat characteristics and species composition on this escarpment. Also in Quebec, Canada, Morisset (1971) and Morisset, Bedard and Lefabvre (1983) studied exposed limestone cliffs in Forillon National Park. They concluded that rare, endemic plants occurring on these cliffs are remnants of an earlier arctic flora that followed the glacial margin at the peak of the last glaciation. These plants occur mainly in the lower parts of the exposed cliff faces, above an embankment of fallen earth (Fig. 4.7).

Table 4.2. *The nature of the topography, the substratum, the dominant plant cover, the maximum elevation, and the number of species with northern or southern affinities along the Eardley Escarpment, Quebec.*

| Site | Topography | Substratum | Dominant plant cover | Maximum elevation (m) | Number of species with northern or southern affinities |
|---|---|---|---|---|---|
| 1 | Low rocky hills | Large marble boulders | Beech, sugar maple, ironwood | 120 | 0 (N)<br>5 (S) |
| 2 | Ravine | East-facing marble wall, shaded, near stream | Sugar maple, red oak, bitternut hickory, white ash | 160 | 0 (N)<br>5 (S) |
| 3 | Series of steep valleys through 15 m marble cliff | Rich organic soil debris from marble cliff; marble boulders and ledges | Bitternut hickory, white oak, hackberry | 180 | 2 (N)<br>9 (S) |
| 4 | 24 m marble cliff with ledges and fissures | Pegmatite seams, calcareous debris and soil | Poison ivy, staghorn sumac, red oak, red maple | 160 | 0 (N)<br>6 (S) |
| 5 | 18 m marble cliff with ledges | Ledges with debris from marble cliff | Red juniper, red oak, white oak | 200 | 2 (N) |
| 6 | 45 m cliff with ledges, 60 m talus | Syenitic gneiss and marble, calceolate veins | Red juniper, red oak, white cedar, staghorn sumac, hackberry | 210 | 6 (N)<br>10 (S) |
| 7 | 15 m cliff above 100 m talus | Syenitic gneiss | Red and sugar maple, ironwood, bitternut hickory, beech | 200 | 3 (N)<br>5 (S) |
| 8 | 45 m cliff, boulder talus, fissures, ledges | Syenitic gneiss, calceolate veins | White cedar, red juniper, common juniper, poison ivy | 210 | 5 (N)<br>11 (S) |
| 9 | 150 m almost barren cliff | Syenite | Red juniper, white cedar, white birch, common juniper | 375 | 1 (N)<br>2 (S) |
| 10 | 30 m marble cliff, ledges | Ledges with debris from marble cliff | Basswood, red maple, red oak | 130 | 1 (N)<br>4 (S) |

*Source:* Data taken from Brunton and Lafontaine (1974).

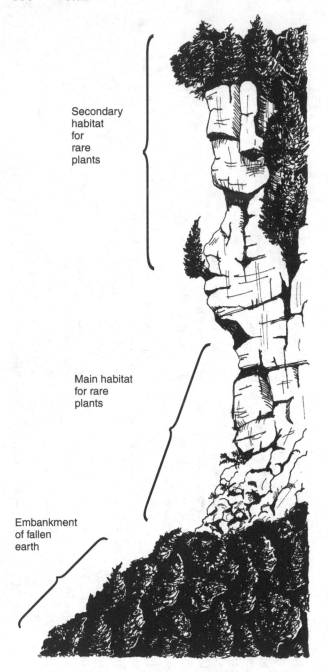

Secondary
habitat
for
rare
plants

Main habitat
for rare
plants

Embankment
of fallen
earth

*Figure 4.*7   The localization of rare plant habitat on cliffs at Forillon National
Park, Quebec, Canada. Original illustration courtesy of C.E. Ryan. Modified
from Morisset *et al.* (1983).

A quantitative study of a large granite cliff in Bon Echo Park, eastern Ontario, Canada, revealed a highly diverse and rich flora, particularly of lichens (Dougan & Associates, 1995). Seven different community types were identified on the cliff face and talus slope. Vertical faces are dominated by twisted *Thuja occidentalis* and *Juniperus communis*, with *Deschampsia flexuosa, Campanula rotundifolia, Pellaea glabella* and two species of *Solidago* in the understory. Some individuals of *Thuja occidentalis* are over 900 years old. Shade-loving species such as *Cryptogramma stelleri* and *Cystopteris fragilis* are common in gullies, and *Polypodium virginianum* and *Woodsia ilvensis* in the talus.

The limestone cliffs of the Niagara Escarpment in southern Ontario, Canada, support a distinct and predictable community of higher plants, mosses, lichens and terrestrial algae (Larson *et al.*, 1989; Cox & Larson, 1993a, 1993b; Gerrath, Gerrath & Larson, 1995). The cliff vegetation remains similar in floristic composition along the 725 km that the escarpment extends from north to south, even though the vegetation of surrounding level ground changes from Carolinian forest in the south to near-boreal forest in the north. The dominant tree species on the exposed cliffs is *Thuja occidentalis*, although *Betula papyrifera* and *Sambucus pubens* are also occasionally present. Individual trees of *Thuja occidentalis* are stunted and deformed, and may reach 1890 years in age; the number of trees pre-dating European settlement is considerable (Larson & Kelly, 1991). A large number of ferns grow on the cliff face, among them *Asplenium trichomanes, A. viride, A. ruta-muraria, Cystopteris bulbifera, C. fragilis, Pellaea glabella* and *Polypodium virginianum*. The cliffs of the escarpment represent the main part of the range for some exceptionally rare plants in North America, including *Asplenium ruta-muraria* and *Phyllitis scolopendrium* (Soper, 1954; United States Fish and Wildlife Service, 1993). Common species include *Poa compressa, Potentilla erecta* and the mosses *Tortella tortuosa* and *Homomalium adnatum* (Larson *et al.*, 1989; Cox & Larson, 1993a, 1993b). In the northern parts of the escarpment a number of grasses with arctic affinities are present, such as *Poa glauca, P. alpina* and *P. canby* (Morton & Venn, 1984, 1987); other species found there are *Draba cana* and *Pinguicula vulgaris* (Fig. 4.8), as well as the western grass, *Trisetum triflorum*. A variety of mostly crustose lichens are also present on the Niagara Escarpment (Yarranton & Green, 1966), and rich communities of endolithic and epilithic micro-organisms were recently discovered and quantitatively sampled (Gerrath *et al.*, 1995; Matthes-Sears *et al.*, 1997; Matthes-Sears *et al.*, 1999). The dark staining of rocks in many locations is actually due to very large and productive communities

*Figure 4.8* *Pinguicula vulgaris* growing on a vertical limestone cliff of the Niagara Escarpment, Ontario, Canada. Photo by P.E. Kelly.

*Figure 4.9* Limestone cliffs of the Niagara Escarpment, Ontario, Canada, showing the intense vertical coloration by epilithic micro-organisms. Photo by D.W. Larson.

of such micro-organisms (Fig. 4.9). The endolithic organisms include 19 taxa of eukaryotic algae and nine taxa of cyanobacteria (Table 4.3), but the group present at the highest frequency by far is fungi. Fungi are predominant in the epilithic zone as well (Fig. 4.10). Like the communities of macroscopic plants on the cliffs, both epilithic and endolithic communities are similar in their composition from north to south along the Niagara Escarpment (Table 4.3; Fig. 4.10).

Part of the work on the Niagara Escarpment has focused on demonstrating the changes that occur in vegetation and environment as one moves from the plateau at the top of the cliff to the face and the talus below (Larson *et al.*, 1989). Figure 4.11 shows canopy profiles across five representative transects to illustrate the sudden change in canopy cover, and its relative consistency from site to site, suggesting distinct and consistent differences between cliffs and surrounding habitat. The distribution of individual plant species across this profile is quite variable (Fig. 4.12), with some species (such as *Polypodium virginianum*) most frequent at the cliff edge, others most frequent on the cliff face (*Pellaea glabella*), and yet others in the talus (*Geranium robertianum*). Lower plants have only been studied on the cliff face, where they appear to occur with about

Table 4.3. *List of endolithic organisms identified from 12 sites along the Niagara Escarpment, Ontario, Canada*

| | Gr | RC | Mi | MN | MC | MR | Me | PV | LH | EL | FL | BR |
|---|---|---|---|---|---|---|---|---|---|---|---|---|
| **Cyanophyta** | | | | | | | | | | | | |
| Chlorogloea sp. | x | x | x | x | x | x | x | x | x | x | x | |
| Chroococcidiopsis sp. | | | | | x | | | x | | | x | |
| Gloeocapsa sp. | x | | | | x | | x | x | x | | | |
| Gloeothece sp. | | x | | | | | | | | | | |
| Nostoc sp. | x | | | | x | | | | | | | |
| Phormidium sp. | x | | | x | | | | | | | | |
| Plectonema sp. | | | | | | | | | x | | | x |
| Schizothrix sp. | | | | | | | | x | | | | |
| Synechocystis sp. | | x | | | x | | x | x | x | | | |
| **Chlorophyta** | | | | | | | | | | | | |
| Chlorella sp. | x | x | x | x | | x | | | | | | |
| Chlorococcum sp. | x | | | | | | | | | | | |
| Chlorosarcina sp. | | | x | x | x | | | | | | | |
| Coccomyxa sp. | x | x | x | | | x | | | | x | x | |
| Gloeocystis sp. | | x | x | | | | | | | | | |
| Klebsormidium flaccidum | x | | | | | | | | | | | |
| Pseudopleurococcus sprintzii | x | x | x | | | | | | | | | |
| Scotiella tuberculata | x | | x | | | | | | | | | |
| Stichococcus bacillaris | x | x | x | x | | x | x | x | x | x | x | x |
| Stichococcus chlorelloides | x | x | x | x | x | | x | | x | x | | x |
| Stichococcus exiguus | | x | | | | | | | | | | |
| Stichococcus minutus | x | | | | x | | | | | x | x | x |
| Trebouxia sp. | | x | | | | | x | x | | | | |
| Trentepohlia sp. | x | | | | | | | | | | | |
| **Chrysophyta–Xanthophyceae** | | | | | | | | | | | | |
| Chloridella neglecta | x | x | x | x | x | x | x | x | x | | x | x |
| Heterococcus sp. | x | | x | x | | | | | | | | |
| Heterothrix sp. | | | | | | | x | | | | | |
| **Bryophyta** | | | | | | | | | | | | |
| Moss protonema | x | | x | | x | x | | | | | | x |

Notes:
Gr = Grimsby, RC = Rock Chapel, Mi = Milton Heights, MN = Mount Nemo, MC = Mono Cliffs, MR = Metcalfe Rock, Me = Meaford, PV = Purple Valley, LH = Lion's Head, EL = Emmett Lake, FL = Flowerpot Island, BR = Bear's Rump Island.

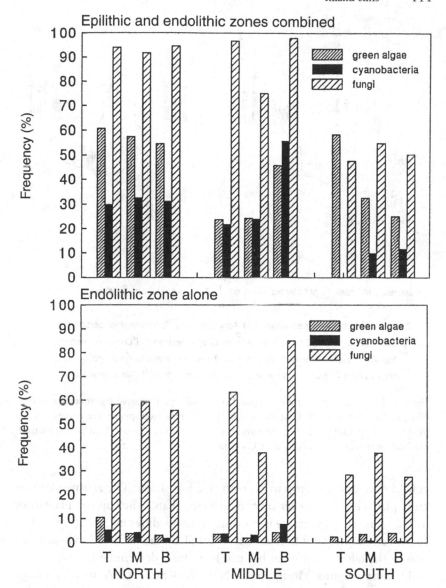

*Figure 4.10* The frequency of green algae, cyanobacteria and fungi on and inside rocks of the Niagara Escarpment, Ontario, Canada. Three sites were surveyed in the north, middle and south regions of the escarpment, and the top (T), middle (M) and bottom (B) sections of the cliff face were sampled. Frequencies were calculated from presence/absence in 25 randomly selected microscope fields. The values shown are means of three samples. The top panel shows data for epilithic and endolithic zones combined, and the bottom panel shows data for the endolithic zone alone. Taken from Matthes-Sears *et al.* (1997), and used with permission.

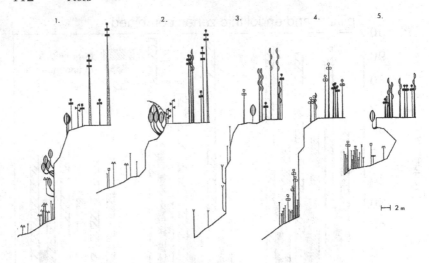

Diameter at breast height classes: | <5 cm, | >5 cm but <20 cm, | >20 cm

Species symbols: •• *Acer saccharum*, Y *Acer spicatum*, ↾ *Amelanchier bartramiana*
↾ *Betula papyrifera*, T *Cornus rugosa*, ⊕ *Fraxinus americana*, ↾ *Ostrya virginiana*,
oo *Populus grandidentata*, oo *P. tremuloides*, ↾ *Prunus virginiana*, ↾ *Quercus rubra*,
ʷ *Rhus typhina*, ↾ *Sambucus pubens*, (|) *Thuja occidentalis*, T *Tilia americana*.

*Figure 4.11* Canopy profile diagrams along five representative transects across the
Niagara Escarpment cliff, Ontario, Canada. Different tree species are represented
by different symbols. Taken from Larson *et al.* (1989), and used with permission of
the National Research Council of Canada.

equal frequency from top to bottom (see Fig. 4.10). The richness of vascular plants is much lower on the cliff face than either on the plateau or
in the talus, and this trend is consistent across different sites (Fig. 4.13).
However, this may not hold true if lower plants are included, whose richness in the talus and plateau has not been fully determined.

The Shawangunk Mountains in New York State, USA, include a large
number of cliffs and rock outcrops composed of limestone, quartz conglomerate and sandstone (Kiviat, 1991). While the cliff faces have not
been sampled, the cliff edges are known to support a forest dominated by
slow-growing pitch pine (*Pinus rigida*). This forest is uneven-aged and
contains trees up to 320 years old (Abrams & Orwig, 1995). A number
of locally or nationally rare species are also present, including *Asplenium
montanum shawangunkense* and *Corema conradii*.

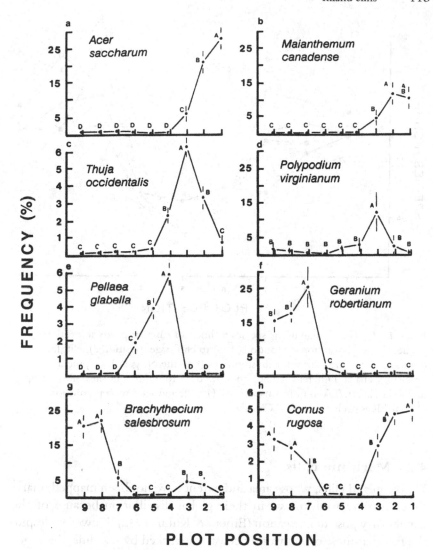

*Figure 4.12* The frequency of eight individual species across the gradient from plateau forest (positions 1–2) to cliff edge (position 3), cliff face (positions 4–6), and talus slope (positions 7–9) on the Niagara Escarpment, Ontario, Canada. Means with the same letter are not significantly different as determined by analysis of variance ($p < 0.05$). Taken from Larson *et al.* (1989), and used with permission of the National Research Council of Canada.

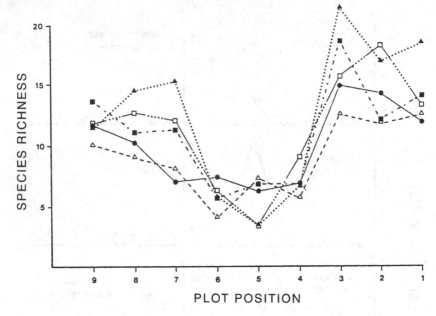

*Figure 4.13* The change in the species richness of vascular plants across the gradient from plateau forest (positions 1–2) to cliff edge (position 3), cliff face (positions 4–6), and talus slope (positions 7–9) on the Niagara Escarpment, Ontario, Canada. Different symbols indicate data from different sites; sites are up to 200 km apart. Taken from Larson *et al.* (1989), and used with permission of the National Research Council of Canada.

## 4.2 Maritime cliffs

Unlike inland cliffs, whose area and extent have not been mapped, maritime cliffs are well known in their global distribution because of the danger they pose to navigation (Emery & Kuhn, 1982). However, despite the fact that most of the world's coasts are backed by sea cliffs, the vegetation of these cliffs has rarely been studied. Information on tropical sea cliffs is especially scarce. Limestone towers in Viet Nam (Fig. 4.14), for example, support lush forests on exposed slopes, but no formal scientific study has been made of these areas. One of the few published papers is a qualitative description of the sea–cliff vegetation in Hawaii (Egler, 1947). Salt-tolerant species are found on spray-swept bluffs up to 100 m tall, but no woody plants, mosses or lichens are among them. On rocky shores and cliffs within the spray zone on the Bermuda Islands, Bermuda cedar (*Juniperus bermudiana*) was common until oyster-shell scale insects

*Figure 4.14* Limestone tower in Viet Nam with lush forest on exposed slopes. Maritime cliffs such as these support a flora that is poorly known, especially in the tropics. Photo courtesy of A. Matheson.

(*Lepidosaphes newsteadi*) caused its dieback in the early 1950s (Challinor & Wingate, 1971; Wingate 1985). Before the outcrops became denuded, this species served as a windbreak for inland habitats. Endemic species of *Peperomia*, as well as ancient trees of Bermuda olivewood (*Cassine laneanum*) and yellow-wood (*Zanthoxylum flavum*), occur on limestone rock ledges on the islands. In the Caribbean and in Asia, volcanic and sedimentary bedrock forms the landmass of many islands, but no studies have been conducted on these outcrops, even though an association between particular species of *Juniperus* and exposed rocky outcrops has been observed (Weber, personal communication).

Maritime cliffs in Australia have been the subject of several investigations. On granite sea cliffs in Victoria (Ashton & Webb, 1977), the lichens *Caloplaca marina* and *Parmelia conspersa* occur in a zone exposed to the highest salt concentrations. Mosses are absent from such sites. In steeply sloped areas exposed to less spray, the drought-resistant shrubs *Leptospermum juniperinum* and *Kunzea ambigua* are common. The authors also report chasmoendolithic algae, including species of *Chlorella*, *Gloeocapsa* and *Stigonema*, many of which are similar to the ones reported from the inland cliffs of the Niagara Escarpment.

Adam *et al.* (1990) report few species restricted to sea cliffs in New

South Wales, Australia. The two species most closely associated with these cliff habitats are the shrubs *Correa alba* and *Westringia fruticosa*; a large number of alien species also occur there. Woody plants are often dwarfed and prostrate due to wind and salt exposure. The authors suggest that genetically distinct cliff and inland populations of certain species such as *Baeckea imbricata* exist. Interestingly, few of the species cited as typical for these sites overlap with the characteristic cliff species in the study by Ashton and Webb (1977), even though the two sites are only a few hundred kilometres apart. Beadle (1981) reports that interesting communities of plants occur on rocky outcrops and headlands along the coast of Australia. *Pandanus peduncularis* and *Casuarina equisetifolia* are especially common and are associated with smaller, less conspicuous chasmophytic plants such as *Myoporum insulare*, *Zygophyllum ammophilum* and *Disphymia australe*. For the most part, however, Beadle observes that in tropical latitudes, the vegetation of sea cliffs and inland cliffs is not that different from the surrounding level-ground communities.

On maritime sandstone and mudstone cliffs in New Zealand, Wilson and Cullen (1986) report few cliff endemics. Species typical of salt-marshes such as *Selliera radicans* and *Samolus repens* occur near the water on the bottom of the cliffs, while others such as *Disphyma australe* and *Blechnum banksii* grow only on the upper parts. A number of species, including *Asplenium obtusatum* and *Poa astonii*, are able to grow in both zones. The vegetation is similar to that of offshore islands. Wardle's (1991) elegant analysis of the vegetation of New Zealand includes a large section dealing with cliffs, bluffs and rock outcrops. Wardle observes that:

> among mainland species that have ranges no larger than a single province, but excluding those restricted by climate to northern districts, at least two-thirds grow only on steep rocky terrain. This is partly because precipitous terrain provides isolation that may encourage speciation, and diverse habitats that buffer against major oscillations in climate and regional vegetation. However, bluffs flanking the valleys and fiords of Westland and Fiordland support no regional endemics other than stragglers from higher altitudes, presumably because they were beneath glacial ice only 14,000 years ago. Cliffs also give refuge to plants that browsing mammals or fire have exterminated from surrounding areas, although their value as refuges for indigenous species is compromised where the surroundings provide an overwhelming seed source of weeds that preempt new surfaces.

Among the interesting plants discussed by Wardle are *Metrosideros excelsa* (pohutukawa) and *M. robusta* (northern rata). Also found abundantly on maritime cliffs of the South Island are *Pachystegia insignis* (Marlborough rock daisy) and *Phormium cookianum* (flax). *Asplenium oblongifolium, A. trichomanes* and *A. lyallii* all occur on wet cliffs in both inland and coastal locations.

Most of the information on sea cliffs in the temperate zone is from the UK (Hepburn, 1943; Goldsmith, 1973a, 1973b; Cooper, 1997; Malloch & Okusanya, 1979; Okusanya, 1979a, 1979b, 1979c), although a few studies have also examined maritime cliffs in temperate Asia (Ohsawa & Yamane, 1988). Cliffs in these areas do not appear to support any woody vegetation (Bates, 1975), but this may reflect more than a millennium of wood harvesting by people rather than the expression of natural plant distribution patterns. Another possibility is that the trees growing on the cliffs may be quite small and inconspicuous. In 1997 we observed gnarled, stunted and grossly misshapen *Taxus baccata* and *Juniperus communis* growing with various species of *Asplenium* and *Phyllitis scolopendrium* on sea cliffs of the Great Orm and the Little Orm of northern Wales. On slate cliffs in north Cornwall, only a small number of species can survive the wind and salt spray on the lower part of the cliffs (Hepburn, 1943). Most of these are saltmarsh species; among the few that are largely confined to cliffs are *Crithmum maritimum* and *Limonium binervosum*. A greater number of species occurs higher up on the cliffs. Goldsmith (1973a) also describes a strong zonation of plant communities on sea cliffs composed of sandstone, shale and conglomerate rock in Anglesey, UK. He emphasizes the relatively high species richness on the most exposed free-faces. In addition to a rich salt-tolerant flora dominated by *Armeria maritima* and *Festuca rubra*, normally salt-intolerant species also occur, such as *Agrostis stolonifera*. The fern *Asplenium marinum* is restricted to shaded rock fissures and often indicates the presence of fresh water; nevertheless, the occurrence of this genus suggests some degree of similarity between maritime and inland cliff floras. Similarities may exist in the ecological processes that control the plant communities on maritime and inland cliffs. This is suggested by the replacement design experiments of Goldsmith (1973b), which showed that species growing abundantly on sea cliffs are also poor competitors in level-ground habitat dominated by more aggressive species. Malloch (1971) and Malloch and Okusanya (1979) have reached a similar conclusion for British sea cliffs in general: 'Sea cliffs represent a fairly hostile environment. It is thus possible that the cliff species, whilst not growing under ideal conditions, are able to tolerate the sea-cliff environment which is too severe for plants capable of growing on similar cliffs inland'. These special and relatively stable conditions on sea cliffs in the UK have resulted in consistent selection pressures on species that can result in a substantial degree of endemism such as that displayed by *Primula scotia*. This high degree of endemism may have once also applied to many plants such as *Beta vulgaris*, *Daucus carota*, *Capparis spinosa*, *Pistacia palaestina* and various species of *Lactuca* and *Allium* as well as to horticulturally important genera such as *Dianthus* and *Campanula* that originally evolved on cliffs or rock outcrops but ended up being used by people in habitats that essentially recreate the open environments typical of cliff faces or talus slopes. This idea which we call the 'urban cliff hypothesis', is developed in Chapter 5.

Coastal cliffs formed of gneiss and granite in south Finland were also surveyed by Makirinta (1985), who found a variety of community types, including open exposed rock outcrops dominated by *Woodsia ilvensis* and *Lychnis viscaria*. Companion species in this community included *Polypodium vulgare*, *Cystopteris fragilis*, *Asplenium trichomanes*, *Poa nemoralis*, *Geranium robertianum* and *Viola tricolor*. This particular assemblage of species appears to be no different in patterns of composition and abundance than communities of plants in similar inland settings in northern Europe.

On the island of Cyprus (Géhu, Costa & Uslu, 1990) a plant community dominated by *Juniperus phoenicea* spp. *lycia* occurs in sites where the roots directly penetrate steeply inclined rocks and cliff faces. Mosses appear to be completely absent from cliffs exposed to salt spray, but lichens, algae, and both halophytic and non-halophytic higher plants are common. *Artemisia arborescens* and *Atriplex halimus* are common on exposed Italian cliffs in the Mediterreanean Sea (Biondi, 1988), and a wide variety of salt-tolerant weedy grasses and forbs occur on cliffs of Atlantic maritime France (Géhu 1964; Bioret *et al.*, 1987). Limestone sea cliffs in Oman, Arabia, are the site of two newly reported shrubby species of *Aloe* (Lavranos, 1995).

Ishizuka (1974) has reported on the vegetation of sea cliffs and coastal screes in Japan. *Juniperus chinensis* var. *sargentii*, *Dianthus superbus*, *Stellaria ruscifolia*, *Draba borealis* and *Sedum ishidae* were all common on cliff faces, while forests of *Acer mono* var. *velutinum*, *Quercus mongolica* var. *gros-seserrata* and *Tilia japonica* occur on plateau forests above these cliffs. On scree slopes, Ishizuka reports *Renoutria sachalinensis*, *Filipendula kamtschatica*, *Artemisia montana* and *Urtica platyphilla*.

No published accounts exist of the vegetation of maritime cliffs in the Americas, even though rocky shorelines exist both on the east and west coasts. Rock outcrops and cliffs of Point Lobos State Reserve, California, support a variety of native and exotic halophytes, salt-damaged lichens, and one of only two remaining natural stands of *Cupressus macrocarpa*. Wind-trimmed trees and the lichen *Ramalina menziesii* are very common at this site (Matthes-Sears, Nash & Larson, 1986, 1987).

## 4.3 Man-made cliffs

Whereas the vegetation of abandoned quarry floors has been extensively studied (especially from the viewpoint of restoration), almost no published information exists on the flora of quarry walls, probably for the same reasons that natural cliff faces have been ignored. The only quantitative studies of quarry wall vegetation are those of Ursic, Kenkel and Larson (1996), Viemann (1997) and Cullen, Wheater and Dunleavy (1998). Ursic *et al.* (1996) studied 18 limestone quarries in southern Ontario, Canada, that had been abandoned for between 18 and 100 years. Species composition is highly variable at recently abandoned sites, but becomes predictable in quarries abandoned for 70 years or more. The resulting community on these man-made cliffs resembles that of the nearby natural cliffs of the Niagara Escarpment in both species richness and composition, with the dominant tree *Thuja occidentalis*, the herb *Geranium robertianum*, ferns such as *Cystopteris bulbifera*, and many lichens and bryophytes.

Hepburn (1943) included a disused slate quarry wall among the sites for his study of sea cliffs in Cornwall, UK, in order to compare the sea-cliff flora with that of nearby rock faces not exposed to salt spray. The quarry face supported several species also found on the sea cliffs, such as *Sedum anglicum* and *Silene maritima*, in addition to ones not present on the sea cliffs, such as *Teucrium scorodonia*, *Cotyledon umbilicus*, ferns of the genus *Asplenium*, and the grass *Brachypodium sylvaticum*.

Quarry walls, like natural cliffs, are colonized by a variety of terrestrial algae, both epilithic and endolithic, but almost nothing is known about these communities. Allen (1971) found a relatively species-poor community of epilithic algae, consisting only of cyanobacteria and not including green algae, in an overgrown limestone and dolerite quarry in the UK. The dominant species there is *Schizothrix calcicola*. In contrast, Gerrath *et al.* (1995) describe a rich cryptoendolithic community from a limestone quarry in southern Ontario, Canada, abandoned nearly 50 years earlier. The most common species is the green alga, *Stichococcus minor*, but other green and blue-green algae are also frequent. Surprisingly, the richness of this community is larger than that of similar cryptoendolithic communities found on the nearby cliffs of the Niagara Escarpment.

In parts of the world where significant amounts of time have passed since settlement altered the original vegetation, many plants and animals have adjusted to the presence of humans and human structures. Thus, throughout Europe, Asia and the Middle East, ancient buildings and walls frequently support a predictable flora of species that would normally be found on cliffs. In fact, many species or genera (*Prunus, Rosa, Ribes, Rubus, Cotoneaster, Buxus, Juniperus, Thuja, Dianthus, Daphne, Sorbus, Hieracium, Saxifraga*) that are endemic to (or that mainly occur on) cliffs and talus slopes in various parts of the world appear to have been exploited by people for horticultural purposes, or these species have opportunistically exploited stone and wooden copies of natural cliff environments created by people. This idea is addressed in Chapter 5.

In some parts of western Europe where cliffs were formed in deposits of loess by the passage of ox-carts over many centuries, unique assemblages of species once occurred. With the advent of more modern means of transportation, and the paving of roadways, these unique assemblages are now being eliminated by natural ecological succession (Baier *et al.*, 1993). In central and southern Germany, a significant effort is being expended to retain these 'Hohlwege' in the landscape. A variety of lichens, mosses and higher plants exploit these areas, including *Polypodium vulgare* and *Campanula rotundifolia*. In contrast to North America where the conservation effort is primary designed to return habitats to their 'presettlement' condition, the effort in Germany is to restore long-abandoned or maturing woodland to its earlier, more disturbed forms. That these forms should take on the appearance of cliff faces is particularly interesting.

A variety of other kinds of disturbances have also lead to the generation of man-made cliffs. Oberdorfer (1977) and Rishbeth (1948) have described some of these communities in Europe and have shown that many species, such as *Sorbus aucuparia, Sedum acre, Asplenium ruta-muraria, Polypodium vulgare* and *Tortula muralis*, occur on walls built in the early part of the second millennium. The majority of these plants are hemicryptophytes, with perennials predominating over annuals (Risbeth, 1948; Lisci & Pacini, 1993a, 1993b), and many of the genera listed also occur on natural cliffs. Some small phanerophytes on these walls have grossly deformed root systems (Risbeth, 1948). The species composition is strongly influenced by the type of wall material, as well as the aspect. For example, *Poa annua* is dominant on the north-facing wall of Trinity College at Cambridge, UK, while *Festuca rubra* and *Sagina apetala* are the dominant species on the south-facing wall (Risbeth, 1948). On the walls of Italian towns, the most common plants that grow in exposed sites on almost any substrate include the perennial *Parietaria diffusa*, the shrub *Capparis spinosa*, and the tree *Ficus carica* (Lisci & Pacini, 1993a; see Fig. 3.15). Ferns such as *Adiantum capillus-veneris, Ceterach officinarum* and *Polypodium vulgare* are found only on damp shaded walls, but the individual microsites where these plants occur reflect the importance of local variation in rock texture, porosity and exposure, as we have already seen for natural rock cliffs. Bryophytes facilitate the establishment of vascular plants by creating germination sites. Grasses predominate on ruins and small arid walls. Vegetated walls are highly susceptible to biotic degradation.

Epilithic and endolithic micro-organisms are common components of the stone and brickwork in historical structures, in particular when masonry is exposed to direct water or condensation, and contribute to the microbial weathering of such structures (Leclerc *et al.*, 1983; Bock & Sand, 1993). These communities may include blue-green algae, red algae, diatoms, green algae and bacteria. For example, epilithic and chasmolithic algal communities colonize the marble of the Acropolis in Athens (Anagnostidis, Economou-Amilli & Roussomoustakaki, 1983).

## 4.4 Summary

The above review of the literature on cliff flora and vegetation shows that cliffs of all types and in all geographical regions have certain characteristics in common. There are also clear trends and differences among different types of cliffs and cliffs in different geographic regions. The flora of cliffs is always very distinct from the flora of surrounding level-ground habitats, and this is true for inland cliffs in all regions of the globe as well as for sea cliffs and artificial cliffs. Often, cliffs are the sites for species that are rare elsewhere for a variety of reasons, including their lack of competitive ability on level ground, their intolerance of disturbance, or their status as glacial relics. Many endemic plants fall into this category. Thus, all cliffs, including sea cliffs and artificial cliffs, can be described as habitats with minimal disturbance by humans and animals that are colonized by non-competitive, disturbance-sensitive plants that grow slowly.

All cliffs are similar in the distribution of life forms that colonize them. Virtually all inland cliffs and man-made cliffs support a sparse vegetation of lower plants including bryophytes, lichens, epilithic algae and endolithic algae or lichens. Annual plants are very rare. Sea cliffs also support lichens and epilithic algae, but bryophytes are less common and no endolithic organisms have been reported, although this may be simply because no one has looked for them. A diverse assemblage of desiccation-tolerant and desiccation-avoidant macroscopic species also occurs. Ferns are widely distributed on inland cliffs, quarry walls and man-made structures, but much less frequent on sea cliffs, especially near the spray zone; like bryophytes, they appear to be unable to tolerate salt spray. Grasses, sedges, small geophytes and cryptophytes are common on all cliffs. Most of these are perennial in their life history but a few are annual plants (often in the form of winter annuals). Shrubs and trees are never very abundant on cliffs, although their relatively large size and perennial life history often make them appear so on casual observation. They are found more frequently on natural inland cliffs than on either sea cliffs or artificial cliffs (although human interference may be partially responsible for their absence from such places). For the trees and shrubs that do occur on cliffs, strikingly similar characteristics have been described from locations around the world. These trees rarely grow very large compared to individuals of the same species on level ground. They are often deformed and twisted, with partial dieback and stem strips. They often grow extremely slowly, and in many cases they grow for very long periods of time, reaching much greater ages than trees on level ground. In areas where the original forest was extensively cut, these trees survived because the cliffs were inaccessible. The significance of cliffs as sites for virgin, or old-growth, forests is discussed further in Chapter 8.

When comparing actual species lists among cliffs in the same geographical region, it is clear that some overlap exists but there are also local differences in the floristic composition. Some species (such as most lichens) appear highly specialized for a certain type of rock substrate, while others (such as many trees, shrubs, herbaceous species and bryophytes) may exist on a variety of rock types in the same region, suggesting that rock chemistry is not a crucial factor in their distribution. The limited evidence available indicates that quarry walls acquire a flora that is very similar to that of natural cliffs, while man-made walls resemble cliff edges and ledges in the predominance of desiccation-tolerant species. Sea cliffs appear to support two distinct groups of plants. The first are salt-tolerant species usually found in saltmarsh habitats, and this group is absent from inland cliffs. The second group is cliff-adapted plants; these usually occur in places that are slightly protected from salt spray, and may be found in saltmarshes and on inland cliffs as well.

Cliffs in different geographical regions (e.g. tropical versus temperate, or Old World versus New World) have few actual species in common. However, some families and genera are found with striking consistency on cliffs around the world. Among them the Cupressaceae, with their representatives such as *Juniperus*, *Cupressus*, *Thuja* or *Taxus*, are often the dominant trees on cliffs. For example, species of the genus *Juniperus* are found on inland as well as maritime cliffs of a variety of materials and on different continents: *J. virginiana* throughout eastern North America on both limestone and granite; *J. ashei* in Texas; *J. communis* in Europe and eastern North America; *J. phoenicea* in France; and *J. bermudiana* in Bermuda. Similar cases could be made for the genera *Polypodium*, *Pellaea*, *Asplenium*, *Adiantum* and *Cystopteris* among the ferns; and *Campanula*, *Geranium* and *Sedum* among the herbaceous vascular plants.

Inland, maritime and artificial cliffs from various parts of the world show many similarities in their vegetation structure and life-form distribution, but distinct local differences in their flora. All support a small array of woody plants as well as a diverse assemblage of vascular and non-vascular, often desiccation-tolerant, species. Commonly encountered genera include *Juniperus*, *Cupressus* and *Thuja* for the trees; *Campanula*, *Geranium* and *Sedum* for the herbs; and *Polypodium*, *Pellaea*, *Asplenium*, *Adiantum* and *Cystopteris* for the ferns. A rich assemblage of lichens, mosses, epilithic algae and endolithic organisms is also present. Cliff vegetation is very sensitive to disturbance and intolerant of competition from neighbouring plants. Woody plants on cliffs are often stunted and deformed, have slow growth rates, and may be very old.

# 5 · *Fauna*

Cliffs worldwide support a variety of protists and animals including a vast array of invertebrates, amphibians, reptiles, birds and mammals. Cliffs in one area of highland in southern British Columbia, Canada, for example, supported 41 per cent (22 of 54) of the faunal species of high conservation value for the region (Sinnemann, 1992). Most studies of cliff fauna, however, have focused on one or a small number of species. This is in striking contrast to the floristic studies reviewed in the last chapter that generally focused on the determination of the composition of the entire plant community. In most faunal studies, the focus of the research is usually not on the habitat but rather directly on the species that occur there. Exceptions to this trend in the faunal literature include Johnson (1986), Reitan (1986), Ward and Anderson (1988), and Camp and Knight (1997). As a result, the organization of this chapter is quite different from the preceding one. This chapter summarizes the information that is presently available in the scientific literature on the ecology and distribution of faunal cliff species and presents it in broad taxonomic groups.

## 5.1 Avifauna

### 5.1.1 General trends

Cliffs appear to support a greater species richness of birds than equal areas within the surrounding habitat, although not all of these birds nest on the cliffs. This difference between adjacent land and cliffs is partly attributable to cliffs being 'permanent habitat edges' characterized by abrupt changes in soil, topography, geomorphology and microclimate combined with local conditions that minimize interspecific competition and predation (Matheson & Larson, 1998). Increased topographic roughness and a greater horizontal and vertical structure to the vegetation appear to be responsible for an increased bird species diversity, especially at the bases of cliffs or in the talus where shrubs are taller and there are alternating patches of vegetation and bare ground. Where vertical and horizontal vegetation structure is more homogeneous, bird species diversity is usually reduced (Ward & Anderson, 1988). Studies that demonstrate this phenomenon include Reitan (1986) in central Norway, Ward and Anderson (1988) in south-central Wyoming, USA, Camp and Knight (1997) in the Mojave Desert, California, USA, and Matheson and Larson (1998) along the Niagara Escarpment, Ontario, Canada. Reitan (1986) placed bird census lines above and below a cliff face and found that the densities of passerine birds were highest at the cliff base, whereas densities on cliffs were higher than in most forest passerine communities. Ward and Anderson (1988) surveyed male birds along talus and plateau transects and found that greater topographic roughness and a more diverse vegetation structure on the cliff sites contributed to a greater species richness and diversity of birds compared with seven control sites away from the cliffs. In Joshua Tree National Park, California, Camp and Knight (1997) also compared the bird diversity of cliff and non-cliff areas and found that, on average, cliff sites supported more bird taxa ($7.67 \pm 0.88$) than non-cliff sites ($4.17 \pm 1.08$) and that 13 bird species were observed exclusively on cliffs compared to only seven bird species at non-cliff sites. They found that cliffs contribute to a unique avian assemblage in this area.

Matheson and Larson (1998) quantitatively examined the distribution and zonation of forest birds along the Niagara Escarpment. Birds roosting in replicated $25 \times 25$ m plots in several cliff zones (edge, face, talus) were compared over four months, and multivariate procedures and analysis of variance were used to determine if habitat variables unique to each zone had an influence on the structure of the bird community. The Niagara Escarpment cliff zones as a whole supported significantly more bird species than would be expected for an equal-sized area of closed forest. While no bird species was restricted to just one habitat zone, many species were found only in the three escarpment zones and not in the level-ground plateau forest. This high forest bird species richness was attributed to greater small-scale topographic heterogeneity associated with talus, face and edge habitats.

Most of the research on cliff-dwelling bird species, however, is not community based but rather focused on two broad taxonomic groups: raptors and sea birds. Inland cliffs and sea cliffs on every continent provide important nesting and perching sites for a broad range of raptors and sea

birds. Relevant research on cliff avifauna is therefore presented in three
sections: raptors, sea birds and a third section on bird species that fall into
neither group but nonetheless have been observed to use cliff environments in some capacity.

### 5.1.2 Raptors

There are many studies on the cliff-nesting members of the
Falconiformes (hawks, eagles, falcons and the New World vultures) and
Strigiformes (owls). Other species of birds in other families normally nest
in level-ground habitats or trees but may select cliffs for nesting when
other sites are unavailable. Most raptors select cliff sites because the relief
is advantageous for hunting or scavenging and topographic relief is positively associated with favourable flight conditions (Janes, 1985). Cliffs
also offer protection from predation on both eggs and young. Scientists
interested in the life cycles of these birds have had to make observations
in and around cliff habitats. These studies offer important insights into
the global distribution of raptors and the use of cliffs by these species.

The cliff-nesting raptor with the broadest range is the peregrine falcon
(*Falco peregrinus*). Seventeen subspecies of peregrines have been identified
worldwide (Weick, 1980) and occupy nesting sites on cliffs in Europe
(Ratcliffe, 1993; Crick & Ratcliffe, 1995; Norriss, 1995), Greenland
(Hovis *et al.*, 1985; Moore, 1987; Meese & Fuller, 1989), North America
(White & Cade, 1971; Grebence & White, 1989; Nelson, 1990;
Rodriguez-Estrella & Brown, 1990), South America (Vasina & Straneck,
1984), Africa (Brown & Cooper, 1987), Australia (Czechura, 1984;
White, Emison & Bren, 1988) and Asia (Flint, 1995). Peregrine falcons
will nest on both inland and sea cliffs (Ratcliffe, 1980, 1993) but generally choose the highest cliffs for nesting, i.e. those with the most com-
manding outlook and protection for eggs and young (Ratcliffe, 1980,
1993). They will only nest on substandard sites such as small crags or
quarries when population levels are high. Peregrine falcons do not
actively build nests but rather scratch a bowl-like depression in the surface
of a soily ledge (Ratcliffe, 1980). Peregrine falcon eyries are often re-used
many times and these ledge sites are usually devoid of vegetation
(Ratcliffe, 1980). Persecution of peregrine falcon chicks and eggs goes
back at least 750 years (see Chapter 7), and Ratcliffe (1980) implies that
nest-site selection may have evolved as a response to human rather than
non-human predation.

In Europe, extensive surveys of peregrine falcon populations have
been conducted in Ireland and the UK, where numbers increased in most

regions between 1981 and 1991. The overall recovery of peregrine falcon levels here probably reflects the elimination of the pesticide DDT which breaks down into DDE, a compound that interferes with egg-shell formation, and HEOD, a toxic breakdown product of the pesticide aldrin (Ratcliffe, 1993; Crick & Ratcliffe, 1995). Territory expansion onto artificial cliffs such as quarries and buildings was also noted, with 13 per cent of the UK population now nesting in quarries. Peregrines may also nest on rocky stream banks or slopes, cathedrals, castles and chimney stacks (Ratcliffe, 1980). In Ireland, peregrine falcons now occupy marginal sites including quarry walls (Norriss, 1995) that were previously only used sporadically.

Numerous observations on peregrine falcon populations have also been made in Greenland where they nest on cliffs along the western coastline (Moore, 1987; Meese & Fuller, 1989). The existence of a peregrine falcon eyrie was also observed to reduce the number of prey species such as Lapland longspur (*Calcarius lapponicus*), northern wheatear (*Oenanthe oenanthe*) and common redpoll (*Carduelis flammea*) in the immediate vicinity. Peregrines were also observed evicting ravens from cliff sites and violently attacking golden eagles (*Aquila chrysaetos*). Similar behaviour was observed in south-western Utah, USA, where peregrines were observed attacking golden eagles and prairie falcons (*Falco mexicanus*) (Hays, 1987). A golden eagle was killed during one such attack after it flew along a cliff face just below a peregrine falcon eyrie (Hays, 1987).

Grebence and White (1989) studied peregrine falcon habitat on cliffs in Utah, USA, along the Colorado, Dolores, Green and San Rafael Rivers. They observed a spatial regularity of nesting sites that could be used to predict additonal nest-site locations. In southern Utah, the distance between nesting sites was approximately 6 to 14 km. Breeding success, expressed as fledgling rates, was on a par with peregrine falcon sites in other areas. In Alaska, USA, White and Cade (1971) surveyed raptor populations in the area of the Colville River and found that peregrine falcons occupied nests exclusively along river courses on high cliffs (Fig. 5.1) and were confined to altitudes lower than 2200 feet a.s.l. (ca. 670 m). Peregrines also occupied no more than 50 per cent of available habitat. The regular spacing observed by Grebence and White (1989) and the presence of unoccupied nesting territory observed by White and Cade (1971) are thought to reduce intraspecific competition.

A long-term survey of peregrine falcons on cliffs was initiated by Nelson (1990) on Langara Island, British Columbia, Canada. Here, peregrine falcon populations have been historically linked to populations of

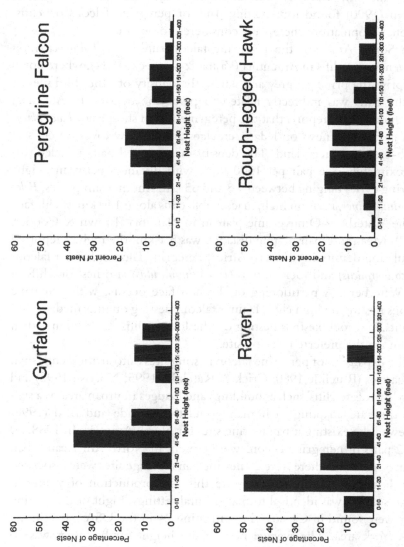

*Figure 5.1*  Nesting height on cliffs for four species of raptors along the Colville River, Alaska, USA. Redrawn from White and Cade (1971).

nesting ancient murrelets (*Synthliboramphus antiquus*), which in turn are preyed upon by introduced Alexandrian rat populations (*Rattus rattus alexandrinus*). A two-year survey of raptor populations along the Rio Yaqui and Rio Bavispe, Sonora, Mexico, by Rodriguez-Estrella and Brown (1990) found four nesting pairs of peregrine falcons on cliffs, although populations there were considered to be rare.

In South America, the peregrine falcon subspecies *Falco peregrinus cassini* nests on cliffs in Argentina (Vasina & Straneck, 1984), where doves comprise the principal prey and where the density of other bird species on the cliffs was indirectly related to peregrine density. In Australia, Czechura (1984) reports that the peregrine falcon subspecies *Falco peregrinus macrocarpus* nests on ledges, crevices and shallow caves on cliffs in south-eastern Queensland. The density of suspected pairs of falcons is approximately one pair per 1500 km$^2$, with distances between neighbouring eyries ranging between 4.8 and 65 km. Another subspecies, *Falco peregrinus minor*, nests on high, inaccessible cliffs along 140 km of cliff face of the Waterberg-Omuverume plateau in Namibia (Brown & Cooper, 1987). Mean interpair nesting distance was 5.1 km, the highest reported population density in southern Africa. Peregrine falcons, Lanner falcons (*Falco biarmicus*) and rock kestrels (*Falco tinnunculus*) also nest on cliffs of the Waterberg. A partitioning of the cliff face occurs, with peregrine falcons nesting on high cliffs, Lanner falcons nesting on intermediate cliff heights, and rock kestrels nesting on the lowest cliffs. Cliffs higher than 100 m are the preferred nesting site.

The expansion of peregrine falcon nesting sites onto artifical cliffs such as quarries (Ratcliffe,1980; Crick & Ratcliffe, 1995; Norriss, 1995) and onto 'surrogate cliffs' such as buildings and bridges in urban areas is a sign that they are adapting to human populations. Cade and Bird (1990) reviewed the existing urban nesting sites for North America. In 1988, 30 to 32 pairs of peregrine falcons were present in North American urban settings, of which there were 24 nesting pairs. Bridge sites were associated with low fledgling success. Despite this, the production of young in urban settings was identical to that in rural settings. Eight of the nesting pairs were found in Canada, the remaining 24 in the USA – five in the New York area. The average nesting site height for buildings was 30 floors. Many hazards are faced by peregrines near urban cliffs that usually are not encountered near natural cliffs. These would include poisoning by pest control agents, lead poisoning, unexpected wind shear, drinking or bathing in contaminated water, tinted windows, wires, aircraft and shooting (Cade & Bird, 1990).

*Figure 5.2* A gyrfalcon nesting pair and two nestlings at an eyrie on a vertical cliff in arctic Alaska. Photo by Frank M. Bond, the Peregrine Fund.

At least six other *Falco* species have been reported nesting on cliffs. Like peregrine falcons, gyrfalcons (*Falco rusticolus*; Fig. 5.2) have a broad range across the northern hemisphere but are primarily restricted to subarctic and arctic latitudes. In North America, White and Cade (1971) report on gyrfalcon populations along the Colville River in Alaska that nest on cliffs adjacent to peregrine falcon populations. Gyrfalcons, with mean weights of 1.2 kg (males) to 1.8 kg (females), are twice as heavy as peregrines, which have mean weights of 0.6 kg (males) to 0.8 kg (females). The contrasting body size affects the type of prey and the tactics used for capture (White & Cade, 1971). The larger body size of gyrfalcons allows them to prey on ptarmigan (*Lagopus mutus*) which peregrines rarely capture as a food source. Gyrfalcons seem more able to tolerate other cliff-nesting bird species in their vicinity than peregrines. The shortest distance between adjacent nests in White and Cade's (1971) survey was 60 m, and occurred between both a gyrfalcon and a raven's nest (*Corvus corax*) and a gyrfalcon and a rough-legged hawk's nest (*Buteo lagopus*). Gyrfalcons nest in abandoned stick nests of ravens on cliffs and may forcibly remove the ravens from these sites. Obst (1994) reports that Gyrfalcons may nest in trees in the western Canadian Arctic but that the mortality rates of young were higher and brood sizes smaller in tree nests compared with nearby cliff nests. Mean internest distances were similar for both tree and cliff nests, at 5.3 and 5.4 km respectively.

Observations on cliffs in Iceland show that raven sites are also preferentially chosen as nesting sites for gyrfalcons. The distance between nesting sites ranged between 9 and 11 km (Woodin, 1980). Similar to White and Cade's (1971) observations in Alaska, Woodin (1980) found that gyrfalcons in Iceland prey mainly on ptarmigan. Long-term monitoring of one gyrfalcon nest showed that ptarmigan comprised 88 to 100 per cent of prey remains during the early nestling stage, although some prey remains were actively removed by the gyrfalcons, presumably to minimize parasitic infection. Fluctuations in prey availability, and late winters and early springs may be important controls of gyrfalcon populations in these environments (Moore, 1987).

Another falcon species, the prairie falcon, nests on cliffs in western North America. Prairie falcons are extremely adaptable to variable cliff conditions and nest in cliff caves, cavities, crevices, ledges and stick nests. Runde and Anderson (1986) reviewed the literature on prairie falcons in the American west and found that the preferred choice for nesting sites varied depending on region. For example, while 43 per cent of surveyed prairie falcon nests in New Mexico were stick nests, none was reported from Colorado or Oregon. Crevices were the preferred nesting habitat for surveyed prairie falcon populations in portions of Colorado and Montana (Runde & Anderson, 1986; Table 5.1). Stick nests were also the preferred habitat in California (Boyce, 1987). Prairie falcons locate nest sites at locations approximately two-thirds of the total height of the cliff, although cliff heights can range between 2.1 and 154.4 m. Nest densities can range between 0.10 and 1.32 nests per km of cliff face (Craig & Craig, 1984). Unlike peregrine falcons, prairie falcons are prone to human disturbance such as mining activity (Bednarz, 1984). Prairie falcons are also prone to disturbance from agriculture because it affects the availabilty of small mammals, their principal prey (Craig & Craig, 1984).

Table 5.1. *Types of eyries used by nesting prairie falcons in the USA*

| State | Caves and cavities[1] (%) | Crevice (%) | Ledge[1] (%) | Stick nest (%) | Reference |
|-------|------|------|------|------|-----------|
| WY | 41 | 24 | 25 | 10 | Runde & Anderson (1986) |
| CO | 14 | 64 | 21 | — | Williams (1981)[2] |
| ID | 60 | — | 19 | 21 | Ogden & Hornocker (1977) |
| OR | 42 | 39 | 19 | — | Denton (1975)[2] |
| NM | 39 | — | 19 | 43 | Platt (1974) |
| UT | 36 | 10 | 32 | 22 | Porter & White (1973) |
| MT | 37 | 37 | 18 | 8 | Leedy (1972) |
| CO–WY | 56 | — | 44 | — | Enderson (1964) |
| WA | 43 | — | 7 | 50 | Decker (1931) |

*Notes:*
[1] Includes a few sites with stick nests present.
[2] Data not reported but stick nests used.

Two African falcons, Taita falcons (*Falco fasciinucha*) and Lanner falcons (*Falco biarmicus*) also preferentially nest on cliffs, while another species, the Saker falcon (*Falco cherrug*), nests on cliffs in the Ukraine, Russia, Kazakhstan and other areas of Central Asia (Flint, 1995). Hartley et al. (1993) studied Taita falcons in Zimbabwe and found that they nest on both high and low cliffs but at heights equalling 75–85 per cent of the total cliff height. Taita falcons are particularly aggressive and Hartley et al. (1993) observed them defending their nests against 11 other species. Taitas, however, will co-ordinate their egg laying and incubating with peregrine falcon behaviour to avoid contact. Taita falcons were observed nesting within 300 m of peregrines with no apparent adverse effect on breeding success. Taita falcons are not dependent on one prey species.

In Namibia, Brown and Cooper (1987) measured average interpair distances of Lanner falcons and compared this data to other previous Lanner falcon studies in southern Africa. Average interpair distances were 4 km on cliffs in Namibia, 2–5 km on continuous cliffs in Transvaal, South Africa, and 1.8 and 3.5 km for two areas in Zambia. Lanner falcon range also extends into the Mediterranean, where Yosef (1991) studied populations of Lanners on cliffs in Israel. Birds known as chukars (*Alectoris chukar*) comprised 43.1 per cent of all prey items. Rodents, bats and invertebrates were also collected. Lanner falcons are particularly successful at fledging young as two different pair fledged at least three young in each of the three years observed by Yosef (1991).

Another group of falcons which frequently nest on cliffs is the kestrels, namely the common kestrel (*Falco tinnunculus*), which is cosmopolitan in the northern hemisphere, and the lesser kestrel (*Falco naumanni*), which is native to the Mediterranean and western Asia. These falcons are very different from other cliff-nesting raptors in that they have low internest site distances and are sometimes colonial birds (Brown & Cooper, 1987). Bustamante (1994) studied a mixed population of these birds on a cliff face in an abandoned quarry in Spain where minimum inter-nest distance was only 6 m. In Namibia, common kestrels are more solitary, with average internest distances of 2.3 km (Brown & Cooper, 1987). Shrubb (1993) also found that, unlike other falcons, kestrels are not obligate cliff dwellers. Only 27 per cent of surveyed common kestrel nest sites in Great Britain were located on cliffs, crags or quarries. Stick nests in trees, tree cavities, buildings and nest boxes make up the other nesting sites. Sixty-five per cent of cliff nests were located on open ledges and 29 per cent were located in holes or crevices. The remainder of the nests were abandoned raven, carrion crow (*Corvus corone*), buzzard (*Buteo buteo*) or jackdaw (*Corvus monedula*) nests. No significant change was observed in the use of cliff nests between 1937 and 1987, although the use of chalk cliffs and quarries by common kestrels declined during this period. Brown and Cooper (1987) found that common kestrel nests are usually located in the bottom half of the cliff face. The kestrels will nest on both inland and sea cliffs (Ratcliffe, 1993).

Some species of eagles and hawks are also cliff dwellers. Golden eagles nest on cliffs and in trees in parts of Europe, Asia and North America. Several subspecies are recognized (Weick, 1980). In the UK, golden eagles nest on both inland cliffs and sea cliffs along the coast of Scotland (Ratcliffe, 1993). The tallest cliffs are the preferred nesting sites and the best sites have been used by many generations. Ledge sites are required that can support the often massive stick nests (Watson, 1997). These nests consist of great heaps of sticks and heather clumps that are visible from a great distance. Breeding pairs may maintain several alternative eyries. These nest sites are often rebuilt in the autumn so that when nesting begins again in March of the following year, the only remaining nest task is the lining of the nest (Ratcliffe, 1993). Because nesting material is added annually to these sites, some nests have been known to reach 6 m in height (Watson, 1997). On canyon walls in southern Idaho, USA, mean internest distance for cliff nests was found to be 4.39 km (Craig & Craig, 1984), while the mean number of young fledged was 1.62 per nesting pair. Black-tailed jack rabbit (*Lepus californicus*) was found to be the principal prey in these areas. Along the east coast of Hudson Bay, Morneau *et al.* (1994) found that nests were located midway up the cliff face and that pair density in this population was 1.04 per 1000 km$^2$. Breeding success was approximately 0.89 to 1.22 fledged young per successful pair. Morneau *et al.* (1994) found this to be lower than at other sites in North America. Golden eagles will also nest on artificial cliffs such as quarry and mine walls in areas that are not traditional nesting territory (Fala, Anderson & Ward, 1985; Watson, 1997).

Other eagles observed nesting on cliffs would include bald eagles (*Haliaeetus leucocephalus*) on cliffs in northern Mexico (Rodriguez-Estrella & Brown, 1990) and black eagles (*Aquila verreauxii*) and booted eagles (*Hieraaetus pennatus*) on cliffs in Namibia (Brown & Cooper, 1987). Despite the fact that black eagles nesting on cliffs of the Warterberg-Omuverume Plateau in Namibia are at the edge of their species range, this population is the densest in the country and comparable to other populations in southern Africa. Booted eagles are expanding their range in southern Africa, largely as a result of an increase in prey availability due to farming practices which have created more open water and larger numbers of seed-eating birds. Booted eagles prefer nests approximately midway up the cliff, whereas black eagles prefer nests more than half way up the cliff face.

A number of hawk species also inhabit cliff faces. Rodriguez-Estrella and Brown (1990) reported five hawk species from northern Mexico. While the two most abundant species in this area are the red-tailed hawk (*Buteo jamaicensis*) and the common black hawk (*Buteogallus anthracinus*), cliffs are the principal habitat for red-tails but for only 3 per cent of the common black hawks. In north-central Oregon, red-tailed hawks and Swainson's hawks (*Buteo swainsoni*) occupy similar cliff habitats (Janes, 1994). Swainson's hawks arrive later than red-tails following migration and usurp some red-tailed hawks from their cliff nesting sites. Core nesting areas are retained, however, especially those with a high density of available perch sites. An abundance of perch sites was found to be an important habitat parameter for red-tailed hawks (Janes, 1994), and red-tailed hawks preferentially include rock outcrops and cliffs in their home ranges (Janes, 1985). Rough-legged hawks are the most abundant cliff-nesting bird of prey in eastern Hudson Bay, Canada (Brodeur *et al.*, 1994). Brodeur *et al.* (1994) found one rough-legged hawk nest every 201 km$^2$, with internest distances of 4.95 km. Most nests were confined to coast-line sites, even though suitable nesting sites were available on cliffs in the interior. Mean brood size was 3.41 nestlings, which Brodeur *et al.* (1994) found to be higher than in other areas. In Europe, Vatev (1987) studied long-legged buzzards (*Buteo rufinus*) on cliff sites in Bulgaria. Three fledglings were typical for each buzzard pair here.

Owls are also known to be cliff nesters but also use a variety of other sites as well. Great horned owls (*Bubo virginianus*) and barn owls (*Tyto alba*) share much of the same range in North America and share similar nest-site characteristics (Andersen, 1996). Interspecific competition is reported between these two species and barn owls may actually choose nest sites on the cliff that protect against predation from great horned owls. In extreme cases, the two species may nest at the same site in the same year (Andersen, 1996). In Zion National Park, Utah, USA, Mexican spotted owls (*Strix occidentalis lucida*) may occupy cliffs in narrow, steep-walled canyons (Rinkevich & Gutiérrez, 1996) but the strict dependence of this species on cliffs has not been demonstrated. The mean ecological density (density in suitable habitat) of owls was 0.48 owls/ km$^2$ and nesting pairs occupy an average area of 50 km$^2$. Rinkevich and Gutiérrez (1996) examined the habitat parameters associated with Mexican spotted owl nesting and found that the absolute humidity and the presence of vegetation strata in canyons, which increases the fre-quency of prey, are the two most important habitat parameters determin-ing Mexican spotted owl abundance in Zion National Park. Herter (1996) also observed the Eurasian eagle-owl (*Bubo bubo*) on cliffs in south-western Germany.

Vultures are another group of raptors which nest on cliffs. In fact, vulture species worldwide use cliffs as nesting sites. In Europe, three prominent cliff-dwelling vulture species are the griffon vulture (*Gyps fulvus*), bearded vulture (*Gypaetus barbatus*) and Egyptian vulture (*Neophron percnopterus*). In Europe, human persecution led to the elimi-nation of griffon vultures from all areas except the Iberian and Balkan Peninsulas and some Mediterranean islands. Griffon vultures have recently been reintroduced into areas such as southern France (Sarrazin *et al.*, 1996). Breeding success rates in introduced individuals were similar to breeding success rates in natural populations. Typically, griffon vultures lay only one egg, and Sarrazin *et al.* (1996) found that specific nesting sites were more likely to be chosen by a nesting pair if the nestling was successfully fledged. Carcasses from local farms are the principal food source for these birds. In Spain, Donázar and Fernandez (1990) found that natural populations of griffon vultures increased in the 1980s as a direct response to a decrease in human persecution. Donázar, Hiraldo and Bustamante (1993) presented a comprehensive study of the ecology of bearded vulture populations in Spain, which have also increased in the last two decades. They found that mean internest distance was almost 1 km and that high cliffs were chosen preferentially for nest sites. The mean number of nests for any one pair was three (range between one and six), while the mean distance between neighbours was 11.1 km. Relief, distance to the nearest nest of a neighbouring pair, altitude and distance to villages were the main factors leading to specific cliff-site selection for nesting. While the risk of mammalian predation on raptors in cliff nests is generally regarded as low, Donázar and Ceballos (1988) report on an unusual incident of a red fox (*Vulpes vulpes*) preying on two Egyptian vulture fledglings. Although the incidents were not observed, the authors speculate that the prey were captured during the initial stages of flight.

In southern Africa, the Cape vulture (*Gyps coprotheres*) and the bearded vulture occupy cliff sites in Namibia and South Africa. Like the European vultures, the Cape vulture in Namibia and the bearded vulture in South Africa and Lesotho are also recovering from human persecution. The Egyptian vulture is extinct in the area (Brown, 1991). The total adult population and numbers of Cape vulture nestlings fledged increased during the 1980s (Brown & Cooper, 1987). Brown (1991) studied bearded vulture populations in Lesotho and South Africa and surmised that the initial population was small and isolated. Estimated carrying capacities were higher than the actual population densities and pairs were found to have a home range estimated at 5000 km². These vultures nest in deep cavities on the cliff face (Brown & Bruton, 1991) and eat carrion ranging in size from small rodents to large ungulates, with 75 per cent of the diet consisting of bone (Brown, 1991).

The black vulture (*Coragyps atratus*) and the turkey vulture (*Cathartes aura*) are the only vulture species in North America. Where their ranges overlap, they are often found roosting together. In Sonora, Mexico, Rodriguez-Estrella and Brown (1990) surveyed raptor populations along Rio Bavispe and Rio Yaqui and found that these two species of vultures were the most abundant raptors in the area, although turkey vultures were the only ones which frequented cliffs. Turkey vultures are very common along the Niagara Escarpment and probably take advantage of the unique wind regime of cliffs to facilitate flying. While nesting on the escarpment face is known, they seem to favour adjacent trees in the talus (Prior, 1990). Wild mammals such as woodchucks (*Marmota monax*), shrews

(Soricidae), unknown bird species and invertebrates such as Coleoptera are the principal food types found in pellets in the area.

### 5.1.3 Sea birds

Many sea birds are adapted to catching fast-moving prey underwater, but unfortunately the morphological adaptations which improve their manoeuvrability and speed in water, such as modified wing area or legs placed further back on the body, hamper their flight and terrestrial mobility. Sea cliffs worldwide support sea bird colonies of one or more species which may utilize the cliff edge, face or talus to avoid predation (Lack, 1968). Sea cliffs are the preferred habitat because of their proximity to the sea and thus access to food supply, the presence of updrafts and air currents along cliffs which facilitates take off and landing, and the reduced pressure from both aerial and terrestrial predation (Buckley & Buckley, 1980).

Many different families of sea birds commonly nest in cliff environments, the most common being the Alcidae (murres, puffins, auklets and murrelets) and the Laridae (gulls and terns). Other sea bird families with species that at least occasionally nest on cliffs include the Sulidae (gannets and boobies), Procellariidae (fulmars, petrels and shearwaters), Diomedeidae (albatrosses), Phaethontidae (tropicbirds) and Phalacrocoracidae (cormorants). Cliff nesting is relatively rare in the latter three families, and most of this discussion will focus on the first five families, particularly the Alcidae.

The thick-billed murre, or Brunnich's guillemot (*Uria lomvia*), and the common murre, or common guillemot (*Uria aalge*), breed on coastal cliffs throughout much of the northern latitudes. Studies on thick-billed murres habitat include those in Canada by Gaston and Nettleship (1981) on Prince Leopold Island, Northwest Territories; Birkhead *et al.* (1985) on Cobourg and Bylot Islands, Northwest Territories; Gaston and Elliot on East Digges Island, Northwest Territories (1996); Gilchrist and Gaston (1997a, 1997b) on Coats Island, Northwest Territories; and Birkhead and Nettleship (1987, 1995) on the Gannet Islands, Newfoundland; those in Alaska on St George Island by Squibb and Hunt (1983) and those in north Greenland on the Carey Islands by Kampp (1990).

Unlike the situation with raptors on inland cliffs, densities of these birds on cliffs can be quite high. A 4 km long 150 m to 250 m high cliff on East Digges Island, Northwest Territories, Canada, was estimated to support 180000 breeding pairs of thick-billed murres (Gaston & Elliot,

1996). Sites as low as 6 m above the high tide line can be occupied all the way to the cliff top (Gaston & Nettleship, 1981; Gaston & Elliot, 1996). Thick-billed murres appear to be limited by habitat rather than food supply and will nest up to 1 km from the shore if coastal cliffs are not available (Kampp, 1990). These murres can forage up to 180 km from the cliff nest (Gaston & Nettleship, 1981).

The limiting factor for thick-billed murres is the number of suitable ledges on a cliff face for nesting (Gaston & Nettleship, 1981). These murres prefer nest sites consisting of very narrow ledges, some as narrow as the length of the bird's foot plus tarsus such that the tail of the incubating bird hangs over the edge (Gaston & Nettleship, 1981). Here they incubate a single egg on bare rock. Complex manoeuvres are therefore required for tasks such as incubation changes or defence. Incubation changes are accomplished by a lengthy manoeuvre involving the transfer of the egg from the foot of one bird to the foot of the other or by the replacement bird leaping onto the egg from above (Gaston & Nettleship, 1981). Defence is especially difficult because thick-billed murres incubate facing the cliff face and have difficulties turning to face predators without dislodging the egg (Gilchrist & Gaston, 1997a, 1997b). Neighbouring birds apparently deter predation as mean breeding success on sites without a neighbour is 7.8 per cent less than on those with one neighbour and 11.2 per cent less than on those with two neighbours (Gaston & Nettleship, 1981). The risk associated with the egg being dislodged by shifting positions or during fights, however, is actually higher than the risk of predation (Gaston & Elliot, 1996). If successfully hatched, thick-billed murre chicks only 15–25 days old and unable to fly, jump directly into the sea followed by the adult male.

Birkhead (1977) on Skomer Island, Wales, UK, Harris, Wanless and Barton (1996) and Harris et al. (1997) on the Isle of May, Scotland, UK, and Birkhead and Nettleship (1987) on the Gannet Islands, Newfoundland, Canada, have also studied common murre habitat. While common murres utilize broader ledges than thick-billed murres for nesting (Birkhead & Nettleship, 1987), Birkhead (1977) found that mean ledge width on Skomer Island was only 0.29 m ± 0.14 m. Common murres also lay their eggs on bare rock ledges or on the tops of rocky stacks (Freethy, 1987) but will occasionally place stones around the eggs (Boekelheide et al., 1990b) or use guano to attach the eggs to the rock (Freethy, 1987). Like the thick-billed murre, they also nest facing the rock face but have a more upright incubating stance (Spring, 1971) and incubate while leaning against the rock face (Boekelheide et al., 1990b).

Common murres have synchronized egg laying and have a higher nesting density and a higher breeding success (Birkhead, 1977). Higher breeding density provides protection from gulls. On Skomer Island, Birkhead (1977) observed great black-backed gulls (*Larus marinus*) threatening murres, grabbing adults off nests and grabbing eggs when they came into view. Common murres also have a high level of site retention and most nest site changes are less than 2 m (Harris *et al.*, 1996).

Black guillemots (*Cepphus grylle*), which are native to the northern latitudes, and pigeon guillemots (*Cepphus columba*), which are native to the North Pacific coast, are alcids which are known for their diversity of nesting sites. While black guillemots prefer nest sites in crevices or small overhangs in fairly low cliffs, they also nest in the talus and have been known to nest on old and ruined buildings (Freethy, 1987). Pigeon guillemots usually choose to nest in crevices on easily accessible, gradually sloping talus (Haley, 1984; Ainley *et al.*, 1990a) but they may also dig holes in steep slopes or cliffs or nest on the substructure of bridges (Haley, 1984).

One of the more well-known members of the Alcidae is the common puffin or Atlantic puffin (*Fratercula arctica*) which nests on steep clay banks as well as cliff faces, edges and talus along the east coast of Canada, the coasts of Greenland, Iceland and the Faeroe Islands, and the west coast of the UK, Ireland and Norway. Puffins feed on a variety of fish species and Harris (1984) estimates that 700000 pairs of puffins nest on cliffs in the UK and Ireland alone. The puffin population of Iceland is several times greater than that of all other countries combined. In the northern portion of its range, rock nesting in crevices in cliffs and talus is more common than burrow nesting in the talus or at the cliff edge, which is more common in the south. Nettleship (1972) studied burrow-nesting puffins at cliff edges on Great Island, Newfoundland, Canada, and found that burrow density increased, fewer eggs disappeared due to predation, and hatching and fledgling success was higher towards the cliff edge.

Razorbills, murrelets and auklets are other groups of the Alcidae which also nest in cliff environments. In the North Atlantic, razorbills (*Alca torda*), which can be solitary or colonial, seek out tiny projections and cracks on cliff faces for nesting. They may also nest in talus and boulder areas but are extremely vulnerable to predation. The introduction of foxes to an island in the Gannet Islands, Newfoundland, Canada, reduced razorbill populations from 1213 to 2 breeding pairs in only nine years (Birkhead & Nettleship, 1995). Murrelets, including Xantus' murrelet (*Synthliboramphus hypoleucus*) (Freethy, 1987), ancient murrelet

(*Synthliboramphus antiquus*) (Gaston, 1992) and Kittlitz's murrelet (*Brachyramphus brevirostris*), nest on cliff environments in the North Pacific (Day, 1995). Xantus' murrelet may nest in crevices or in the corners of caves in cliffs (Freethy, 1987) while Kittlitz's murrelet nests in talus slopes in remote areas of Alaska and the Russian far east (Day, 1995). On Reef Island, British Columbia, Canada, Gaston (1992) found that while the majority of ancient murrelets nested in living tree roots, stumps or logs, 6 per cent of the birds were found to be nesting in burrows in rock crevices.

The group of species commonly referred to as auklets and auks nests on cliffs and in talus in coastal areas throughout the North Pacific. In some areas, these species exhibit broad ecological overlap. For example, the whiskered auklet (*Aethia pygmaea*), least auklet (*Aethia pusilla*), crested auklet (*Aethia cristatella*) and parakeet auklet (*Cyclorrhynchus psittacula*) all nest in crevices primarily in unvegetated talus throughout the Bering Sea and the Sea of Okhotsk (Day & Byrd, 1989). The size of the crevices used, however, varies in relation to the size of the bird, with least auklets using the smallest crevices and crested auklets the largest (Bédard, 1969; Day & Byrd, 1989). However, the least auklet, because of its body size, can utilize a broader range of nest sites if necessary (Bédard, 1969). Other auklets include the rhinoceros auklet (*Cerorhinca monocerata*) (Wilson & Manuwal, 1986) and Cassin's auklet (*Ptychoramphus aleuticus*) (Ainley *et al.*, 1990b). Wilson and Manuwal (1986) studied three populations of rhinoceros auklet along the Washington coast and found that 3.9 per cent, 4.9 per cent and 26.8 per cent of the birds nested on steep rocky slopes and cliffs. These birds have developed a nocturnal visitation pattern in response to pressure from predation and kleptoparasitism. Cassin's auklet, which is typically densely colonial, occupies natural crevices and hollows in talus slopes off the California coast, USA (Ainley *et al.*, 1990b).

The little auk or dovekie (*Alle alle*) may be the most numerous Atlantic alcid species, with approximately 12 million breeding pairs throughout much of the Arctic (Evans & Nettleship, 1985). It nests in three primary habitats: eroded crevices on cliff faces, amongst hollows on talus slopes, or on rocky slopes. Evidence gathered by Stempniewicz (1995) from Spitsbergen, however, suggests that coastal cliffs were its original nesting habitat and it has since expanded to flatter ground. Evidence for this would include the presence of egg-shell pigmentation (unimportant in nest burrows) and two brood patches for only one egg (originally two eggs may have lessened the impact of gull predation on cliffs). Fifteen-day-old chicks also possess plumage similar to other cliff nesters (e.g.

(a)                                          (b)

*Figure 5.3*   (a) Kittiwakes occupying a variety of nest sites from minuscule ledges to small grassy banks on sea cliffs facing the North Atlantic, Newfoundland, Canada. (b) A single adult kittiwake occupying a nest site on the cliffs in (a). Photos by P.E. Kelly.

razorbills and murres), test their wings outside (a sign of intent to leave the cliff colony), and have well-developed belly and flank feathers for protection when striking the ground (Stempniewicz, 1995). While some populations still nest on cliffs, many younger populations have moved onto rocky mountain slopes. Stempniewicz (1995) believes this shift in nesting habitat has occurred because it lengthens the prelaying period, females gather more energy for egg production, and egg size increases, thus increasing chick growth and development. Also, there were no previous competitors for these breeding sites.

Cliff-nesting species are also found in the Laridae; the kittiwakes, gulls and terns (Fig. 5.3a, b). Black-legged kittiwakes (*Rissa tridactyla*) nest on sea cliffs throughout much of the northern hemisphere (Cullen, 1957; Furness & Barrett, 1985; Chapdelaine & Brousseau, 1989; Cadiou, Monnat & Danchin, 1994; Birkhead & Nettleship, 1995). Cadiou *et al.* (1994) found that the nest sites of black-legged kittiwakes are actively sought in advance by squatters or failed breeders which occupy unattended nests or nests with chicks to assess the environmental quality of a nest site. Adaptations to cliff nesting in the black-legged kittiwake would include sharp claws and mud collecting. Cullen (1957) compared black-legged kittiwakes with ground-nesting gulls and found that anti-predator behaviours such as an aggressive upright posture, predator attacks and alarm calling have been lost or reduced in cliff-nesting kittiwakes. While kittiwakes are unaffected by terrestrial predators (Birkhead & Nettleship, 1995), aerial predation from gulls can still result in a significant loss of kittiwake chicks (Chapdelaine & Brousseau, 1989).

During the breeding season, cliff and talus environments provide predatory gulls like the herring gull (*Larus argentatus*), glaucous gull (*Larus hyperboreus*) and lesser black-backed gull (*Larus fuscus*) with a significant proportion of their food supply. Gulls often nest on the cliff face itself (Gaston & Elliot, 1996; Gilchrist & Gaston, 1997a) or above the colonies and, in the case of glaucous gulls on Coats Island, Northwest Territories, Canada, may obtain greater than 85 per cent of their diet from thick-billed murre eggs or chicks (Gilchrist & Gaston, 1997b). Some cliff-nesting gulls show nesting behaviour that has evolved as a direct response to the rock face. In the Galápagos Islands, Ecuador, nesting swallow-tailed gulls (*Creagrus furcatus*), like thick-billed murres, face the cliff to protect nestlings from falling and place their feet and wings between the chick and the abyss (Burtt, 1993). Chicks peck at the wall to initiate feeding. While they are generally regarded as predators in sea bird communities, gulls can also be the victims of predation themselves, as is the case with the yellow-footed gulls (*Larus livens*) in Baja California, Mexico. Ravens were observed attacking young and eggs at yellow-footed gull nest sites in talus and along beach-berms (Spear & Anderson, 1989).

Terns generally do not nest near cliffs, but Scolaro, Laurenti and Gallelli (1996) observed South American tern (*Sterna hirundinacea*) colonies nesting at cliff edges along the Argentina coastline. These colonies are apparently founded by randomly nesting pairs which then attract other nesters. The sedentary behaviour of tern chicks until the fledgling stage was seen as an adaptation to the cliff environment. The brown noddy (*Anous stolidus*), another member of the Laridae, was observed nesting on cliffs along the coast of Puerto Rico (Burger & Gochfeld, 1985). The only member of the Sulidae to prefer nesting on cliff faces is the North Atlantic gannet (*Sula bassana*) which nests on cliffs that have horizontal ledges wide enough for the accretion of nest material. These nests begin as a ring of seaweed, other vegetation and flotsam cemented with excreta, which forms a base to which a new cup is added annually. The North Atlantic gannet is morphologically suited to cliff nesting because it has high aspect-ratio wings, small pectoral muscles and it is heavy (up to 3.6 kg), and thus has difficulty taking off from flat surfaces (Nelson, 1978a). Adaptations to the cliff environment would include nest cementing with excreta, the inability to retrieve eggs, the extreme cling-ing ability of young, the passive begging of young (anti-falling behavi-our) and the black plumage of young (attack inhibiting) (Nelson, 1978b). Other Sulidae include the brown booby (*Sula leucogaster*), which nests on cliff edges (Burger & Gochfeld, 1985), and the red-footed booby (*Sula sula*) and Peruvian booby (*Sula variegata*), which will occasionally nest on cliff faces (Nelson, 1978b) in tropical regions. In the Peruvian booby, chick mortality is high, however, as nests often break away and fall off (Nelson, 1978b). In the Pelagic cormorant (*Phalacrocorax pelagicus*) of the Phalacrocoracidae, nesting can occur on cliffs. On the Farallon Islands, California, USA, colonies of these birds use their excrement to cement nests onto cliff ledges. The shape of the colonies is dependent on the availability of ledges (Boekelheide *et al.*, 1990a). Successful breeding in these birds is dependent on the availability of juvenile rockfish, which dominate their diet (Boekelheide *et al.*, 1990a).

Some fulmars and petrels, members of the Procellariidae, may nest on cliffs but are typically species that burrow into softer substrates. The northern fulmar (*Fulmarus glacialis*) has a circumpolar distribution, including the coasts of Greenland (Falk & Møller, 1997), Alaska (Hatch, 1989), Russia (Vyatkin, 1993) and the UK and Ireland (Cramp, Bourne & Saunders, 1974). This species nests on level ledges on coastal cliffs and rocky breakdown but may also nest well inland (Warham, 1990). In the western Bering Sea, Vyatkin (1993) found that fulmar colonies nest from 20 m above sea level to the top of the cliffs at 300–400 m. Cramp *et al.* (1974) report that apart from cliffs, fulmars may nest in hollows on steep grassy slopes, broken areas at the cliff edge and on the tops of walls, castles and ruined buildings. Nest sites are usually associated with collections of stones and debris (Warham, 1990). In north-eastern Greenland northern fulmar colonies, Falk and Møller (1997) found that nest sites were no more than 3 m from each other.

Some petrel species nest on cliff faces and talus, although most nest on steep, often grassy, slopes. Most are smaller species and choose crevices bounded on all sides by rock to prevent takeover by larger petrels (Warham, 1990). Examples include the Cape petrel (*Daption capense*) of the southern hemisphere (Weidinger, 1996) and the snow petrel (*Pagodroma nivea*) of Antarctic waters which nest in colonies on cliffs, and the white-headed petrel (*Pterodroma lessoni*) and Georgian diving petrel (*Pelecanoides georgicus*) which nest in talus burrows in the southern hemisphere (Warham, 1990).

Other cliff-nesting sea birds include the wedge-rumped storm petrel (*Oceanodroma tethys*) and Wilson's storm petrel (*Oceanites oceanicus*) of the Oceanitidae (Warham, 1990), some members of the Phaethontidae, the tropicbirds (Haley, 1984) and the Diomedeidae, the albatrosses. Cliff-nesting albatrosses include the grey-headed albatross (*Diomedea chrysostoma*), yellow-nosed albatross (*Diomedea chlororhynchos*), black-browed albatross (*Diomedea melanophris*), light-mantled sooty albatross (*Phoebetria palpebrata*) and the sooty albatross (*Phoebetria fusca*) (Weimerskirch, Jouventin & Stahl, 1986). Although all five species occupy cliffs of the Crozet Islands in the south-western Indian Ocean, there appears to be no evidence of competition for nest sites. Available habitat does not seem to limit the population. Also, the five species have varying foraging areas (Weimerskirch *et al.*, 1986). The light-mantled sooty albatross prefers to nest solitarily and places its nests well away from others (Hosking & Lockley, 1984).

### 5.1.4   *Other bird species*

Other cliff-nesting birds include the raven, rock wren (*Salpinctes obsoletus*), cliff swallow (*Hirundo pyrrhonota*), Waldrapp ibis (*Geronticus eremita*), and many species of *Columba*, including *Columba livia* (rock dove, or common pigeon). Ravens are commonly associated with nesting pairs of raptors on both inland and sea cliffs (Ratcliffe, 1993) and many raptor species will use abandoned raven stick nests as nest sites (White & Cade, 1971; Boyce, 1987). In the Mojave Desert, 49 per cent of prairie falcon nests were built in abandoned raven or other raptor nests (Boyce, 1987), although ravens are known to use nest sites repeatedly over many generations (White & Cade, 1971). Gyrfalcons often rely heavily on raven nests for their own use, thus ravens become important modifiers of the gyrfalcon environment on cliff faces (White & Cade, 1971). In the UK, raven nests are often lined with sheep's wool (Ratcliffe, 1993). Ravens are also extremely hardy birds whose nesting patterns are rarely disrupted, even during severe springs or winters (Ratcliffe, 1993).

Both ravens and gyrfalcons nest in the lowermost sections of cliff faces (see Fig. 5.1), usually under an overhang, and both species have been observed nesting in close proximity to each other. Gyrfalcons have also been observed displacing ravens from their nests and in some cases killing them (White & Cade, 1971). In southern Idaho, mean internest distance for ravens was 7.48 km, which was greater than for the other three species of raptors surveyed along the same cliffs (Craig & Craig, 1984). Along the Colville River in Alaska, White and Cade (1971) found that the average internest distance for ravens was 21.1 miles (range 3.5–102 miles), also greater than for the other raptor species along these cliffs. In Alaska, White and Cade (1971) found that ravens are 'functional raptors' and feed partly on carrion (often stolen from raptor nests) and partly on live prey.

The rock wren is a cliff nester native to rocky, arid habitats of western North and Central America. Camp and Knight (1997) and Ward and Anderson (1988) reported that the rock wren was one of the most abundant species in their studies of avian cliff communities in California and Wyoming. The rock wren nests in crevices or holes in cliffs, talus slopes and boulders, where it builds a small foundation of stones upon which the nest is built (Rumble 1987; Merola, 1995). This remarkable feat is accomplished by the female rock wren, which can move stones weighing over 6 g despite an approximate body weight of only 16 g! (Merola, 1995). In the UK, winter wrens (*Troglodytes troglodytes*) also occupy cliffs and steep banks, where the male birds build nests for roosting (Ratcliffe, 1993).

The coastal limestone cliffs of Morocco provide the last refuge for another cliff-dwelling species: the Waldrapp ibis. Once common throughout Central Europe, North Africa and the Middle East, this species was decimated by uncontrolled hunting, pesticide use and the loss of habitat off cliffs which was largely converted to farmland (Manry, 1993). Approximately 350 individuals remain in the wild. These birds build bulky nests on rocky ledges along the seacoast but then travel inland approximately 19 km to feed on beetles and crickets picked off sand dunes. Today they are largely protected by the isolation of these coastal cliffs. Reintroduction efforts are currently being considered in non-native cliff habitat in Spain (Manry, 1993).

The cliff swallow is a cliff-nesting bird species which nests inland along rivers and freshwater bodies. These birds build nests on natural cliffs, canyons, bridges and road culverts near mud supplies essential to nest construction (Brown & Brown, 1990). Colonies of cliff swallows have thus expanded eastward through the Great Plains, USA, by following river courses and colonizing bridges. In dense colonies, cliff swallows often practise brood parasitism, whereby eggs are carried to other nests. Brown and Brown (1990) estimate that 22–43 per cent of nests in large colonies have an alien egg. This provides a greater chance that all of an individual's eggs will be hatched.

Other common cliff-nesting bird species found by Ward and Anderson (1988) in Wyoming were the brewer's sparrow (*Spizella breweri*), green-tailed towhee (*Piplio chlorurus*), vesper sparrow (*Pooecetes gramineus*), sage thrasher (*Oreoscoptes montanus*) and horned lark (*Eremophila alpestris*). The lark sparrow (*Chondestes grammacus*), western meadowlark (*Sturnella neglecta*) and vesper sparrow occurred in significantly greater densities on rock outcrops. In Joshua Tree National Park, the mourning dove (*Zenaida macoura*), black-chinned hummingbird (*Archilochus alexandri*), common raven and black-throated sparrow (*Amphispiza bilineata*), along with the rock wren, accounted for over half of all birds observed at cliff sites (Camp & Knight, 1997). Along the Niagara Escarpment, the American goldfinch (*Carduelis tristis*) and house wren (*Troglodytes aedon*) were found at both sites in all three cliff zones but not in the plateau, while the common grackle (*Quiscalus quiscula*) was found in the talus and edge zones at both sites but not in the plateau (Matheson & Larson, 1998). In central Norway, the four most dominant passerines surveyed around cliffs were the chaffinch (*Fringilla coelebs*), hedge accentor (*Prunella modularis*), common chiffchaff (*Phylloscopus collybita*) and the European robin (*Erithacus rubecula*). These four species constituted approximately 50 per cent of the cliff birds encountered. The highest density of birds was observed at the cliff base, where the European pied flycatcher (*Ficedula hypoleuca*) rather than the European robin was one of the four dominant species (Reitan, 1986).

In the UK, other cliff-nesting birds observed would include the stock dove (*Columba oenas*), jackdaw, carrion crow (*Corvus corrone*) and the chough (*Pyrrhocorax pyrrhocorax*) (Ratcliffe, 1993). Stock doves nest in isolated pairs or small groups in rock crannies or holes in the cliff face, while jackdaws nest in colonies in cliff fissures and crevices where ravens and peregrine falcons are the primary predators (Ratcliffe, 1993). In summer, the jackdaws are often seen on neighbouring hill slopes looking for insects. Carrion crows are usually tree nesters but nest on cliffs in Wales. Sea cliffs on the west coast of Wales and Ireland are also the home of the chough. The chough feeds on insect larvae and ants from short turf or recently burned heath. Agricultural developments have consequently led to a population decline in this species (Ratcliffe, 1993). The common pigeon or rock dove is probably the most familiar example of a once highly endemic cliff species that took advantage of anthropogenic cliffs and the enormous food supply adjacent to human settlements. It is tempting to consider here and below the question of whether the use of cliffs by palaeohumans had the effect of offering new opportunities to certain species of plants and animals that were normally present as highly endemic populations spatially restricted to locations with cliffs. Among the various species of *Columba* that still occur in India, for example, there are several species that are morphologically and behaviourally similar to the rock dove, but they are still present only as highly endemic and, in some cases, rare species. It is possible that *Columba livia* and other animals such as *Rattus rattus* and *Mus musculus* at one time used cliffs as their sole or principal habitat type.

Some birds like the great reed warbler (*Acrocephalus arundinaceus*) in the Shetland Islands, UK (Riddiford & Potts 1993), and the house sparrow (*Passer domesticus*) in the Isles of Scilly, UK (Penhallurick, 1993), have been observed foraging or nesting in cliff habitats on rare occasions. The house sparrows were observed nesting in sea cliffs in the soil at the top of the cliff face underneath overhanging vegetation. Nearby grasses were used to line holes in the cliff for nests (Penhallurick, 1993).

## 5.2 Mammals

Cliffs, talus and associated caves are also important habitats to wildlife. In the Blue Mountains of Oregon and Washington, USA, Maser *et al.* (1979) describe the important characteristics of these habitats and the various mammals which use the cliff environments. They also point out that the abrupt, relatively stable boundary defined by the cliff edge provides a secure habitat for wildlife and increases species diversity. Maser *et al.* (1979) list mice (Cricetidae), woodrats (Muridae), coyotes (Canidae), bobcats and pumas (Felidae) as users of shallow cave habitat in cliffs for shelter, reproduction and nesting. Thirty-three mammal species are listed as users of cliffs, 36 species are listed as users of talus, and 20 species are listed as users of caves in cliffs in Wyoming for feeding, reproduction or both (Table 5.2). Pikas (*Ochotona princeps*) are listed as the only obligate user for they are dependent on the availability of talus for their survival.

The suitability of cliff faces as viable habitat for mammals is scale dependent. Small-scale cracks, ledges, crevices and the trees rooted within them provide ample opportunities for the movement of small mammal species such as rodents. Despite the obvious difficulties imparted by the vertical orientation of cliffs, which select against large body weight and size, there are some larger mammal species for which cliff environments also provide optimal habitat (Fig. 5.4). A synopsis of the available literature on the use of cliff environments by both small and large mammals is presented below.

Table 5.2. *Summary of vertebrate use of cliffs in western USA.*

| | | | |
|---|---|---|---|
| *Amphibians* | | Great horned owl | fr |
| Woodhouse toad | f | Burrowing owl | f |
| Pacific treefrog | f | Townsend's solitaire | fr |
| | | Loggerhead shrike | f |
| *Reptiles* | | Starling | fr |
| Western fence lizard | fr | Gray-crowned rosy finch | fr |
| Side-blotched lizard | fr | Black rosy finch | fr |
| Western skink | f | Pine siskin | f |
| Striped whipsnake | f | Lesser goldfinch | f |
| Gopher snake | f | Green-tailed towhee | f |
| Night snake | f | Sage sparrow | f |
| Western rattlesnake | f | | |
| | | *Mammals* | |
| *Birds* | | Little brown myotis | fr |
| Canada goose | r | California myotis | fr |
| Turkey vulture | r | Small-footed myotis | fr |
| Goshawk | f | Western pipistrelle | fr |
| Red-tailed hawk | r | Big brown bat | fr |
| Swainson's hawk | f | Spotted bat | fr |
| Ferruginous hawk | r | Western big-eared bat | f |
| Golden eagle | fr | Pallid bat | fr |
| Prairie falcon | fr | Least chipmunk | fr |
| Peregrine | fr | Yellow pine chipmunk | fr |
| American kestrel | fr | Yellow-bellied marmot | fr |
| Chukar | fr | Columbian ground squirrel | r |
| Rock dove | r | Mantled ground squirrel | fr |
| Barn owl | fr | Canyon mouse | fr |
| Screech owl | fr | Deer mouse | fr |
| Poorwill | f | Pinyon mouse | fr |
| Common nighthawk | fr | Bushy-tailed woodrat | fr |
| Black swift | fr | Porcupine | r |
| Vaux's swift | fr | Coyote | fr |
| White-throated swift | fr | Red fox | f |
| Say's phoebe | fr | Black bear | f |
| Gray flycatcher | f | Marten | f |
| Barn swallow | fr | Fisher | f |
| Cliff swallow | fr | Short-tailed weasel | f |
| Steller's jay | f | Long-tailed weasel | fr |
| Black-billed magpie | f | Wolverine | fr |
| Clark's nutcracker | f | Badger | f |
| Common raven | fr | Striped skunk | f |
| Dipper | r | Spotted skunk | fr |
| Canyon wren | fr | Lynx | f |
| Sage thrasher | f | Bobcat | fr |
| Western bluebird | f | Mountain goat | fr |
| Mountain bluebird | f | Big-horned sheep | fr |
| Rock wren | fr | | |

*Notes:*
f = feeding; r = reproduction; fr = feeding and reproduction.
*Source:* Modified from Maser *et al.* 1979.

*Figure 5.4* A raccoon family climbing a 260-year old inverted eastern white cedar on a cliff face on the Niagara Escarpment. Raccoons use cracks, fissures, ledges and trees to scale the cliff face successfully. Photo by P.E. Kelly.

Table 5.3. *Total number of small mammals captured in plateau plots, cliff edge plots, cliff face plots and talus slope plots along the Niagara Escarpment*

| Species name | Capture location | | | | |
| --- | --- | --- | --- | --- | --- |
| | Plateau | Edge | Face | Talus | Total |
| White-footed mouse (*Peromyscus leucopus*) | 21 | 21 | 10 | 16 | 68 |
| Deer mouse (*Peromyscus maniculatus*) | 4 | 11 | 17 | 14 | 46 |
| Raccoon (*Procyon lotor*) | 4 | 2 | 1 | 3 | 10 |
| Red squirrel (*Tamiasciurus hudsonicus*) | 5 | 5 | 2 | 2 | 14 |
| Eastern chipmunk (*Tamias striatus*) | 0 | 0 | 2 | 2 | 4 |
| Southern red-backed vole (*Clethrionomys gapperi*) | 0 | 0 | 0 | 1 | 1 |
| Total | 34 | 42 | 32 | 38 | 153 |

*Note:*
Sampling took place over the summer months June–August. All animals were livetrapped and released within 12 hours of capture.

### 5.2.1 Small mammals

The majority of small mammals which use cliffs are rodents, although we will also discuss the use of cliffs by bats. Small mammals on or near cliffs have been live-trapped by Matheson (1995), Johnson (1986) and Ward and Anderson (1988) in Ontario, Canada, Utah, USA, and Wyoming, USA, respectively. Matheson (1995) explored the use of the 25 m high Niagara Escarpment cliff face, edge, talus and plateau by small mammals at two sites. More species of small mammals were caught in live traps placed within 25 × 25 m plots on the cliff face than in the plateau (Table 5.3). The white-footed mouse (*Peromyscus maniculatus*), deer mouse (*Peromyscus leucopus*), red squirrel (*Tamiasciurus hudsonicus*) and raccoon (*Procyon lotor*) were caught in both locations but the eastern chipmunk (*Tamias striatus*) was only caught on the cliff face. These five species plus the southern red-backed vole (*Clethrionomys gapperi*) were captured in the talus. Habitat partitioning was evident amongst the two species of mice as the white-footed mouse was most frequently caught on the plateau and cliff edge (21 total captures in each habitat) and least frequently caught on the cliff face (10 captures). The deer mouse was most frequently

*Figure 5.5* Heavily used animal trails occur along cliff edges and faces of the Niagara Escarpment. These often include obvious latrines such as the base of this inverted eastern white cedar along the Niagara Escarpment. Photo by D.W. Larson.

caught on the cliff face (17 captures) and least frequently caught in the plateau (4 captures) (Table 5.3).

Matheson (1995) concluded that the benefits must outweigh the costs of using the cliff face, including increased risk of falling and lower food availability, and that small mammals are using the Niagara Escarpment cliff face for its non-consumptive properties, i.e. it acts as a refuge from predation. The cliff represents 'enemy-free space' (Jeffries & Lawton, 1984) where small mammals can retreat to reduce or eliminate the risk of predation. The discrete boundaries offered by the cliff face as a landscape element allow small mammals to exploit several different habitats within close proximity to each other. The cliff environment is ideal for these highly mobile mammals, which can feed in the adjacent plateau forest but retreat to the cliff face for safety from predation. While on the cliff face, animals defecate on ledges and trees in spatially distinct latrines (Fig. 5.5; Table 5.4). These allochthonous inputs, small though they may be, provide important small-scale contributions to the nutrient pool on the cliff face (Matheson, 1995).

Table 5.4. *Dry weight of mammal faeces collected in seven cliff-face sites along the Niagara Escarpment*

| Site | Location | Number of samples | Site total (g) | g/m² |
|------|----------|-------------------|----------------|------|
| S1 | South Site | 12 | 1963.69 | 2.38 |
| S2 | South Site | 7 | 1132.69 | 1.81 |
| S3 | South Site | 27 | 4656.6 | 7.44 |
| N1 | North Site | 2 | 18.24 | 0.03 |
| N2 | North Site | 0 | 0.00 | 0.00 |
| N3 | North Site | 6 | 57.50 | 0.08 |
| N4 | Purple Valley | 9 | 4327.47 | 3.82 |

*Note:*
The amount of faeces per square metre was calculated by dividing the site totals by the surface area of each site.

Winter mammal track data (Table 5.5) were also collected by Matheson (1995). Small mammals show a preference for using the cliff edge and talus as open corridors for movement during the winter months. Trails along the cliff edge seem to have a common pathway, whereas trails in the plateau have seemingly random orientations (Matheson, 1995). The increased frequency of use of these habitats is probably a function of snow depth, which increases with increasing distance from the cliff edge (Bartlett *et al.*, 1990).

In south-eastern Utah, USA, Johnson (1986) trapped small mammals on five butte summits and bases and at a nearby control site. These cliffs range in height from 27 to 104 m. A significant relationship was found between cliff height and the number of species found on the summits. The high cliffs were found to be significant barriers to small mammal movement. While the species richness of mammals at the control sites and the bases of the buttes was nearly identical, far fewer species were found on the butte summits. Sixteen different mammal species were trapped at the control site, whereas only five different mammal species were trapped on the butte summits, the number of species at each butte varying between two and five. The canyon mouse (*Peromyscus crinitus*) was the only species trapped on all five butte summits. The Colorado chipmunk (*Eutamias quadrivittatus*) and the bushy-tailed woodrat (*Neotoma cinerea*) were trapped on four of the five summits. The deer mouse and the pinyon mouse (*Peromyscus truei*) were also trapped. Mammalian richness was also correlated with microrelief. There was no species–summit

Table 5.5. *Mammal track encounter rates in plateau, edge and talus locations along the Niagara Escarpment*

| Animal | Plateau | Edge | Talus | $\chi^2$ | $p$ value |
|---|---|---|---|---|---|
| Coyote | 1.0 | 0.5 | 1.8 | 0.8 | >0.05 |
| Mouse | 0.5 | 4.2 | 1.0 | 4.2 | >0.05 |
| Chipmunk | 2.0 | 2.0 | 1.5 | 0.1 | >0.05 |
| Squirrel | 5.8 | 22.3 | 27.0 | 13.5 | <0.05 |
| Porcupine | 0.0 | 3.0 | 0.0 | 6.0 | <0.05 |
| Vole | 0.0 | 0.5 | 0.0 | 1.0 | >0.05 |
| Raccoon | 0.0 | 1.5 | 4.5 | 5.3 | >0.05 |
| Shrew | 0.7 | 0.0 | 0.0 | 1.3 | >0.05 |
| Total | 9.97 | 34.00 | 35.80 | 15.7 | <0.05 |

*Note:*
Each value is the total number of tracks observed along 250 m transects 12 hours after three separate snowfalls. $p$ value greater than 0.05 indicates no significant difference.

area relationship. Johnson (1986) concluded that cliffs do not provide barriers to the movement of some species, in particular the canyon mouse which can exist indefinitely with no free water, is omnivorous, a good climber, and prefers rocky habitat. Cliffs may be the preferred habitat for this species. The bushy-tailed woodrat prefers shelter in vertical crevices and clefts on cliffs, which may have provided the access route to the summit for this species. The deer mouse and bushy-tailed woodrat were also identified by Ward and Anderson (1988) and Maser *et al.* (1979) as cliff dwellers.

In Wyoming, USA, Ward and Anderson (1988) found a greater abundance and diversity of small mammals on cliff sites than in talus and plateau control sites 150 m away from the face. Ward and Anderson (1988) found that cliffs were disproportionately important as wildlife habitat. The six most common species captured on cliff sites were the deer mouse, least chipmunk (*Eutamium minimus*), Wyoming ground squirrel (*Spermophilus elegans*), bushy-tailed woodrat, white-tailed prairie dog (*Cynomys gunnisoni*), and the northern grasshopper mouse (*Onychomys leucogaster*). The six common species at the control sites were the same as on the cliff face except for the inclusion of the silky pocket mouse (*Peroganthus flavus*). The bushy-tailed woodrat was absent at the control site. This abundance and richness of small mammals were attrib-

uted to increased topographic roughness and a greater diversity of vegetation on cliff sites (Ward & Anderson, 1988).

Another small mammal native to western North America, the pika is
an obligate talus dweller and is diagnostic of a cool, mesic, rocky habitat
(Hafner, 1993). Pikas live in talus environments and nowhere else and
have a maximum dispersal distance of only 3 km (Hafner, 1993). They
occupy large, continuous talus habitats and very small isolated talus fragments scattered throughout the western USA (Udall, 1991; Hafner,
1993; Peacock & Smith, 1997). Pikas are individually territorial, do not
hibernate, and define their territory by constructing 'hay piles' amongst
the rocks (Peacock & Smith, 1997). These hay piles are stores of vegetation that are used as food supply during the winter months.

Other rodents observed in talus environments include the pack rat or
eastern woodrat (*Neotoma floridana*). This species is found predominantly
in cliffs, caves and rocky areas in the eastern USA. Nests are constructed
of twigs, bark, other vegetation and assorted shiny objects (Feldhamer,
Gates & Chapman, 1984). The small shrew (*Sorex minutus*) nests in talus
slopes in the Krkonose Mountains of the Czech Republic (Růžička &
Zacharda, 1994) and feeds on the large numbers of arthropod species that
inhabit the stony debris. It is quite likely that cliff and talus environments
worldwide are habitats for many endemic rodent species. One such indication comes from a study conducted of the wild populations of rodents
in Libya (Ranck, 1968). In this work it was found that wild populations
of *Rattus rattus, Mus musculus,* and *Gerbilis campestris* were almost entirely
confined to cliffs and their associated talus slopes along the
Mediterranean coast. The author pointed out that these wild populations
were not commensal with any known populations of people and he speculated that such steep rocky areas might be the preferred areas for such
species in the absence of contact with humans.

Other small mammals commonly associated with cliffs include bats
because of their obvious preference for cave and crevice habitats (i.e.
Lüth, 1993). Adam, Lacki and Barnes (1994), however, have found that
bats also spend a disproportionate amount of time foraging outside their
night roosts in and around cliffs. The activities of Virginia big-eared bats
(*Plecotus townsendii virginianus*) in Kentucky, USA, in and around 15 to 60
m high sandstone/limestone cliffs were monitored using radio transmit-
ters (Adam *et al.*, 1994). Foraging activity was concentrated in and around
night roost areas along the cliffs and in the neighbouring forest habitat,
particularly at the cliff top. Adam *et al.* (1994) and Lacki, Adam and

Shoemaker (1993) conclude that cliffs are important components of the habitat for Virginia big-eared bats. The latter authors proposed the use of buffer strips adjacent to the cliff edge and in the talus to protect vulnerable bat populations such as the Virginia big-eared bat.

The above literature gives greater weight to an idea that we call the 'urban cliff hypothesis'. We believe that many of the rodents and other cliff-dwelling organisms that are now commensal with humans (many horticulturally important plants, birds and mammals) come from taxa that were once endemic to cliffs and therefore much less common. The distant ancestors of *Rattus* and *Mus* species that are now viewed as pests in cities and towns worldwide could very well have been taxa that were highly specialized and endemic to cliffs and talus slopes. When increases in human population size made it necessary for *Homo sapiens* to leave the protection of caves and rock shelters as habitat and food storage areas, new buildings were constructed of stone and wood. The idea here is that perhaps the rats, mice, pigeons, bats and many plants have simply followed us from one kind of cliff habitat of natural origin to others constructed of stone, wood, glass and steel by humans. Even though it is probably impossible to test this idea without finding truly wild populations of these species and then testing their genetic relationship to commensal populations, other workers long ago have noticed this trend. Audubon reported in the 1846 folio edition of *Viviparous Quadrupeds of North America* some observations made by a civil engineer working in central Pennsylvania, USA:

> In April 1831, when leading the exploring party which located the portage railroad over the Alleghany Mountains, in Pennsylvania, I found a multitude of these animals living in the crevices of silicious limestone rocks on the Upper Conemaugh river, in Cambria county, where the large viaduct over that stream now stands. The country was then a wilderness, and as soon as buildings were put up the rats deserted the rocks, and established themselves in the shanties, to our great annoyance . . . We have on various botanical excursions, explored these mountains at different points, to an extent of seven hundred miles, and although we saw them in the houses of settlers, we never observed any locality where they existed permanently in the woods, as they did according to the above account.

Analysis of the ecological structure of human dwellings, however, suggests that we may now live in caves and rock shelters that are cement and glass copies of dwellings made of sandstone and dolomite during the Palaeolithic period.

## 5.2.2 Large mammals

Large mammals which use cliffs include ungulates such as mountain goats and bighorn sheep (Artiodactylidae), foxes (Canidae), langurs (Cercopithecidae), rock-wallabies (Macropodidae) and other rock mammals of Australia. Mountain goats and sheep primarily use cliffs as escape terrain and foraging sites. Haynes (1992) found that mountain goats (*Oreamnos americanus*) in north-western Wyoming, USA, spend most of their lives within 0.4 km of cliff terrain, thus their home ranges are comprised almost entirely of cliffs and nearby slopes. Smith (1986) obtained similar results in Alaska, USA, but also noted that winter habitat preference varies. He found that cliffs were not as preferable as old-growth stands for winter habitat in Alaska. Gilbert and Raedeke (1992) found that cliffs comprised 65 per cent of all winter goat locations in the Cascades, Washington, USA, and Geist (1971) noted that 52 per cent of mountain goats in mid-winter in the Palliser Range, Alberta, Canada, were located on sheer cliffs. These different findings may reflect climatic differences between the three regions.

Mountain goats also have a diurnal cycle of cliff occupancy. Mountain goats in the Palliser Range move off cliffs at midday and move towards mountain slopes to feed on alpine fir (*Abies lasiocarpa*). Before dusk, they return to the cliffs to locate a resting place. Mountain goats will also run to cliff terrain when threatened, flattening themselves against the rock (Geist, 1971). Wherever possible, they will position themselves under overhangs. This response can even be initiated by the noise of a low-flying aircraft. Mountain goats are very competent climbers, even on snow-covered and ice-covered cliffs. In the event that they slip, they spread their legs apart and flatten themselves against the rock and claw for footholds as they slide.

Stone's sheep (*Ovis dalli stonei*) (also native to western Canada), on the other hand, jump along cliffs and if they fall they leap to other ledges or footholds. Stone's sheep are more generalized mountaineers which roam extensively over mountain terrain. Their escape response is similar to that of the mountain goat, as is their diurnal cycle, although Stone's sheep are generally more widely dispersed at nightfall but congregate on the cliff faces by morning. Stone's sheep apparently also use cliffs as breeding habitat. Geist (1971) noted that unreceptive ewes head towards cliff faces to escape the advances of rams. Frequently, they run towards narrow ledges or back into crevices. These tactics are usually not successful as the ram often dislodges the ewe from her position and breeding takes place on the cliff face.

Kelly (1980), Graham (1980) and Wakelyn (1987) have also reported that cliffs are important as escape terrain in bighorn sheep (*Ovis canadensis*) as a response to predators such as bobcat (*Lynx rufus*) (Kelly, 1980) or humans (Graham, 1980). Wakelyn (1987) found that the distribution of Rocky Mountain bighorn sheep (*O. canadensis canadensis*) in Colorado was controlled by the extent and distribution of cliffs or slopes greater than 60°. Cliffs are high-visibility habitats and thus preferred by bighorns because they can detect predators visually and communicate danger visually (Wakelyn, 1987). It was also found that rugged escape terrain is the most critical Rocky Mountain bighorn sheep habitat component, and 95 per cent of all bighorn sheep activity in north-eastern Utah occurs within 300 m of cliff escape terrain.

If surprised, the desert bighorn (*O. canadensis mexicana*) generally escapes by climbing uphill, but if surprised from above it is capable of taking headlong flights down steep escarpments (Graham, 1980). Injuries and falls do occur but the desert bighorn has an amazing capacity to recover from broken bones and persist for years, even though the accidents may lead to malformed appendages or organs (Allen, 1980). Cliff habitats are also beneficial in that they allow for smaller group sizes, greater foraging efficiency and lower heart rates due to the perception of greater safety (Wakelyn, 1987).

In Europe, exposed cliff edges, ledges and talus slopes of large cliffs are the home of the chamois or moufflon (*Rupicapra rupicapra*). Native to the Alps, this species has been introduced elsewhere in Europe where it has established populations on cliffs where there has never been a large herbivore (Herter, 1993; Lüth, 1993). The chamois often find shelter in caves or under overhangs at the cliff base (Herter, 1996). In the upper Danube Valley, this has led to extensive grazing and trampling damage (Herter, 1993, 1996; Stärr *et al.*, 1995) as the xerothermic vegetation on the cliffs is the preferred food. Grasses such as *Sesleria alba* are the preferred food in the summer but more woody plants are eaten in the winter. Selective removal of *S. alba*, eutrophication and trampling have all led to a change in the community composition of these cliffs (Herter, 1993, 1996). Chamois in Baden-Württemberg, Germany, have in part contributed to the addition of ten plant species to the endangered species list and 17 plant species to the threatened species list between 1983 and 1994 (Stärr *et al.*, 1995). Wild pigs, badgers and hares also contribute to trampling damage at cliff edges in these areas (Herter, 1996).

Other ungulates such as mule deer (*Odocoileus hemionus*) and pronghorns (*Antilocapra americana*) make use of talus slopes. In Wyoming, USA, these two species use talus slopes throughout the year, but for varying purposes. In winter, south-facing talus slopes are important foraging sites as they are free of snow and provide protection from winds, while in the summer, these two species seek shelter under overhangs to escape the heat (Ward & Anderson, 1988).

Australia has a relatively large number of mammal species specializing in rock habitats and many of these species occupy cliff-face, cliff-edge and talus environments. Australia has 21 marsupial and rodent species which are specialist rock dwellers, approximately 11 per cent of the Australian species belonging to these groups (Freeland, Winter & Raskin, 1988). In Arnhemland in north-western Australia, there are many quartz sandstone cliffs, escarpments and rock outcrops that rise up to 100 m above the surrounding landscape which are home to species such as the northern quoll (*Dasyurus hallucatus*) (Begg, 1981b), large rock rat (*Zyzomys woodwardi*) and common rock rat (*Zyzomys argurus*) (Begg, 1981c; Freeland *et al.*, 1988), sandstone antechinus (*Parantechinus bilarni*) (Begg, 1981a; Freeland *et al.*, 1988), rock ringtail possum (*Pseudocheirus dahli*), black wallaroo (*Macropus bernardus*), and nabarlek or miniature rock-wallaby (*Peradorcas concinna*) (Freeland *et al.*, 1988). Other Australian rock mammals include the scaly-tailed possum (*Wyulda squamicaudata*), central rock rat (*Zyzomys pedunculatus*), long-tailed dunnart (*Sminthopsis longicaudata*) and fat-tailed antechinus (*Pseudoantechinus macdonnellensis*) (Freeland *et al.*, 1988). Other species such as the echidna (*Taccyglossus aculeatus*) are not usually rock dwellers but in western Arnhemland they are found most frequently on rock escarpments (Freeland *et al.*, 1988). Such a clear exploitation of cliffs appears to be a direct result of the surrounding wet–dry tropical climate where long periods may go by with little rain. Near these rock 'islands', however, moisture availability is high, and this leads to a high plant species diversity and richness and thus a broad assortment of plant morphologies and plant secondary chemistries which in turn support an unusually large fauna of specialized rock-dwelling mammals (Freeland *et al.*, 1988). Relatively few of these species have been studied to date.

Begg (1981a, 1981b, 1981c) studied the reproduction of the sandstone antechinus, northern quoll, large rock rat and common rock rat of the Arnhem Land plateau. The habitats favoured most by the sandstone antechinus were talus slopes where the boulders and high plant diversity provide suitable nest sites (Begg, 1981a). Talus slopes were also preferred by the northern quoll, a principal predator of the sandstone antechinus, especially during the early wet season (Begg, 1981b). Rock crevices were the second most favoured habitat for the northern quoll, as they were for the female antechinus. The rock rat habitats are partitioned, with the common rock rat preferring talus slopes and the large rock rat preferring closed forest at the cliff top (Begg, 1981c). Some large rock rat individuals have been captured in rock crevices and talus.

The most frequently studied Australian rock species are the rock-wallabies and wallaroos. The brush-tailed rock-wallaby (*Petrogale penicillata*) (Short, 1982), yellow-footed rock-wallaby (*P. xanthopus*) (Copley, 1983; Copley & Robinson, 1983; Robinson *et al.*, 1994), black-footed rock-wallaby (*P. lateralis*) (Pearson, 1992), allied rock-wallaby (*P. assimilis*) (Horsup & Marsh, 1992; Horsup, 1994), northern wallaroo (*Macropus robustus woodwardii*) (Croft, 1987) and black wallaroo (Press, 1989) live primarily on rock outcrops and escarpments in the arid regions of Australia. The distribution of these species is largely determined by the availability of rocky outcrop and cliff habitat. Consequently, they have a patchy distribution throughout their ranges but occur in high densities on these discrete sites (Short, 1982).

While all these species require rock outcrops, they inhabit different home ranges throughout Australia and occupy varying components of the cliff environment. Brush-tailed rock-wallabies occur throughout much of mainland Australia but only inhabit boulder areas with mazes of subterranean passageways and cliff faces on isolated rock stacks, where they seek shelter in caves. Cliffs without ledges and overhanging cliffs are usually uninhabited (Short, 1982). Black-footed rock-wallabies are native to central Australia and require deep overhangs or caves for shelter and prefer rocky outliers with talus slopes. They have been observed in colonies in areas ranging from small granite outcrops less than one hectare in area all the way up to extensive gabbro talus slopes (Pearson, 1992). The allied rock-wallaby is an inhabitant of northern Queensland and shelters in talus slopes during the day but forages into the surrounding countryside at night. This pair-bonding species exhibits a strong fidelity for its shelter and home range (Horsup, 1994). Yellow-footed rock-wallabies live on cliff faces and in gullies in south-western Australia. In the Northern Territory, sandstone rocklands are home to the black wallaroo, which seeks shelter in the talus slopes at the base of heavily eroded sandstone cliffs (Press, 1989). This species is also found in abundance in the monsoon forests of the region. The northern wallaroo also occupies steep rocky habitats in the Northern Territory (Croft, 1987).

Rock-wallabies shift food preferences as the seasons shift between the wet and dry seasons. While some of these species such as the allied rock-wallaby are well adapted to the wet–dry tropics by showing generalist feeding strategy (Horsup & Marsh, 1992), yellow-footed rock-wallabies prefer grasses during the wet season and coarser browse plants during drought (Copley & Robinson, 1983). The rocky habitat provides a refuge from climatic extremes and is an important source of pockets of nutrient resources (Pearson, 1992).

The cliff and talus environments inhabited by rock-wallabies provide escape territory from a variety of native and introduced species. The dingo (*Canis familiaris dingo*), for example, is a primary predator of wallabies but it cannot maintain attack velocity in talus environments (Croft, 1987) and cannot access most cliff faces. Rock-wallabies also use caves in cliffs as protection from wedge-tailed eagles (*Aquila audax*) (Short, 1982; Pearson, 1992). Explorers and hunters in the early part of the century used to hunt rock-wallabies right off cliffs (Copley, 1983). Hunting led to a drastic reduction in rock-wallaby populations in the twentieth century. Feral cats have also been known to kill rock-wallabies (Kinnear, Onus & Bromilow, 1988) and feral goats can evict rock-wallabies from the most accessible shelters (Copley, 1983). Species such as feral rabbits also compete with rock-wallabies for food supply (Pearson, 1992).

Introduced European red fox (*Vulpes vulpes*) populations, however, are the principal threat to the long-term survival of wallaroo and rock-wallaby populations (Copley, 1983; Kinnear et al., 1988; Pearson, 1992). Foxes are very agile and reportedly are capable of hunting wallabies right off cliff faces (Short, 1982). Introduction of the fox has led to the decline and extinction of populations and the reductions in their range and distribution (Kinnear et al., 1988). A fox control programme implemented between 1982 and 1986 in the central wheatbelt region of Western Australia led to population increases of 138 and 233 per cent in two populations of black-footed rock-wallabies after four years of static or declining numbers (Kinnear et al., 1988).

Foxes are also common predators in coastal cliff sea bird communities (Maccarone & Montevecchi, 1981; Squibb & Hunt, 1983; Haley, 1984; Evans & Nettleship, 1985; Jones, 1992; Gilchrist & Gaston, 1997a). For example, Maccarone and Montevecchi (1981) observed a fox climbing down a nearly vertical cliff face on Baccalieu Island, Newfoundland, Canada, to obtain murre eggs. Sixteen eggs were stolen in only 95 minutes. Foxes also kill and cache petrel, kittiwake, puffin and murre chicks and adults for winter survival (Maccarone & Montevecchi, 1981).

Feral cats, dogs, rats and mice are also predators in sea bird colonies around the globe (Haley, 1984).

Another unusual cliff-dwelling species is the Hanuman langur (*Presbytis entellus*), an arboreal monkey native to southern Asia. At two sites in north-western India on the north-west Indian plain and in the Himalaya foothills, Hanuman langurs use cliffs as surrogates for trees. While they are not obligate cliff dwellers, the langurs use the cliffs for sleeping, sunbathing, escape and as lookouts (Vogel, 1976). Ledges on the upper half of the cliff face are used for sleeping and in the mornings the cliff edge is used for sunbathing and as a lookout. The cliffs also provide excellent escape terrain from predators for langurs can move up or down cliffs as quickly as they move through trees (Vogel, 1976). They move on the cliff face by using small grooves in the rock as handholds and narrow ledges to land onto or from which to jump. They bounce off smooth rock faces to change direction and can climb up crevices by bracing against the two sides.

### 5.2.3   Past use of cliffs by mammals

Bateman (1961), Churcher and Fenton (1968), Churcher and Dods (1979), Guilday (1971) and Riley (1993) have identified the remains of a wide variety of cliff fauna from crevice caves in eastern North America. Mammals such as whitetail deer (*Odocoileus virginianus*) and beaver (*Castor canadensis*) (Bateman, 1961), short-tailed shrew (*Blarina brevicauda*), smoky shrew (*Sorex fumeus*), big brown bat (*Eptesicus fuscus*), long-eared bat (*Myotis keenii*), red-backed vole (*Clethrionomys gapperi*), meadow vole (*Microtus pennsylvanicus*), red squirrel (*Tamiasciurus hudsonicus*) and wapiti (*Cervus canadensis*) (Churcher & Fenton, 1968), striped skunk (*Mephitis mephitis*), varying hare (*Lepus americanus*), cottontail rabbit (*Sylvilagus floridanus*) and the extinct large pika (*Ochotona* sp.) (Churcher & Dods, 1979), little brown bat (*Myotis lucifugus*), woodland deer mouse (*Peromyscus maniculatus*) and muskrat (*Ondatra zibethicus*) (Churcher & Fenton, 1968; Churcher & Dods, 1979) have been found in crevice caves along the Niagara Escarpment. Riley (1993) reports a similar array of faunal remains from escarpment crevice caves including marten (*Martes americana*), porcupine (*Erethizon dorsatum*) and the extinct large pika. While some of these species represent the remains of prey items and do not or have not used the cliff environment for feeding or reproduction, the variety and abundance of faunal remains do indicate that the cliffs of the Niagara Escarpment may have been important to many mammal species in the past. Hafner (1993) also used fossil pika remains to identify areas with cool, mesic, rocky habitats during the Pleistocene period in the western USA.

## 5.3 Amphibians and reptiles

Herrington (1988) examined the use of 183 talus slopes and non-talus control sites by amphibians and reptiles in Oregon and Washington, USA, and found that talus areas are extremely important habitat types for the herpetofauna of these two states. More than 60 per cent of the amphibians and reptiles that occur in these states were found to utilize talus habitats. Four species of frogs, 14 species of salamanders, three species of lizards and ten species of snakes were observed in the talus. Four species of salamander – *Plethodon elongatus, P. larselli, P. vandykei* and *P. stormi* – were restricted to talus areas although all the salamander species found were capable of completing their entire life cycle within the talus habitat. Twenty herpetofaunal species used the talus as protection against adverse weather conditions, while eight species used the talus for reproductive activities. Snake species congregate in the talus, which they use as hibernacula. Only five of the herpetofaunal species were considered occasional users of the talus. Talus altered by human activities seemed to affect the abundance of herpetofauna which occupied them. Only 42 per cent of the talus slopes surveyed were considered unaltered yet yielded 73 per cent of the total number of individuals.

Along the Niagara Escarpment, Bogart, Cook and Rye (1995) found three species of ambystomatid salamanders and several associated hybrids. The yellow-spotted salamander (*Ambystoma maculatum*) was particularly widespread and inhabited a wide variety of habitats, particularly wetter areas. From their distribution and the rate of hybridization, Bogart *et al.* (1995) found that these salamanders may be remigrating into northern sections of Ontario. In Utah, Johnson (1986) found that some reptiles were capable of negotiating cliffs on buttes as five of the 11 species of reptiles surveyed in a control area were found to have established populations at the top of buttes. One species, the eastern fence lizard (*Sceloporus undulatus*) was found on all five butte summits. The remaining four species were the western whiptail (*Cnemidophorus tigris*), bullsnake (*Pituophis melanoleucus*), tree lizard (*Urosaurus ornatus*) and side-blotched lizard (*Uta stansburiana*). The presence of the bullsnake was a surprise, although bullsnakes are reported to be excellent climbers. Reptilian species richness was not correlated with cliff height or microrelief and there was a 'nearly' significant relationship between reptilian species number and butte summit area. Considering these five species are primarily insectivorous and arachnivorous, Johnson (1986) speculated that the distribution of ants and spiders might affect the distribution and abundance of reptiles.

## 5.4 Invertebrates

Except for the flying insects, most invertebrates are much less mobile than vertebrates and as a consequence their presence on cliffs suggests the ecological role of *in situ* herbivore, microcarnivore or detritivore. This role in autochthonous nutrient cycling is quite different from that of many of the birds, small mammals and large mammals described above, which for the most part bring nutrients to cliffs in the form of dead prey, excrement, or live young which die. Oettli (1904) attributes the large masses of rich organic debris in cracks and crevices in limestone cliffs in Switzerland to the activities of earthworms and isopods that feed on dead and decaying plant material. While studies of the invertebrate communities in cliff environments are not as numerous as the number of studies on the vertebrate components of the cliff ecosystem, the distribution and ecology of some invertebrate communities in cliff environments have been investigated. For example, the entire arthropod community of talus slopes in the Krkonoše Mountains in the Czech Republic was investigated by Růžička and Zacharda (1994), who left pitfall traps in the slopes for up to a year. Species of Diptera (flies), Aphidinea (aphids), Opiliones (harvestmen), Coleoptera (beetles) and Araneae (spiders) were the most abundant. Twenty-three species of spiders and 31 species of beetles were collected. The most abundant species was a rhagidiid mite (*Evadorhagidia oblikensis corcontica*). Lauterer (1991) found 50 species in the insect family Psylloidea on limestone cliffs in the Pavlovské Hills in the southern Czech Republic when only 14 species had previously been described.

In Baden-Württemberg, in south-western Germany, ladybugs (*Coccinella septempunctata*) overwinter at the top of talus slopes, while glacial relict species such as the beetle *Nebria castanea* are found at the toe of talus slopes (Lüth, 1993). On cliff faces, some insects such as *Machilis germanica* and *Cryphia domestica* feed on epilithic lichens (Lüth, 1993). The larvae of *Cryphia domestica* also use the lichen to build sheaths in which they hide during the day. The larvae of the endangered butterfly species *Parnassius apollo* eat *Sedum album*, which grows on the face, and the moth *Apamea platinea* lays its eggs on small stunted tufts of the grass *Festuca pallens* (Lüth, 1993). In the same area, the rare grasshopper

*Podisma pedestris* was found on cliffs in 1990 after it was thought to have been eliminated from the region (Herter, 1996). In Nepal, India, Bhutan and China, the Himalayan honeybee (*Apis laboriosa*) builds its nests in the open on sheer cliff faces at elevations up to 3500 m. While most open-nesting honeybees are native to the tropics or subtropics, the Himalayan honeybee can forage in ambient temperatures at least 5–6°C lower. It is the largest honeybee in the world and its large size is seen as an adaptation to survival on open cliff faces in the Himalayas (Underwood, 1991).

Cliffs are home to unique assemblages of spider species, as discovered by Růžička (1990), Koponen (1990), Lüth (1993), Snazell and Bosmans (1997), Lindqvist (1964), and Buddle and Larson (unpublished). In Germany, the jumping spider (*Salticus scenicus*) hunts insects on cliff faces and the walls of buildings (Lüth 1993). In the Czech Republic, Růžička (1990), Růžička and Zacharda (1994), Růžička, Hajer and Zacharda (1995) and Růžička (1996) found unique communities of spiders living on talus slopes. The stony debris in the talus was found to have a voluminous subterranean environment with a stable microclimate that supported different spider communities at different depths and thus was a well-defined underground ecosystem (Růžička, 1990, 1996; Růžička & Zacharda, 1994; Růžička *et al.*, 1995). While the temperature of the upper surface of the talus overheats and fluctuates appreciably during the day and during the year, several metres down into the talus the temperatures are cool but stable all year, with a high relative humidity (Růžička, 1990; Růžička *et al.*, 1995). In northern Bohemia, Czech Republic, the surface of the talus was found to be the home of thermophilous spiders such as *Echemus angustifrons* and *Callilepis schuszteri* (Růžička *et al.*, 1995). Psychrophilous spiders such as *Lepthyphantes tripartitus* and *Micrargus apertus* were common at a few metres depth, while the troglobitic spider *Porrhoma egera* and the predatory mite *Poecilophysis spelaea* were the dominant species deep in the talus (Růžička *et al.*, 1995). In southern Moravia, Czech Republic, *Pholcus opilionoides* was the dominant species at the surface, *Lepthyphantes improbulus* was dominant at depths between 0.5 and 5.0 m and *Porrhomma egeria* was dominant at depths greater than 5.0 m (Růžička, 1996). While spider diversity decreased downward from the surface (Růžička, 1996), new subspecies of spiders with morphological adaptations such as depigmentation and elongated appendages were found in the deep talus debris (Růžička, 1990; Růžička & Zacharda, 1994). Shallow subterranean environments were shown to be just as important as karst caves for the underground evolution of invertebrates (Růžička, 1996).

Table 5.6. *Number of specimens of different species of invertebrates at three cliff sites in Forillon National Park, Quebec, Canada*

| | Cliff Sites | | | |
|---|:---:|:---:|:---:|:---:|
| Species | I | II | III | Total |
| **Amaurobiidae** | | | | |
| *Callobius bennetti* (Blackwall) | 2 | 7 | 14 | 23 |
| *Coras montanus* (Emerton) | 1 | — | — | 1 |
| *Wadotes calcaratus* (Keyserling) | 19 | 20 | 23 | 62 |
| Family total (3 species) | 22 | 27 | 37 | 86 |
| **Gnaphosidae** | | | | |
| *Gnaphosa muscorum* (L. Koch) | 1 | — | — | 1 |
| *Zelotes fratris* Chamberlin | 4 | 1 | 5 | 10 |
| Family total (2 species) | 5 | 1 | 5 | 11 |
| **Clubionidae** | | | | |
| *Clubiona canadensis* Emerton | — | 3 | 1 | 4 |
| **Liocranidae** | | | | |
| *Agroeca ornata* Banks | 6 | — | — | 6 |
| **Thomisidae** | | | | |
| *Ozyptila* sp. | — | 3 | 1 | 4 |
| **Salticidae** | | | | |
| *Neo nelli* Peckham & Peckham | 2 | 2 | 3 | 7 |
| **Lycosidae** | | | | |
| *Pardosa lapidicina* Emerton | 22 | — | — | 22 |
| *P. mackenziana* (Keyserling) | — | 10 | 6 | 16 |
| *Trochosa terricola* Thorell | — | 1 | — | 1 |
| Family total (3 species) | 22 | 11 | 6 | 39 |
| **Agelenidae** | | | | |
| *Cicurina brevis* (Emerton) | — | 3 | — | 3 |
| *Cryphoeca montana* Emerton | 2 | 5 | 9 | 16 |
| Family total (2 species) | 2 | 8 | 9 | 19 |
| **Hahniidae** | | | | |
| *Neoantistea magna* (Keyserling) | — | 1 | 2 | 3 |
| **Theridiidae** | | | | |
| *Robertus riparius* (Keyserling) | 2 | — | — | 2 |
| *Theridion sexpunctatum* Emerton | 1 | 1 | — | 1 |
| Family total (2 species) | 3 | 1 | — | 3 |
| **Mimetidae** | | | | |
| *Ero* sp. | — | 1 | — | 1 |

Table 5.6 (*cont.*).

| Species | Cliff Sites | | | |
|---------|:---:|:---:|:---:|:---:|
| | I | II | III | Total |
| Linyphiidae: Linyphiinae | | | | |
| *Bathyphantes pallidus* (Banks) | — | 1 | 2 | 3 |
| *Lepthyphantes alpinus* (Emerton) | — | — | 1 | 1 |
| *L. intricatus* (Emerton) | 1 | — | 9 | 10 |
| *L. zebra* (Emerton) | 14 | — | — | 14 |
| *L. turbatrix* (O.P. -Cambridge) | — | — | 1 | 1 |
| *Meioneta?* sp. | — | — | 2 | 2 |
| *Microneta viaria* (Blackwall) | — | — | 1 | 1 |
| *Oreonetides vaginatus* (Thorell) | — | — | 1 | 1 |
| *Porrhomma terrestris* (Emerton) | 1 | — | — | 1 |
| Linyphiidae: Erigoninae | | | | |
| *Ceraticelus laetabilis* (O. P. -Cambridge) | — | 2 | — | 2 |
| *Diplocentria bidentata* (Emerton) | — | — | 1 | 1 |
| *Gonatium crassipalpum* Bryant | — | 2 | — | 2 |
| *Islandiana flaveola* (Banks) | — | — | 1 | 1 |
| *Metopobactrus prominulus* (O.P. -Cambridge) | — | — | 1 | 1 |
| *Sisicottus montanus* (Emerton) | — | — | 2 | 2 |
| *Tapinocyba minuta* | — | 1 | — | 1 |
| *Tunagyna debilis* (Banks) | — | — | 2 | 2 |
| *Walckenaeria atrotibialis* O. P. -Cambridge | — | 1 | 1 | 2 |
| *W. castanea* (Emerton) | — | 1 | 1 | 2 |
| *W. exigua* Millidge | — | 4 | 2 | 6 |
| *Zornella cultrigera* (L. Koch) | — | 1 | 1 | 2 |
| Family total (21 species) | 16 | 13 | 29 | 58 |

*Source:* Modified from Koponen (1990).

Koponen (1990) used pitfall traps to collect spiders at the base of three different sections of a long cliff face (labelled I, II and III in Table 5.6) at Forillon National Park, Quebec, Canada. A total of 39 species of spiders were caught at the three cliffs, 21 of which belonged to the aerial web spider family (Linyphiidae). Twenty-one of the 39 species of spiders were found at only one of the three cliff faces. Five species, or 12.8 per cent of the species found, were subarctic-alpine in their distribution, and only occur south of Forillon National Park in mountainous areas of New Hampshire, USA. Five other thermophilous species with a southern range were found. Forillon National Park represented the northern limit for these species. Cliffs with many arctic–alpine plant species were also found to support more spiders with northern affinities (Koponen, 1990). At Shakespeare Cliff, Dover, UK, Snazell and Bosmans (1997) also discovered disjunct populations of spiders in the talus of these vertical chalk faces. Two new species, *Minicia marginella* and *Zodarion vicinum*, were discovered for the UK, *Z. vicinum* having only previously been recorded from central and southern Italy.

Buddle and Larson (unpublished) arranged three transects at randomly selected distances along a 700 m section of Niagara Escarpment cliff near Milton, Ontario, Canada. Four yellow pitfall traps, half-filled with water, detergent and a salt solution, were then positioned at the top of the cliff in the plateau forest, on small ledges on the cliff face directly, and between rocks on the talus slope at the bottom of the cliff. On the cliffs, small rocks and litter were used to make small bridges to the edges of the pan pitfall traps when solid rock prevented them from being immersed in the surface. There were 36 traps at each transect and the contents of the traps were collected four times in the late summer and late fall before sub-zero temperatures occurred. This design was used to sample arthropods, including spiders, along the cliffs of the Niagara Escarpment and to compare the taxonomic pattern observed there with talus slopes and level-ground woodland. For the spiders, a total of 222 individuals in 9 families were trapped. The funnel-web spiders (Agelenidae) and the aerial-web spiders (Linyphiidae) represented 85 per cent of the individuals trapped. The latter family was by far the most common on the cliff faces. An analysis of variance revealed that there was a significant effect of habitat type on the distribution of Linyphiidae. Figure 5.6 illustrates the frequency distribution of the different families of spiders found. While identifications were only made to genus in this study, the results suggest some factors that influence the structure of the arachnid community on cliffs.

Snails and slugs (Mollusca) in cliff environments have been reported by Baur and Baur (1990) in Sweden, Lawrey (1980) in Virginia, USA, and Nekola *et al.* (1996) along the Niagara Escarpment in Wisconsin, USA. The rock-dwelling snails *Chondrina clienta* and *Balea perversa* were reported from stone walls and stone piles on Öland, Sweden (Baur & Baur, 1990). *Chondrina clienta* was also reported from quarry faces, karst areas and fissures in limestone on the same island. Each species has specific habitat preferences and occurs in different densities depending on the structure of substrate. Baur and Baur (1990) found that *Chondrina clienta*

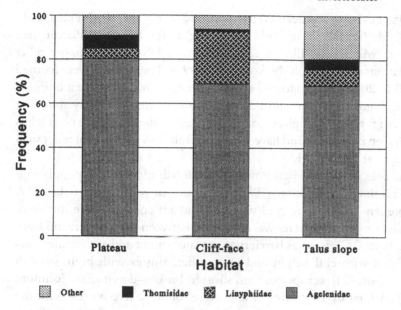

*Figure 5.6* Percentage frequency of Thomisidae, Linyphiidae, Agelenidae and other spider families trapped along the plateau, cliff face and talus slope of the Niagara Escarpment. Unpublished data from C. Buddle and D.W. Larson.

is the only species discovered on exposed vertical surfaces because it can survive sudden changes in temperature. It grazes exclusively on lichens, and large numbers may exert a high grazing pressure on the rock surface (Baur & Baur, 1990). Another cliff snail species (*Chondrina avenacea*) has been observed by Lüth (1993). This snail lives exclusively on endolithic lichens on cliffs in south-western Germany. The slug *Pallifera varia* in Virginia, USA, was found to graze on lichens on talus slopes (Lawrey, 1980). The genus *Aspicilia* was found to be the most important food source for these slugs. The common lichens *Xanthoparmelia cumberlandia* and *Huilia albocaerulescens* were actively avoided by *Pallifera varia*. These slugs were found to make food choices based on lichen chemistry rather than simply grazing the most frequently encountered lichen species (Lawrey, 1980).

Nekola *et al.* (1996) found 41 taxa of land snails at the base of cliffs along the Niagara Escarpment in Wisconsin, USA. At undisturbed sites, mean diversity of snails was 20 taxa per site. These taxa include both ubiquitous and glacial relict species. Since the cliff base is a popular site for hiking, these species were found to be vulnerable to disturbance. A full intact canopy was deemed important for these species (Nekola *et al.*, 1996). Algific talus slopes (cold producing) in the Paleozoic Plateau area of Iowa, Minnesota, Illinois and Wisconsin, USA, also support relict populations of land snails (Nekola *et al.*, 1996). These slopes were formed 15 000 to 20 000 years ago and support ice caves which create a buffered habitat. Eight rare species found on these slopes were formerly abundant at the height of the last glaciation. On the associated maderate cliffs (cliffs which seep cold water and have minor cold air flow), six glacial relict snail species were also found.

The role of cliffs in determining the distribution of ant populations was examined by Johnson (1986) on buttes in south-eastern Utah. A comprehensive sampling grid was used and ant colonies were surveyed. The diversity of ant species was compared between the plateaus and bases of the buttes. Cliffs act as barriers to ant movement and ant richness was correlated with cliff height and microrelief. Buttes with high, smooth cliffs supported fewer species than shorter, broken-down cliffs (Johnson, 1986). A total of 16 species of ants were found at five buttes, with the number of species at any one butte varying between six and eight. These 16 species belonged to the families Dolichoderinae, Formicinae and Myrmecrinae. The contrasting distribution of ant species in this environment may reflect the dispersal mechanisms and flight distances of the individual species, especially during nuptial flights. Species–area relationships were absent and it was the cliff and not the cliff top that determined the ant species composition.

In cliff environments, ant populations (Hymenoptera) have been shown to be restricted by the presence of ant lion larvae, one of their principal predators (Gotelli, 1996). In central Oklahoma, ant lion larvae live in dense aggregations at the base of sandstone cliffs <2 m from the cliff base. Rainfall and disturbance were the two principal factors influencing ant lion larvae numbers (Gotelli, 1993). Rainfall, in particular, makes the soil impenetrable and the ant lions succumb to high temperatures on the surface. Ants appear to use a chemical cue to avoid ant lion larvae aggregations. This in turn affects the distribution and abundance of ant populations in the talus. Other types of invertebrate communities studied on cliffs include the protistan microbe community (Lynn & Olson, 1990), invertebrate communities in raptor nests (Philips and Dindal, 1977) and invertebrate communities in madicolous habitats, i.e. habitats created by the movement of thin sheets of water over a surface (Sinclair & Marshall, 1986). Lynn and Olson (1990) collected water samples from rock depressions and small springs at 12 sites along the Niagara Escarpment and found 63 genera of protists. The most common protists were the colpodid *Colpoda*, the hypotrichs *Urostyla* and *Euplotes*, the testate amoebae *Arcella* and *Difflugia*, the oligotrich *Halteria* and the peritrich *Vorticella*.

Philips and Dindal (1977) found that raptor nests are important habitat islands for invertebrates on cliff faces. Three groups of invertebrates were identified, namely the parasitic fauna, such as blackflies, bloodsucking *Protocalliphora* larvae, ticks and mites; the animal saprovores such as hide or carpet beetles (Dermestidae) and skin beetles (Trogidae); and the humus fauna such as springtails (Collembola). The parasitic fauna are found on the raptors themselves and their prey, the animal saprovores are associated with decomposition, and the humus fauna are associated with decomposition of nest material. Nest parasites such as *Protocalliphora*, blackflies, the tick *Ornithodoros concanensis*, the mite *Ornithonyssus bursa* and the Mexican chicken bug (*Haematosiphon inodorus*) can cause death in cliff-dwelling raptors.

Ticks are also principal ectoparasites in sea bird colonies, particularly the hard Ixididae and the soft Argasidae (Hoogstraal, 1978). Boulinier and Danchin (1996) found *Ixodes uriae* in 22 British sea bird colonies. The prevalence and intensity of these ticks were higher in sea bird colonies which were decreasing in number. Boulinier and Danchin (1996) believe this to be a cause and effect relationship whereby the ectoparasite population levels influence breeding success and population trends within the sea bird colonies. Dispersal is probably accomplished at the end of the breeding season when newly fledged sea birds mix with other colonies (Boulinier & Danchin, 1996). While all of the above ectoparasites are not themselves specifically associated with cliffs, their hosts are. As a consequence, these animals are inevitably having an impact on the flow of energy and matter in these communities.

Along the Niagara Escarpment, Sinclair and Marshall (1986) found over 70 species of arthropods in madicolous habitats. These species were divided into four categories. Thirteen of the species were restricted to the madicolous habitat (wet rocks). These include species in the Trichoptera, Diptera, Thaumaleidae, Ceratopogonidae, Chironomidae and Acari families. Another 15 species were common in the madicolous habitat but able to survive in thicker films of water. Thirty-six other species spend only part of their life cycle in this habitat and are more commonly found in streams and brooks. The final ten species were considered accidental occurrences and do not breed in these habitats. Sinclair and Marshall also reported species that used madicolous sites for short parts of their life cycles. It was concluded that rock outcrops and cliffs support a diverse array of invertebrates in a variety of trophic levels, but no conclusions could be made about the functional roles of these species because distribution data were lacking.

## 5.5 Summary

While there is a paucity of literature on faunal cliff communities, the life cycle, behaviour and habitat requirements of many individual species in cliff environments have been well documented. This chapter has shown that cliffs worldwide support a wide diversity of fauna. Cliff cavities, crevices and ledges provide suitable nesting sites for a large number of species. Raptors, sea birds and other cliff-dwelling avifauna utilize cliffs as relatively safe nesting and roosting sites but venture into the surrounding habitat for their food supply. For aerial hunters, cliff faces provide excellent starting points for hunting and foraging trips, as well as vantage points for prey detection. Flight, however, is not a mandatory requirement for faunal access to cliff environments as the small-scale heterogeneities on the cliff are often linked to form complex networks of pathways that facilitate navigation. Cliffs are not barriers to the movement of small mammals and reptiles and for some species the small-scale heterogeneities are actually preferred habitat. Cliff faces also represent a zone of enemy-free space for those species that are capable of utilizing it. Others are capable of completing their entire life cycle without leaving the cliff environment. While cliffs appear to be hostile environments for many organisms, for the key groups described in this chapter they are heavily exploited refuges. It is possible that certain taxa such as the rats, mice and rock doves that are now cosmopolitan and commensal with humans, were once highly endemic to cliffs in their original wild state. This possibility gives rise to the 'urban cliff hypothesis', namely that humans in palaeolithic times exploited cliff, talus and cave environments that were already home to a wide variety of endemic plants and animals that were tolerant of high degrees of stress and disturbance, and that humans still exist in environments that copy the characteristics of these original rock shelters.

A variety of other vertebrates as well as invertebrates are associated with cliffs and some are highly adapted to the unusual conditions that exist on bare or fractured rock. Although comprehensive surveys have not been carried out, the scattered reports that have been published show that salamanders, spiders, snails and a restricted array of insects are very common on cliffs. It would be very useful if careful and comprehensive sampling of individual cliffs could be carried out in the future in order to determine if cliffs select for these particular groupings on the basis of their functional ecological characteristics.

# 6 · *Controlling processes*

While inaccessibility has protected cliffs from significant amounts of disturbance, it has also limited the amount of experimental work that deals with questions of the genesis and maintenance of cliff communities. Most previous studies have inferred mechanisms of community or ecosystem function from descriptions of the cliff biota. In this chapter, literature on the growth of individual species and populations, the establishment of patterns of relative abundance, and the development of species composition is briefly reviewed. Ideas about how physical factors influence the biota of cliffs are presented first, followed by a discussion of the control of communities through biotic interactions.

## 6.1 Bedrock composition

There are three aspects of geology and geomorphology that influence the biotic communities of cliffs: bedrock composition, structural heterogeneities, and erosion. Bedrock composition falls into three large categories: (1) hard siliceous rocks, mainly of igneous origin but also including some sedimentary rocks such as sandstones; (2) hard calcareous rocks, mainly of sedimentary origin but also including igneous or metamorphic rocks such as basalt and marble; (3) unconsolidated or indurated materials such as sand, gravel or loess. It is generally known that siliceous rocks produce acid soils that select for an array of plant species commonly called calcifuges ('lime avoiders'). Conversely, calcareous rocks produce chalky soils with neutral to high pH values that select for a different array of plants known as calcicoles ('lime seekers') (Fitter & Hay, 1987). The low pH values of soils derived from acid rocks cause the accumulation of toxic levels of $Fe^{2+}/Fe^{3+}$ or $Al^{3+}$ ions. In contrast, the neutral to elevated pH values of limestone soils cause accelerated rates of nutrient loss so that plants growing in these areas are nutrient depleted. This link between rock chemistry, soil chemistry and plant species composition, richness or growth has been well established for level ground (Etherington, 1981; Wentworth, 1981; Jeffries, 1985). However, a survey of the available literature suggests that the relationship is much less clear on vertical cliffs. This is not entirely surprising if one considers that the normal process of soil formation is also absent on cliffs.

Rohrer (1982, 1983) found that higher plant and bryophyte species composition is similar on rocks that differ greatly in their chemistry, and Rune (1953) shows that serpentine rocks support lichens and other plants that normally grow on both calcareous and siliceous substrates. Karlsson (1973) and Lundqvist (1968) also argue that rock chemistry has little effect on cliff vegetation, and the same conclusion is reached in the comprehensive but little-known work of Oettli (1904). Faegri (1960) reports that some species of *Asplenium* are much more dependent on the accumulation of organic matter than on rock chemistry, and thus grow equally well on limestone and on granite. *Taxus baccata* and *Phyllitis scolopendrium* show the same lack of affinity for rocks of a particular chemical composition. Many species of *Rhododendron* that commonly occur in acid habitats such as bogs or heaths are also abundant on limestone cliffs in China (Cox, 1945). Walker (1987) and Young (1996) observed that the flora of cliffs in the south-eastern USA is similar despite the fact that the cliffs are composed of both acid sandstones and basic limestones. Likewise, Cowles (1901) studied the vegetation of sandstone cliffs near Chicago at Starved Rock (Fig. 6.1) and found an assemblage of species very similar to that reported by Larson *et al.* (1989) for the limestone-based Niagara Escarpment (Fig. 6.2). He also notes that the vegetation of sandstone and limestone ravines in the Chicago area is the same, and concludes that exposure and atmospheric conditions, rather than rock material or soil, control the development of plant communities. Cooper (1997) found that there were general correlations between vegetation composition and rock chemistry, but that elevation and distance from the sea were much more important factors controlling the structure of the vegetation on sea cliffs and inland cliffs than the characteristics of the rock itself. Cooper also points out that rock chemistry co-varies with small-scale rock structure. Therefore, the two effects must be separated before any claims of a direct influence of rock chemistry on plant distribution can be made, and few studies to date have done this.

A number of studies emphasize that local variation in the nature of cracks, crevices, ledges and solution hollows may be of much greater importance than the overall variation in chemistry imposed by the bed-

*Figure 6.1* Vertical sandstone cliffs at Starved Rock Park, Illinois River, Illinois, USA. These cliffs were studied by Cowles (1901) and have many of the same species that occur on the cliffs of the Niagara Escarpment. Photo courtesy of M. Ursic.

rocks themselves. Hora (1947) reports that few cliff species show clear site selection based on the chemistry and pH of the substrate, but that there is enormous variability of pH values among microsites only a few centimetres away from each other. Lundqvist (1968) and Oettli (1904) report that on acid rocks, calcareous materials go into solution and make the acid surfaces more alkaline, while on alkaline rocks, humus materials acidify the limey soils and make them more neutral. This is supported by

*Figure 6.2* Vertical limestone cliffs of the Niagara Escarpment support a similar array of species to that shown in Fig. 6.1. Photo by D.W. Larson.

evidence from Cox and Larson (1993a, 1993b) that the pH between rocks of large-block talus at the base of limestone cliffs is very high initially, but drops as low as 4.6 once organic matter accumulates to a significant degree. Lüth (1993) points out that siliceous rocks such as gneiss may contain traces of calcium, which accumulates in crevices; for this reason crevices in siliceous rock often support calcicoles. Likewise, calcifuges can grow on limestone that is superficially leached of calcium (Lüth, 1993). Rodwell (1992) shows that some soils sampled on limestone cliffs in the UK had 700 mg calcium per gram of sample, but still had pH values between 6.8 and 7.5. All of these results together show that there is considerable overlap in the chemistry of the microsites between cliffs formed of different rock types.

While the majority of papers seem to argue that there is no or only limited control of vegetation by rock chemistry, there are also a number that show evidence to the contrary. For example, Jarvis and Pigott (1973) report that *Lychnis viscaria* only shows good growth on limestone cliff faces that are high in phosphorus. In Baden-Württemberg, Germany, Lüth (1993) suggests rock type as an explanation for the pronounced differences in the cliff and talus flora between the Schwäbische Alb (calcareous rock) and the Odenwald and Black Forest (siliceous rock). Lichens as a group seem to be controlled more by rock type or chemistry than any other group of organisms. Boyle, McCarthy and Stewart (1987) show that lichen communities can be good predictors of substrate chemistry, and others have shown that granite rocks have a very different lichen flora from limestone rocks in the same geographical area (Yarranton & Green, 1966; Brodo, 1973; Larson, 1980; Lawrey, 1984). What has not been shown, however, is whether these distinctive distributional characteristics of some lichen communities reflect the chemistry of the rocks or some other property such as small-scale weathering patterns (Larson, 1980). Until more research is done comparing the lichens with other groups that occur on rocks, this question will remain unanswered. What seems to be clear is that many workers *expect* rock chemistry to exert a large influence on the species composition of cliffs and rock outcrops, and accordingly there is the temptation to interpret equivocal results that way. Thus, Hofman, Nowak and Winkler (1974) found that for 27 lichen species whose distributions were determined on the basis of substrate specificity (calcareous versus siliceous), 22 showed very little evidence of being substrate specific whereas five others did. These authors concluded that lichens show strong substrate specificity, but their conclusions are actually contradicted by their results.

The available literature also indicates that there are few relationships between cliff fauna and the geological composition of cliffs. Růžička (1990) lists rock type as an important factor affecting the abundance and type of spider species found on talus slopes in the Czech Republic, but the nature of the control of the fauna by the rock chemistry was not shown. Most raptor species use habitat parameters other than rock type as cues for choosing nest sites (Runde & Anderson, 1986). The bearded vulture (*Gypaetus barbatus*) is an interesting exception to these general findings. In South Africa, bearded vultures preferentially nest on basalt and sandstone cliffs (Brown & Bruton, 1991). At these locations, the vultures come in contact with filmy concentrations of iron in caves and on ledges and then spread the iron oxide through their feathers by preening. The resultant rufous discoloration of lighter feathers on underparts camouflages the birds on the cliff face. Microscopic analysis of the feathers indicates that the iron oxide acts as a protective coating which reduces wear. Also, the iron oxide concentrates on the portions of feathers where wear occurs, thus acting as a repair mechanism. The iron oxide may also help to control ectoparasites (Brown & Bruton, 1991).

## 6.2 Heterogeneities

The second aspect of cliff geology and geomorphology that is important to community organization is the degree to which physical heterogeneities are present, and their spatial scale (Oettli, 1904; Wunder & Möseler, 1996; Cooper, 1997). As a general rule, cliffs formed of unconsolidated materials offer few large-scale heterogeneities compared with cliffs formed of solid rock. This is due to their relatively rapid erosion. Cliffs composed of igneous rocks such as basalt and andesite tend to provide more habitat than softer sedimentary rocks (Maser et al., 1979). However, the number and types of heterogeneities formed can be similar in many different substrates. The presence of heterogeneities can change from the top to the bottom of a cliff face, although not consistently between different cliffs. For example, Morisset et al. (1983) found the greatest density of variable microsites in the lower part of the free-face, where most of the rare plants occurred. Conversely, Nuzzo (1995) found that most of the fracture lines and stable rock that provided habitat for *Solidago sciaphila* were present on the upper sections of the cliffs.

The number, size and spacing of crevices, ledges, pockets and fracture planes strongly influence the density and diversity of the flora of cliffs (Oettli, 1904). This can be seen clearly on sandstone cliffs in Tasmania, where colonies of higher plants are concentrated along joint crevices while sheer faces are covered only by mosses and lichens (Coates & Kirkpatrick, 1992). Davis (1951), Morisset et al. (1983) and Brunton and Lafontaine (1974) have all pointed out that the greater the heterogeneity of a rock face, the greater the variety and density of plant species. Oettli (1904) also states that the wide diversity of plant life forms on cliffs is explained by the presence of a wide diversity of microsites in very close proximity. Davis (1951) has argued that cliffs lacking structural non-conformities would be colonized only by endolithic and epilithic microbes, and indeed many cliffs formed of largely unweathered rock have vast expanses of free-face that support little vegetative cover. Closer examination, however, shows that a wide variety of microscopic organisms is present on such faces and is controlled by topographic heterogeneity on an equally small scale (Matthes-Sears *et al.*, 1997).

All workers have observed that unclaimed sites are generated very slowly on cliffs and their colonization is very difficult; for this reason, very few therophytes (annual plants) are found on cliffs. Almost all cliff plants are at least short-lived perennials or annual plants that take on a perennial habit when growing in cracks and crevices. *Geranium robertianum* is the best example of this kind of organism on the cliffs of the Niagara Escarpment. The plant occurs as an obligate annual plant on talus slopes, but individual rosette-shaped colonies on cliffs have persisted for many years.

The fracturing of rock greatly increases its water-holding capacity, and this can lead to locally favourable moisture conditions in arid climates. For example in Australia, highly fractured sandstone outcrops support a diverse community of rock mammals and vegetation in the otherwise species-poor and seasonally dry tropics (Freeland *et al.*, 1988). Water availability is usually high within 50 m of these rocky areas. It is unlikely these types of communities would have evolved in rock without such capabilities. Zohary (1973) also reports that rock outcrops in desert regions of the Middle East have a much greater supply of water than the surrounding level-ground habitat, and believes that this explains the high diversity of plants found in these areas relative to other level-ground rocky terrain. When cliff edges form overhangs, the ability of the rock to supply and retain water as well as buffer fluctuations in temperature is decreased. No evidence has been published on this topic, but our own observations support the idea that overhanging rocks (Fig. 6.3) are less vegetated both on top and underneath than cliffs with a solid foundation.

Fracture lines and crevices caused by weathering or rockfall are common at the cliff edge as well as on the face. In some settings, the edges of cliffs can form small islands or peninsulas of rock (Fig. 6.4). These fractures can be very shallow and small scale, or else extend all the way to the rocky pediment that underlies the entire cliff, providing plant roots with access to water deep within the rock. Crevices narrower than 5 cm are those most favourable for plant growth, since they retain detritus (Oettli, 1904). The larger crevices on the Niagara Escarpment provide rooting space for trees such as *Thuja occidentalis* and *Juniperus virginiana,* and small shrubs such as *Acer spicatum* (Larson *et al.*, 1989). Similar microsites

*Figure 6.3* Overhanging rocks support far fewer macroscopic organisms than vertical rocks. In many cases these overhangs are the product of weaker rocks underlying stronger ones. Photo by P.E. Kelly.

*Figure 6.4* Small rock island separated from the main body of cliff along the Niagara Escarpment. Photo by D.W. Larson.

on cliffs in south-western Germany support *Amelanchier ovalis* and *Cotoneaster integerrimus*; smaller fissures may support cushion plants (Lüth, 1993). On low-elevation cliffs in Switzerland, species that are particularly dependent on crevices include *Potentilla caulescens, Laserpitium siler* and *Globularia cordifolia* (Oettli, 1904). While some tree root systems penetrate deep into the rock (Lüth, 1993), others are relatively shallow: root excavations of cliff-face *Thuja occidentalis* showed that roots expand primarily in a horizontal direction along bedding planes parallel to the cliff face, as well as following vertical fissures both upward and downward from the point of insertion (Matthes-Sears & Larson, 1995). On limestone cliffs and steep slopes in Israel, the majority of woody species exploit pre-existing cracks and crevices, but some such as *Pistacia lentiscus* are apparently able actively to penetrate solid rock with their roots (Oppenheimer, 1956, 1957). Both active penetration and growth of roots into pre-existing fissures can provide plants with a substantial degree of drought avoidance, as they get access to water that seeps along small cracks and crevices within the rock. Larger fissures and cracks are also exploited by mammals, which use them as den sites (Fig. 6.5), and smaller crevices are used by earthworms and isopods (Oettli, 1904). Variations in exposure, however, can greatly modify the microhabitat conditions among otherwise similar crevices and result in different plant communities colonizing them. For example, on cliffs in south-western Germany, cracks and crevices on moist, shady walls support communities dominated by *Asplenium viride* and *Cystopteris fragilis,* while those on dry, sunny and exposed walls support *Hieracium/Draba* communities (Herter, 1996).

For species that are unable to penetrate rock strata, desiccation tolerance is an alternative survival strategy. This may explain why many of the genera that recur on cliffs in many parts of the world are poikilohydric, and why poikilohydric species make up a relatively large proportion of the flora of most cliffs (Ashton & Webb, 1977; Dougan & Associates, 1995; Porembski *et al.*, 1996). For example, *Polypodium virginianum* and *Asplenium ruta-muraria* are both found along cliff edges of the Niagara Escarpment, and related species are found in similar locations in other countries, as shown in Chapter 4. Apart from poikilohydric plants, a small number of other specialists can survive exclusively on superficial water supplied by rainfall. In central Europe these include succulents such as *Sempervivum tectorum* and *Sedum album*, winter or spring annuals such as *Leontodon incanus* and *Erophila* sp., and xerophytes with small, rolled, hairy or waxy leaves such as *Dianthus* sp. (Oettli, 1904; Lüth, 1993). They are able to colonize shallow soil accumulations on cliff edges, chimney

*Figure 6.5* Vertical and horizontal fracturing of bedrock along cliffs results in habitat that is specifically exploited by vertebrates. Here, intense use of cliff edges by mammals is shown by dark movement trails in snow (A). Similarly, mammal use of crevice and karst caves is intense (B). Photos by D.W. Larson.

tops, and isolated boulders that are disconnected from the deep crevice system and have little internal water supply. The predominance of desiccation-tolerant or desiccation-avoidant plants along cliff edges reinforces the idea that these habitats are ecologically very different from the surrounding communities.

Spatial heterogeneities are equally important for the fauna of cliffs. The abundance and species richness of animals are greater on cliff faces compared to surrounding level ground (Ward & Anderson, 1988). Peregrine falcons in Utah show preference for individual facies of sandstone rock with protruding ledges and overhangs (Grebence & White, 1989). Crevice caves formed from collapsing caprock or from flaking sheetrock from the free-face are exploited by a variety of vertebrates but rarely birds (Maser *et al.*, 1979). Short (1982) found the number of ledges to be one of the most important habitat variables for brush-tailed rock-wallabies on cliffs in Australia. More rock-wallabies are found on cliffs with more caves, more crack or chimney routes onto the cliff, and more sheltered but shorter ledges. The availability of larger, more easily detected feeding roosts is also important to bats. Shelter width, distance to the back of the shelter, entrance width and height are important to the Virginia big-eared bat (Lacki *et al.*, 1993).

Bird species diversity and richness are also greater on cliffs because of the vertical structure provided by the cliff face (Ward & Anderson, 1988; Camp & Knight, 1997; Matheson & Larson, 1998). Maser *et al.* (1979) observed that small mammals and forest birds use shallow and deep solution and crevice caves for roosting or overwintering, while most raptors exploit ledges and areas of moderately overhanging rock (Fig. 6.6). The number, distribution and availability of nesting, roosting and perching sites for raptors are a reflection of the morphology and number of fractures, ledges, overhangs and caves on a particular cliff face. Morphological preferences for these habitat parameters vary amongst species. For example, 97 per cent of prairie falcons (*Falco mexicanus*) in the western USA (Runde & Anderson, 1986) and 53 per cent of cliff-nesting golden eagles along Hudson Bay, Canada, choose nest sites with overhangs (Morneau *et al.*, 1994). Rough-legged hawks along Hudson Bay, Canada, however, choose nest sites lacking overhangs (Brodeur *et al.*, 1994). Other species such as the glaucous-winged gull (*Larus glaucescens*) in south-east Alaska show no relationship between the habitat attributes of nest sites and breeding success (Murphy *et al.*, 1992). Some species may expand nesting sites to less desirable and less secure cliffs when population levels expand. The recovery of peregrine falcon populations since the 1970s in the UK was accompanied by an expansion of nesting sites to small and unsuitable rocks, broken banks and sites on the ground (Crick & Ratcliffe, 1995).

Nest sites, however, are not the only locations on cliffs required by raptors. Reproductive performance in red-tailed hawks (*Buteo jamaicensis*) in Oregon, USA, is positively correlated with the number of perch-related features in the area from which they seek prey (Janes, 1994). Vasina and Straneck (1984) found that one peregrine falcon pair requires at least five cliff locations: one as a nest site, a second as an eating and plucking ledge for the transference of prey, a third as a plucking and resting ledge for the male (which is also used for sunning), a fourth as the female's sleeping roost, and a fifth as the male's sleeping roost. Other ledges are required for caching uneaten prey (Vasina & Straneck, 1984). Some species such as prairie falcons (Runde & Anderson, 1986) and bearded vultures (Donázar *et al.*, 1993) rotate eyrie and nest sites and have a low rate of reoccupancy in consecutive years. Bearded vultures may require as many as three other nests to be used in alternating years. This rotation reduces the risk of parasitic infestation (Runde & Anderson, 1986; Donázar *et al.*, 1993).

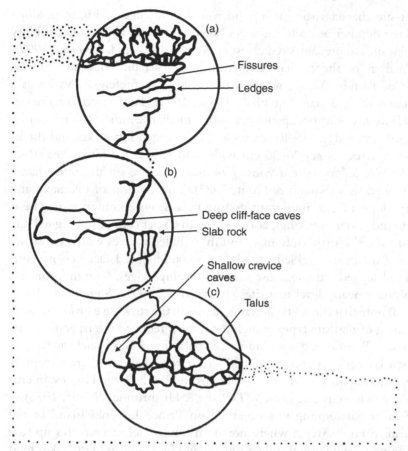

*Figure 6.6* Illustration of cliff features that influence wildlife. (a) Cliff edges include table rocks and ledges, as well as both horizontal and vertical fracture lines. These areas can be used by raptors for roosting, nesting and perching. In (b), deep solution caves are shown. These can be prime habitat for bats. (c) shows loose slab rock, shallow crevice caves and accumulated rock in talus that can be used by birds such as swifts for roosting and reproduction, as well as by lizards, snakes, amphibians, mice, woodrats, coyotes, bobcats and pumas for shelter, feeding and reproduction. Modified from Maser *et al.* (1979).

In sea birds as with raptors, low stepped cliffs are often more vulnerable to predation, and high, vertical cliffs limit predator access and are more favourable for nesting (Chapdelaine & Brousseau, 1989; and see Fig. 5.3). Certain horizontal strata may produce more ideal nesting conditions, thus producing nesting in bands across the cliff face (Gaston & Nettleship, 1981; Squibb & Hunt, 1983). Ideal nesting conditions are often associated with the very specific dimensions of cliff-face ledges.

Nest-site characteristics such as burrow depth, ledge width, ledge slope and the number of walls can affect breeding success and the accessibility of nest sites to predators (Gilchrist & Gaston, 1997b; Harris *et al.*, 1997).

Analysis of the habitat characteristics of sea bird nest sites in the Pribilof Islands, Alaska, revealed that ledge size preference varies as a function of bird size (Squibb & Hunt, 1983). Red-faced cormorants (*Phalacrocorax urile*) occupy ledges with a median width >40 cm; northern fulmars, ledges 30–40 cm wide; black-legged kittiwakes and thick-billed murres, ledges 20–30 cm wide; and red-legged kittiwakes (*Rissa brevirostris*), ledges 10 cm wide. Common murres, on the other hand, occupy all sites (Squibb & Hunt, 1983). The amount of overhang and ledge slope are also important nesting factors: no common murre nests are found under overhangs, compared with 63 per cent of red-legged kittiwake nests. Northern fulmars and thick-billed murres which construct no nest use horizontal ledges, while nest-building red-faced cormorants and red-legged kittiwakes use seaward-sloping ledges. Common murres also prefer nearly level nest sites (Harris *et al.*, 1997). Squibb and Hunt (1983) note that the subtle heterogeneities in the rock have produced partitioning of nest site types, which has in turn reduced interspecific competition. Partitioning also influences behaviour. Birds such as murres which lay eggs on vulnerable cliff sites rarely abandon them except in times of extreme disturbance, while razorbills, which tend to nest in crevices, will leave the egg often (Tschanz & Hirsbrunner-Scharf, 1975).

Similar partitioning was observed on Prince Leopold Island in the Canadian High Arctic, where northern fulmars select nest sites up to a metre wide, while thick-billed murres and black-legged kittiwakes tend to nest on sites less than 15 cm wide (Gaston & Nettleship, 1981). Interestingly, in thick-billed murres, breeding success is lower on narrow ledges – their preferred nesting habitat (Gaston & Nettleship, 1981; Birkhead *et al.*, 1985). Further research on thick-billed murres by Gilchrist and Gaston (1997a), however, revealed that nest sites on broad ledges associated with a higher short-term survival rate are not the same nest sites associated with long-term survival. Narrow ledges allow surviving chicks to leap directly into the sea while accompanied by a parent without harassment by neighbours. Murres nesting in the centre of dense groups on broad ledges are not as successful because the chicks have difficulty reaching the cliff-ledge edge. Survival is dependent on the parent quickly reuniting with the chick in the water. Delays in reaching the edge result in the chick missing the parent in the water; unaccompanied chicks are then mobbed by other murres and drowned or taken by gulls (Gilchrist & Gaston, 1997a).

Talus slopes represent an extremely heterogeneous and unstable habitat for most large organisms, including trees and mammals. However, for smaller organisms that are highly mobile, neither the heterogeneities nor the instability are a problem (Fisher, 1952; Guariguata, 1990; Pérez, 1994; Fig. 6.7). A wide variety of disturbance-tolerant vegetation as well as spiders, amphibians, insects and small mammals such as pika exploit talus slopes as habitat (Moss & Nickling, 1980; Herrington, 1988; Koponen, 1990; Cox & Larson, 1993a, 1993b; Hafner, 1993; Růžička et al., 1995; Peacock & Smith, 1997). Some organisms, such as salamanders, nest deep within the talus slope (Herrington, 1988). The snow bunting (Plectrophenax nivalis) is a bird species that nests in the talus despite the presence of predators nesting on adjacent cliffs (Meese & Fuller, 1989). This behaviour is a trade-off between the threat of predation and the availability of ideal nesting sites.

The size of the individual particles in the debris field is an important factor since it determines both the size of the spaces available for colonization and the stability of the surface. The block size in the talus is strongly influenced by the type of rock forming the cliff face and the rate of cliff erosion. At one extreme, rapidly eroding sandstone will form an unstable sandy talus with very small particle size that is suitable habitat for small organisms such as ant lions (Gotelli, 1996). At the other extreme, hard dolostone caprock may form a stable talus consisting of large blocks that provides habitat for much larger organisms. The distribution pattern of talus organisms is thus closely related to rock size and the inverse of the rate of particle movement (Pérez, 1994). In Germany, Herter (1996) distinguishes a number of different talus plant communities based on block size and stability, ranging from patchy Galeopsis-communities in sunny, unstable talus with small-sized debris to bryophyte-rich Gymnocarpium-communities in shady, stable, blocky talus. Rumex-communities predominate on unstable, sunny talus with medium-sized debris. In the coarsest talus, where there is no accumulation of humus between the individual rock particles, lichens, mosses and pioneer plants may predominate (Lüth, 1993). Shade-tolerant trees can also occur in such stable talus because their seedlings can germinate and grow in the dim light deep within the fissures until the crown of the trees can emerge through the tops of the talus blocks. We believe that such a mechanism applies to the widespread occurrence of Thuja occidentalis on talus slopes of the Niagara Escarpment, and to various Juniper species that grow on talus elsewhere. The abundance and type of spiders found in talus are also in part a function of the size and shape of the talus debris and the way the stones have accumulated (Růžička, 1990).

*Figure 6.7* Talus slope of the Niagara Escarpment illustrating the large patches of broken, unstable rock. Such slopes include large air spaces which could represent significant and well-protected habitat for vertebrates such as pika and rattlesnakes. Photo by P.E. Kelly.

Among the more conspicuous and interesting characteristics of the woody and herbaceous vegetation of talus slopes is the apparent selection for taxa with the capacity to reorient their growth following the displacement of the shoot by rockslide. Plants that lack the ability to bend or reorient their growth are simply removed from the ground by rockslides. Fisher (1952) discusses this behaviour for a variety of talus species in New Zealand, and similar adaptations are evident in *Geranium robertianum* and *Daucus carota* growing on talus slopes of the Niagara Escarpment and the Mediterranean Alps in southern France, respectively. Plants with tough and flexible roots are also capable of enduring in unstable talus environments, and many of these plants have runners which help stabilize the rock (Lüth, 1993).

## 6.3 Erosion

The third aspect of geology or geomorphology that is important to the ecology of cliff organisms is the rate at which the materials forming the cliff are eroded. The removal of rocky debris occurs mainly by the interaction of gravity with water, although wind can also play a role in moving small particles (Hétu, 1992) and can have a significant sculpting effect if the erosion continues uninterrupted for hundreds of thousands of years. Cliffs adjacent to large bodies of water or rivers are continually scoured at the base and in some instances no talus slope is present. The abrasive action of water can be so extreme that microhabitats are removed almost as rapidly as they are formed. Cliffs and talus slopes exposed to such stresses do not represent habitat for macroscopic organisms, and only epilithic mosses, lichens and algae can be expected to occur (Hedderson & Brassard, 1990). If significant undercutting of cliff faces occurs, erosion at the base can also accelerate the mass wasting from the free-face, leading to a rapid retreat of the entire cliff edge as discussed in Chapter 2. In some cases, the absence of woody vegetation from sea cliffs is probably not due to salinity, but rather the result of the high rate of mass wasting compared to inland cliffs not exposed to water. This will be especially true for cliffs formed from indurated or unconsolidated materials. Sea cliffs formed of hard rock, on the other hand, often support trees even when they are directly exposed to salt spray. For example, along the west coast of North America from California to Alaska, *Cupressus macrocarpa* (Fig. 6.8) and *Pseudotsuga menziesii* both occur on cliffs and rock outcrops that are directly exposed to salt spray, but very resistant to mass wasting.

For cliffs not exposed to a significant erosive force, the stability of

*Figure 6.8* Sea cliffs at Point Lobos State Reserve, California, USA, support the last natural stand of *Cupressus macrocarpa* in the USA. The trees grow abundantly in this reserve despite the continuous exposure to salt spray. Total annual precipitation at the site is high because of drenching fogs that occur almost daily. Photo by D.W. Larson.

microhabitats and the amount of time that organisms have to exploit them are much greater. Trees growing for centuries on ledges, in cracks and in crevices have been reported by Larson and Kelly (1991) and Kelly *et al.* (1994). For these populations, recruitment occurs in widely separated pulses (Barden, 1988; Larson, 1990). The maximum life span of these trees is nearly 1900 years, but even at this age the death of trees appears to be caused by rockfall rather than by senescence (Briand,

*Figure 6.9* A stunted *Juniperus communis* on an inland cliff in the Lake District, UK. Photo by A. Charlton.

Posluszny & Larson, 1993). Although the age structure of cliff forests has not been determined by other workers, the scattered reports of large and apparently old and deformed woody trees on cliffs in a variety of locations (Sealy, 1949; Jackson & Sheldon, 1949; Davis, 1951; Hotchkiss *et al.*, 1986; Donahue, 1996) suggest that cliffs may support ancient forests in settings where the current rates of weathering and erosion are low. Figure 6.9 shows vertical granite cliffs in the Lake District, UK, that support deformed and apparently slow-growing *Juniperus communis* that are potentially very much older than similar sized trees in the surrounding landscape. Larson *et al.* (1999) have recently shown for cliffs in eastern North America, central England and Wales, Germany and France that ancient woodlands occur generally on rock cliffs composed of materials that undergo slow mass wasting.

The susceptibility of rock to erosion and mass wasting can also influence the distribution of sea bird colonies. On the east-facing cliffs of Prince Leopold Island, Northwest Territories, Canada, Gaston and Nettleship (1981) observed that thick-billed murres are concentrated in the uppermost 100 m of the cliff, even though there would be a thermal advantage to nesting lower down. Rockfall activity in the lowermost portions of the cliff was considered the likely explanation. Rockfalls were observed that destroyed entire ledges, although new nesting sites were created as a result. The authors conclude that rockfall is a major cause of egg mortality at this site. On Bylot Island, Northwest Territories, Canada, breeding success of thick-billed murres is only 36.4 per cent, largely due to increased mortality from rockfalls and ice-falls onto breeding ledges. Eggs or young not directly hit are often displaced by panicking birds (Birkhead et al., 1985). The preference of nest sites with adjacent rock walls may be an adaptation to rockfall activity at this site.

## 6.4  Light, temperature and water

Few people have made quantitative measurements of light, temperature or water on cliffs, and even fewer have tried to show the quantitative links between the physical environment and the organization of the biota. Karlsson (1973) tried to interpret the autecological characteristics of a large array of cliff-dwelling plants in northern Sweden, but the large number of species and sites made it impossible for him to measure precisely the physical properties of the microsites where the plants were found. He reports values for temperature and water on cliff edges, ledges and talus slopes, but many of these measurements were made by directly feeling the soil with the hand. Such subjective evaluation of the factors that might control the vegetation or the fauna of cliffs may be better than having no information, but readers must remain cautious when statements about the dryness or coolness of cliffs are made without using a reliable standard of measurement. Maycock and Fahselt (1992), for example, report vegetation differences between dry cliffs and wet cliffs near Sverdrup Pass, Northwest Territories, Canada, based on a visual assessment of soil water content. They then attempt to make conclusions about the environmental control of vegetation patterns based in part on the subjectively measured environmental conditions. Such conclusions should be viewed with some scepticism.

Photosynthetic organisms on cliffs usually have a predictable access to saturating levels of photosynthetically active radiation, suggesting that competition for light is minimized. An exception to this is the vegetation of caves and deep crevices, where distinct zonations of plants may exist based on the amount of light available. For example, Zehnder (1953) describes the algal vegetation of a tropical cave, where green unicellular algae and some filaments colonize the entrance area but cyanobacteria of the genus Chroococcus and very few green algae are found in the deeper, darker zones. Besides direct sunlight, conditions that are typical for cliffs include rapidly moving air, rapidly fluctuating temperatures, and a lack of abundant well-developed soil. The human interpretation of such exposure is that cliffs are hot and dry; however, this is contrary to the available evidence. For example, *Thuja occidentalis* trees on cliffs do not show stomatal closure even during periods of intense drought (Matthes–Sears & Larson, 1991), and small trees given a continuous supply of water over a two-year period show no significant increase in shoot elongation (Fig. 6.10). This suggests that a predictable amount of water is available in at least some cliffs, probably from the current year's precipitation that has percolated through the bedrock. Endolithic algae, which grow within the rock (Fig. 6.11), can also exploit this water supply (Gerrath *et al.*, 1995), but epilithic lower plants cannot and may be more dependent on atmospheric moisture conditions. Thus, epilithic algae dominate moist and shaded rock surfaces in tropical and temperate zones, but only lichens can grow on faces that receive much radiation and are dry. Surfaces that are both shaded and dry are uncolonized or at most able to support species of *Lepraria* (Jaag, 1945; Zehnder, 1953).

The availability of water in rock differs from that in soil not only quantitatively, but also qualitatively. Rock substrate is essentially impenetrable to macroscopic organisms and cannot be moved except by mass wasting or large-scale phenomena such as earthquakes. Thus, plant roots and animal eggs or larvae must find microhabitats on cliff faces where their water-absorbing structures can exploit the natural flux of water from source to sink through individual crystals or small solution pockets. In other words, water in rocks cannot be easily foraged for by macroscopic organisms. This is in striking contrast with soil, where organisms of all phyla can forage through an almost fluid-like matrix of small solid particles, gases and liquids. The exposure of cliffs in summer to topographically buffered supplies of water and radiation in turn buffers the growing season temperatures and evaporation rates (Dahl, 1951). In the measurements of lichen thallus temperatures in various habitats by Lange (1953), lichens on rock faces consistently heated up less above air temperature than lichens on level ground, and the reasons for this include radiation exposure as well as the high conductivity of the rock, as discussed in Chapter 3. Nevertheless, lichens of south-facing limestone cliffs in a submediterranean region of Europe are able to survive exposure to laboratory-imposed temperatures of at least 90–100 °C (Lange, 1953). Low temperature resistance may explain why tropical and temperate cliffs differ drastically in their macrovegetation but not their algal vegetation.

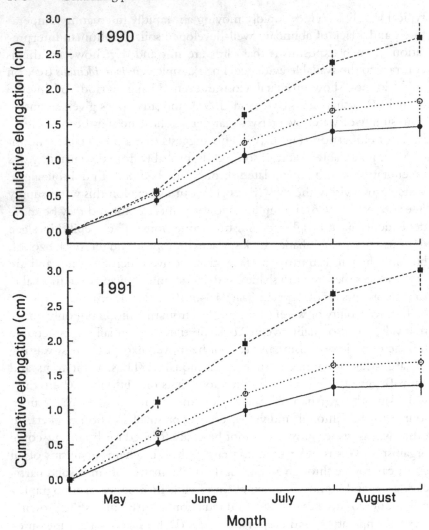

*Figure 6.10* Mean cumulative stem elongation for *Thuja occidentalis* growing on limestone cliffs over a two-year period (1990, 1991). Plants supplemented with water (open circles) or nutrient solutions (squares) grew only slightly more than control plants (solid circles) supplied with nothing. From Matthes-Sears *et al.* (1995), and used with permission of the University of Chicago Press.

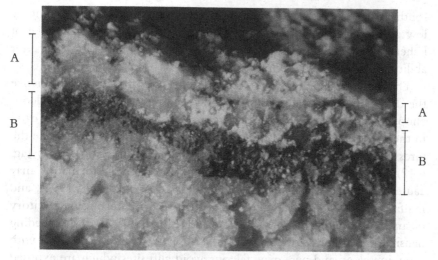

*Figure 6.11* The endolithic algal zone within limestone rocks of the Niagara Escarpment. The depth of the zone varies between 2 and 5 mm depending on the translucency of the rock to photosynthetically active radiation. Cells from epilithic algae, fungi and lichens are predominant in the surface zone, A, while true endolithic organisms are present in zone B. Photo by D.W. Larson.

Minimum temperature in the habitat imposes a northern limit on vascular plants, but lower plants tolerate almost any low temperature by simply becoming inactive (Zehnder, 1953).

Despite the above suggestion that cliffs provide some moderation of hostile environmental conditions, the productivity of photosynthetic organisms that occur on cliff faces and cliff edges is very low. Individual rosettes of cliff species stay rooted in small pockets for long periods of time (Davis, 1951). Tree species such as *Thuja occidentalis* have low but variable growth rates (0.05 mm radial growth per year; Kelly, Cook & Larson, 1992) and low short-term photosynthetic rates that do not change substantially through the growing season (Matthes-Sears & Larson, 1990). Even lower rates occur in the poikilohydric fern *Polypodium virginianum* along cliff edges (Gildner & Larson, 1992a, 1992b). This fern balances its carbon budget over the year by maintaining low but even rates of photosynthesis throughout the growing season. The fronds are green and appear physiologically active even in winter; however, experiments using stable isotope tracers and *in-situ* photosynthesis measurements found no evidence for significant photosynthesis during this period (Matthes-Sears, Kelly & Larson, 1993). Lichens on south-facing cliffs may dry out completely, reaching water contents as low as 2 per cent of dry weight (Lange, 1953). The predominance of lichens over algae in this exposed microhabitat may be due to their better ability to survive such complete desiccation.

The light, temperature and water regime of cliff faces also influence the distribution and behaviour of cliff-dwelling raptors. Prairie falcons choose eyries with overhead protection to maximize radiant heating early in the breeding season and minimize exposure to the hot sun while the birds are roosting later in the season (Runde & Anderson, 1986; Fig. 6.12). Inappropriate protection from rain, snow and ice formation may cause nest abandonment in golden eagles (Morneau *et al.*, 1994), and rainfall levels on cliff faces are an important factor determining territory occupancy in peregrine falcons in Ireland (Norriss, 1995). The breeding density of bearded vultures on cliffs decreases as the number of days with snow increases, and peregrine falcons avoid cliff sites which are exposed to wet northwesterlies and receive no direct sunlight during spring (Norriss, 1995). Breeding success in peregrine falcons is reduced in more exposed nest sites where eggs often end up in pools of mud (White *et al.*, 1988). Rinkevich and Gutiérrez (1996) found that a very specific micro-climate provides the habitat structure for Mexican spotted owls (*Strix occidentalis lucida*) in Utah which allows them to avoid physiological stress. These owls prefer canyons with narrow and deep walls in part because of a higher absolute humidity and no extensive direct exposure to the sun. Light levels also seem to initiate daily behavioural responses in raptors. The appearance of sunlight on the cliff face initiates 30 minutes of preening and sunning in peregrine falcons in Argentina, immediately followed by a hunt. Hunting is reinitiated just before sunset and stopped at twilight (Vasina & Straneck, 1984).

The proximity of open bodies of water is an important factor determining the occupancy of cliff sites by birds. All Taita falcon cliff nests studied by Hartley *et al.* (1993) in Zimbabwe were situated close to permanent water, and the long-legged buzzard nests studied by Vatev (1987) in Bulgaria were located within 500 m of moving or standing water. Mexican spotted owls also prefer canyon nest sites where open water is available (Rinkevich & Gutiérrez, 1996).

While the shape and dimensions of nesting cracks and ledges are the principal determinants of breeding success in sea birds, microclimatic conditions may have an effect on the timing and success rate of breeding. Excessive winter snowfall or a delayed spring melt may hinder successful hatching as eggs laid directly on snow may slip off the cliff or

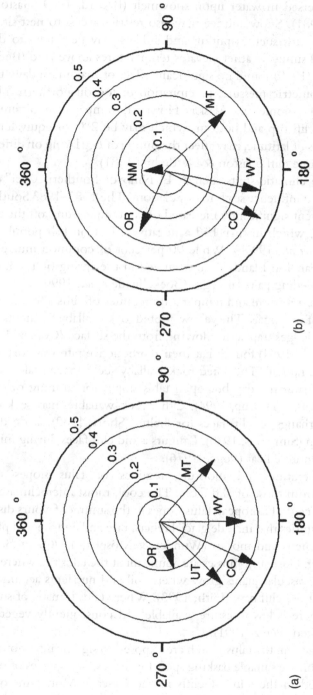

*Figure 6.12* Mean aspect of (a) prairie falcon eyries and (b) prairie falcon nest cliffs in five USA states. The vector lengths are proportional to the degree of concentration (*r*) about mean aspect. Redrawn from Runde and Anderson (1986).

become immersed in water upon snowmelt (Bédard, 1969; Gaston & Nettleship, 1981). Snow and ice may also restrict access to nest sites in crevices, and birds such as puffins and auklets may be forced to delay breeding until sunny weather elevates temperatures above zero (Bédard, 1969). Hatch (1989) found no significant effect of maximum daily temperature, barometric pressure, precipitation or fog on northern fulmar attendance in colonies in Alaska. However, temperature differences between the cliff top and bottom, which may be 2°C or equivalent to several degrees of latitude, may affect the timing of egg laying on different parts of the same cliff (Gaston & Nettleship, 1981).

Large-scale climatic shifts such as El Niño or Southern Oscillation events may also influence sea bird populations. The 1982–1983 Southern Oscillation event significantly increased ocean temperatures off the coast of California, which in turn had a negative effect on fish populations (Boekelheide et al., 1990b). While 90 per cent of common murre cliff sites on the Farallon Islands were occupied in the spring of 1983, only half of the breeding pairs laid eggs (Boekelheide et al., 1990b).

The unique moisture and temperature regimes of cliffs also affect the distribution of mammals. The yellow-footed rock-wallaby frequents cliffs because rock ledges trap water flowing from the surface (Copley, 1983). Robinson et al. (1994) found that their reproductive rate was correlated with effective rainfall. The allied rock-wallaby seeks shaded talus slopes for shelter during the day but open talus slopes for sunning on cool, winter afternoons (Horsup, 1994). Other rock-wallabies may seek caves or rocky overhangs on cliff faces for shelter (Short, 1982), as do desert bighorn sheep (Simmons, 1980). Langurs avoid cliff faces during midday to avoid the intense heat (Vogel, 1976).

The temperature and moisture regimes of talus slopes differ significantly from those of cliff faces. The cool, moist microclimate that characterizes central European talus slopes in the summer favours distinct plant communities that include glacial relicts, rare and endangered plants, and many lichens and mosses (Wunder & Möseler, 1996). In Baden-Württemberg, Germany, mosses are abundant at the talus toe where cold air exits, and vascular plants grow where soil and nutrients accumulate, particularly at the cliff base (Lüth, 1993). Where talus is made of smaller fragments, there is less moisture available and consequently vegetation cover is reduced (Pérez, 1991).

Rock surfaces in the talus which are exposed to significant amounts of direct radiation are suitable basking spots for snakes. Ten species of snakes were observed in the talus of cliffs in the Cascade Mountains of the north-west USA. The talus is an important habitat in the reproductive biology of these snakes and several species use the rocky habitat as hibernacula (Herrington, 1988). Populations of salamander species in the talus are dependent on the unique thermal and moisture regimes of these habitats, and many salamanders can complete their entire life cycle there. Adverse conditions such as reduced rainfall or elevated surface temperatures can negatively affect the surface activity of these species (Herrington, 1988). Cold and wet conditions may also negatively affect talus-nesting Atlantic puffins. In Newfoundland, puffins fledged only 27.7 per cent of their young under cooler, wetter conditions compared to 50.5 per cent under normal conditions (Nettleship, 1972). Death was primarily due to chilling.

## 6.5 Wind

Wind speed adjacent to cliffs is higher than in surrounding habitats, as shown in Chapter 3. This is evident from the growth form of cliff-face and cliff-edge trees (Fig. 6.13). Species such as *Didymo panax pittieri* in Costa Rica become stunted and deformed near the crests of ridges because of shortened internode lengths (Lawton, 1982; Fig. 6.14). Similarly, *Pinus longaeva* at high altitudes in the White Mountains of California is stunted, multi-stemmed, and extremely ancient (>4000 years) (Fig. 6.15), while trees in protected valleys usually show more productive growth, larger dimensions and less age (<1000 years) (Schulman, 1954). Tranquillini (1979) and Grace (1977) report a variety of negative effects of wind on photosynthetic rates, allocation patterns, and growth rates for trees growing on steep slopes. However, they fail to make the connection between declines in productivity and increases in longevity (Schulman, 1954; Larson & Kelly, 1991).

The presence of accelerated wind speeds will cause woody vegetation on cliffs to move more violently in crevices and solution pockets. It is thus expected that windstorms will cause not only more treefall events but also more rockfall events from cliffs. The literature on the impact of severe storm events on ecological communities is growing (Foster & Boose, 1992), but it is still too sparse to permit conclusions to be drawn about the impact of windthrow on the physical and biological structure of cliffs. The gross deformations of stems of *Thuja occidentalis* on cliffs of the Niagara Escarpment may be due to windthrow that is caused by extreme storm events, but the role of rockfall cannot be discounted.

Wind is also a factor in the distribution of animal species on cliffs.

*Figure 6.13* Cliff-edge trees have a more obvious wind-trimmed appearance than cliff-face trees that are more clearly influenced by rockfall. In this case, a 250-year-old tree has had its major axis removed by wind and the current growth axis is comprised of lateral branches on the leeward side of the main axis. Photo by P.E. Kelly.

Some spider species in the Czech Republic are only found on windy talus slopes because the winds provide them with an abundant food supply (Růžička *et al.*, 1995). Maser *et al.* (1979) state that the higher the cliff face, the more suitable it is for wildlife. As cliff heights increase, there are more predictable upward-flowing air currents or thermals, conditions which are necessary for many species of raptors. For this reason, prairie falcons (Maser *et al.*, 1979) and bearded vultures (Donázar *et al.*, 1993) only select nest sites on higher than average cliffs. Cape vultures, which are common components of cliffs in Namibia, also respond to wind exposure on cliffs by selecting nesting sites to the lee of prevailing winds (Brown & Cooper, 1987). During nesting season, 84 per cent of occupied nests occur in such locations, which are thought to offer the greatest protection for vulnerable nestlings in the face of 75 km/h winds.

Wind is also an important abiotic factor in coastal sea bird colonies. High wind speeds can either be a detriment or an advantage, depending on the species. Murre eggs laid on ledges with a calm wind regime survive significantly longer than eggs laid on exposed ledges because, under windy conditions, gulls are better able to reach narrow ledges using

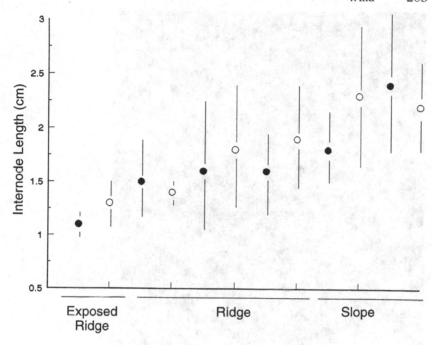

*Figure 6.14* Mean internode lengths and 95 per cent confidence intervals for branches on the windward (solid circles) and leeward (open circles) sides of the crowns of a tropical shrub, *Didymopanax pittieri*, grouped into three position classes on a mountain in Costa Rica. Redrawn from Lawton (1982).

gliding flight (Gilchrist & Gaston, 1997a). Morphological differences give glaucous gulls a predatory advantage over dovekies during windy conditions. Dovekies only leave their breeding colonies during calm weather (Stempniewicz, 1995). Extreme updraughts can also dislodge birds such as gannets from nests (Nelson, 1978a). On the other hand, Falk and Möller (1997) found no correlation between wind speed and the number of attending birds during June and July at northern fulmar colonies in north-eastern Greenland.

The prevailing wind direction can also influence the distribution of sea bird colonies. For example, northern fulmars in Alaska had a lower average attendance on days with northerly winds, and Hatch (1989) surmised that their feeding grounds lay in this direction. In Antarctica, snow petrels nest on cliffs in the lee of prevailing south-west winds (Warham, 1990). Mammals such as desert bighorn sheep seek relief on cliffs from desiccating winds on hot days (Simmons, 1980), while Stone's sheep escape to sheltered cliffs during the winter to avoid storms (Geist, 1971).

*Figure 6.15* *Pinus longaeva* growing in the White Mountains of California, USA. Stunted and wind-trimmed trees of this species occur on rocky slopes and cliffs above 3000 m elevation and show extensive strip-bark growth. Photo by P.E. Kelly.

## 6.6 Nutrient availability and food supply

Cliffs have the appearance of being nutrient depleted because they have only a sparse vegetation cover with low productivity. However, this conclusion applies to cliff faces as a whole but not to individual microsites. The cliff is composed of scattered small-scale habitats capable of supporting high productivity interspersed amongst much larger areas of exceptionally low productivity. To give a sense of scale to this variation in productivity, many temperate, boreal or tropical forests have annual productivity values in the range of 400–1000 g m$^{-2}$ yr$^{-1}$, and 100 g m$^{-2}$ yr$^{-1}$ is considered low. In comparison with this, Matthes–Sears *et al.* (1997) report maximum values of approximately 23 g m$^{-2}$ yr $^{-1}$ before respiratory losses were considered, and individual *Thuja occidentalis* on cliffs of the Niagara Escarpment may grow at rates of 0.01 g yr $^{-1}$. With stem densities of about 1000 trees per hectare (Kelly & Larson, 1997a), we can determine that each tree has access to approximately 10 m$^2$ of open cliff face. If we divide the observed lower limit of annual productivity by this area, we calculate a lower limit of annual production of 0.001 g m$^{-2}$ yr $^{-1}$. This makes these cliffs one of the least productive ecosystems on the planet.

Despite this exceptionally low productivity, individual plants growing on cliffs show no strong evidence of nutrient limitation (Matthes–Sears & Larson, 1991, 1995). This is true even when cliff trees are compared to the same species growing rapidly in swamps (Fig. 6.16) or when comparisons are made among microhabitats and species (Matthes–Sears & Larson, 1991; Cox & Larson, 1993a, 1993b; Hora, 1947). A number of factors may explain this. Soil availability of cliffs is patchy, but where cliff soils occur, the concentration of nutrients such as Ca, Mg, P and K may be even higher than in soils on level-ground forest at cliff edges (Young, 1996). Plants on cliffs are also heavily colonized by mycorrhizal fungi (Matthes–Sears *et al.*, 1992). This may explain why the force-feeding of extra water and nutrients to cliff-face trees had only a marginal impact on the growth of eastern white cedar (Matthes–Sears *et al.*, 1995). The growth rates of obligate cliff-dwelling ferns are also unaffected by the addition of nutrients (Coates & Kirkpatrick, 1992). These results suggest that plant productivity on the cliff face is normally low. Plant and animal size and population growth on cliffs may be regulated on the basis of the available flux of nutrients, and growth rates adjusted such that tissue nutrient concentrations are not deficient and the survival of plant parts is ensured (Ingestad & Lund, 1986). Matthes–Sears and Larson (1999) have recently shown that the factors that influence the growth rates of trees are not necessarily the same factors that influence their survival. Small *Thuja occidentalis* were planted in pots that differed greatly in soil volume and nutrient level. Nutrient supplements increased growth rates of trees but did not influence their survival, yet larger volumes of soil available for root systems had a positive impact on survival but not on growth.

Pockets of high nutrient availability on cliff faces include sites that receive direct inputs from the excrement and guano of small mammals and bird populations (Oettli, 1904). Small ledges in areas of water seepage

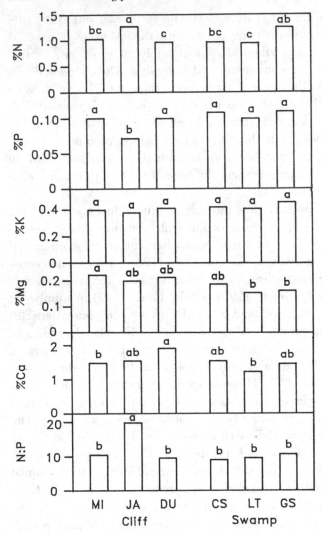

*Figure 6.16*  Mean tissue nutrient concentrations of *Thuja occidentalis* growing at three cliff and three swamp sites. Means with the same letter are not statistically different at $p < 0.05$. These graphs show that cliff-face *Thuja occidentalis* are not limited by nutrients when compared to swamp-grown trees. Data taken from Matthes-Sears and Larson (1991), and used with permission of the University of Chicago Press.

also permit higher levels of primary production. Small mammals are likely to be significant sources, rather than sinks, of nutrients on cliff faces by providing small-scale sources of nutrient input to the cliff face for resident plant species (Matheson, 1995). Cliff faces and talus below nesting birds represent nutrient-rich habitats, particularly in the Arctic, an otherwise nutrient-poor environment (Batzli 1975; Odasz 1994). Plants growing closest to recent guano deposits in Svalbard, Norway, have the highest leaf nitrate reductase activity. The effective assimilation of nitrate may be a competitive advantage (Odasz, 1994). Mosses, lichens and algae are often abundant below old golden eagle nests (Ratcliffe, 1993). New plant associations with coprophilous characteristics were also observed by Guitián and Guitián (1989) in coastal sea bird colonies in Spain. Peregrine falcon guano kills plants such as ling heather (*Calluna vulgaris*) on cliffs in the UK, although these are soon replaced by bilberry (*Vaccinium myrtillus*) (Ratcliffe, 1980). Sea birds are unimportant contributors to the nutrient budget of coastal waters (Bédard *et al.*, 1980), but affect the structure of plant communities on cliffs (Wootton, 1991). On sea cliffs in Washington, USA, some species are unaffected by guano from overhead sea bird colonies, whereas others either increase or decrease in response to the input. Lichens such as *Xanthoria* and *Caloplaca* may increase, while others such as *Physcia* may be eliminated. *Xanthoria* and *Caloplaca* may either be responding positively to the input or expanding into free space vacated by *Physcia*. Overall, four of the 18 taxa examined are positively influenced by guano inputs (Wooton, 1991).

Bird species' distribution and nesting site selection are not directly influenced by nutrient availability on cliffs because prey is generally captured away from the cliff face and brought back to nesting sites for consumption. The factors that limit plant productivity on cliffs do not apply to mobile animals. Cliff faces chosen for nesting, however, need to be located close to the available food supply. For example, Egyptian vultures in Spain minimize the energy investment in finding and carrying food to the nest by locating their nests close to feeding places, even if this means increased human contact (Ceballos & Donázar, 1989). In Namibia, Lanner falcons monopolize nesting sites on cliffs above open country while peregrine falcons choose nesting sites on cliffs above ground-level woodlands and thickets. Rock kestrels monopolize sites on the lowest sections of the face (Brown & Cooper, 1987; Fig. 6.17). The ground-level hunting practices of Lanner falcons exclude aerial-hunting peregrine falcons from nesting sites above open country. Declines in the

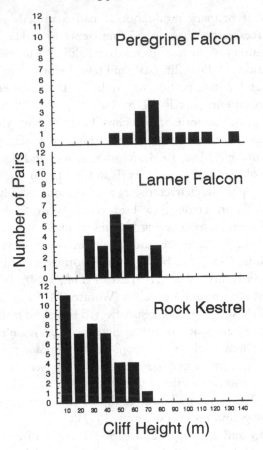

*Figure 6.17* Cliff heights selected by nesting pairs of three raptors on cliffs in Namibia. Redrawn from Brown and Cooper (1987).

Cape vulture (*Gyps coprotheres*) population have also been attributed to a decline in food supply resulting from severe bush encroachment into savanna regions (Brown & Cooper, 1987). Similarly, the occupancy rates of peregrine falcons on cliffs in Ireland are greatest where the cliffs are in close proximity to open farmland (Norriss, 1995). Peregrine falcons may choose less desirable nesting sites and have higher nesting densities where food supplies near cliffs are plentiful (Brown & Cooper, 1987). For example, the distance between nests decreases on cliffs near popular racing pigeon routes (Norriss, 1995).

Species such as the bearded vulture have very specific foraging requirements that confine the distribution of their nesting sites to cliffs in areas of irregular topography. While the amount of open land available for foraging is positively correlated with productivity, the highest productivity occurs in the most rugged terrain (Donázar et al., 1993). This is directly related to the feeding habits of these birds. The majority of the bearded vulture food supply is bones, and rocky areas such as talus are used as ossuaries where bones are dropped and broken on exposed rock to facilitate ingestion. Areas with high relief also tend to have significant slope winds which blow these rocky areas free of snow, thus facilitating feeding (Donázar et al., 1993).

Populations of cliff-nesting birds respond primarily to fluctuations in the populations of their off-cliff prey. Peregrine numbers on cliffs on Langara Island, British Columbia, Canada, for example, gradually declined after the introduction of rats to the island led to a decline in the population of ground-nesting ancient murrelets (Synthliboramphus antiquus), the principal food supply of the falcons (Nelson, 1990). Golden eagle density has been observed to decline in response to deer management strategies in Scotland which reduced the amount of deer carrion available through increased culling (Watson, Payne & Rae, 1989). Sea bird colonies are in direct competition with humans for fish resources in some areas (Freethy, 1987), and some sea birds such as the common murre have actually altered their feeding behaviour to avoid areas with heavy gill-netting (Boekelheide et al., 1990b). In contrast, increases in capelin (Mallotus villosus) and sand lance (Ammodytes) stocks have been linked to the expansion of black-legged kittiwake populations in the Gulf of St Lawrence (Chapdelaine & Brousseau, 1989).

Some cliff-nesting bird species do forage on cliffs, but this is the exception rather than the rule. Gulls prey on other cliff-nesting birds (see Chapter 5) and a large portion of some gull diets is comprised of the young and eggs of sea birds such as the common murre and thick-billed murre (Birkhead, 1977; Gilchrist & Gaston, 1997a). The great reed warbler (Acrocephalus arundinaceus), which is not a cliff-nesting bird species, has been observed abandoning its traditional habitat to forage on thrift (Armeria maritima) and sea campion (Silene maritima) on a vertical cliff face on Fair Isle, Scotland, UK (Riddiford & Potts, 1993). After 14 days, its claw tips were worn to the toe. Australian rock mammals such as the black-footed rock-wallaby forage in talus environments that provide refuge to plants from climatic extremes (Pearson, 1992). In the Canadian Rockies, bighorn sheep often forage on cliff faces in March when the snow has crusted over on the adjacent slopes. Cliffs are the only habitat at this time of year where forage is readily available (Geist, 1971).

## 6.7 Aspect

It has been widely assumed that aspect is an important factor influencing the plant communities present on cliffs even though work designed to specifically test for this effect often produces negative results (Orwin, 1972). While a number of studies have indeed found an influence of aspect on cliffs (Hepburn, 1943; Goldsmith, 1973a; Ashton & Webb, 1977; Fuls et al., 1992) and on artificial walls (Risbeth, 1948), in most cases aspect is not the primary controlling factor but only of secondary importance. Karlsson (1973) showed that on the cliffs of Sarek National Park in northern Sweden, aspect plays very little role in regulating the distribution patterns of the abundant plants. However, it appears to be much more important for many of the rare species typical of more southern or northern sites. A similar example is described by Pigott and Pigott (1993), who have shown that Tilia cordata (small-leaved lime) is confined to steep, north-facing granite cliffs in the southern portion of its range in France. All other aspects and exposures incur such a high demand on the transpiration stream that seedlings die very quickly. However, in the central and northern portions of its range where the climatic conditions do not impose as much stress from evapotranspiration, T. cordata shows no such aspect dependence. Considering the effect of aspect on radiation supply on cliffs (as discussed in Chapter 3), we believe that most species that occur mainly on cliffs will have their distribution only slightly limited by aspect. The greatest impact of aspect ought to be the extension of the growing season for plants on south-facing cliffs in the northern hemisphere (and north-facing cliffs in the southern hemisphere), which receive their seasonal peak in direct beam radiation in winter.

Vascular plants and endolithic and epilithic algae on the Niagara Escarpment are fairly consistent from site to site even when aspect is changing (Larson et al., 1989; Matthes-Sears et al., 1997). The most likely explanation for this is that the highly heterogeneous surface of cliffs creates variation in the microenvironmental conditions that is far greater than the differences imposed by a simple change in aspect. This heterogeneity is most prevalent on the smallest scale; as a consequence, it is easily understood why small plants such as algae show no clear dependence on the overall aspect of a cliff. Ferris and Lowson (1997) have shown that endolithic micro-organisms in the pores of the crystalline limestone of the Niagara Escarpment are essentially hiding from intense illumination. If true, this would suggest that aspect differences among cliff faces have little to do with the distribution of these organisms since all aspects are exposed to amounts of radiation in excess of what the organisms need to survive. The likelihood of aspect effects should increase with increasing size of the plants and should therefore be greatest for trees, but in the literature there is a surprising lack of data that address this point. In our own studies along the Niagara Escarpment, cliff aspect has shown no impact on the number, shape, size or age of *Thuja occidentalis* (Larson & Kelly, 1991; Kelly *et al.*, 1994). However, preliminary data relating individual tree growth rates to aspect show indeed that trees on north-facing cliffs have higher growth rates than trees on cliffs with other aspects. This result is consistent with the general preference of this species for microhabitats and sites that remain cool and moist for extended periods of time (Walker, 1987; Archambault & Bergeron, 1992).

Cliff-face aspect is a contributing factor to the distribution of many raptor nesting sites. In most habitats, nesting raptor populations prefer very specific cliff orientations which can significantly influence territory occupancy (Norriss, 1995). Southwest-facing to east-facing cliffs (225° to 90°) are the most popular nesting sites. Prairie falcons in the western USA use southwest-facing to southeast-facing cliffs (Runde & Anderson, 1986; Boyce, 1987) because southerly exposures moderate cold temperatures during incubation and brooding in the early spring, and minimize exposure to inclement weather (Runde & Anderson, 1986; Morneau *et al.*, 1994). Golden eagles and red-tailed hawks in southern Idaho also prefer to nest on southwest-facing cliffs (Craig & Craig, 1984), as do golden eagles along Hudson Bay, western Quebec, Canada, which nest on southwest-facing to south-facing cliffs in valleys (Morneau *et al.*, 1994). Rough-legged hawks in north-western Quebec prefer nests which face south (Brodeur *et al.*, 1994), and peregrine falcons on cliffs in west Greenland nest on west-facing to southwest-facing cliffs at cliff heights between 60 and 120 m (Moore, 1987). Prairie falcons in Colorado prefer east-facing cliffs and similar observations have been made for gyrfalcons and peregrine falcons in Alaska (Williams, 1984), as well as for peregrine falcons in many urban environments (Cade & Bird,

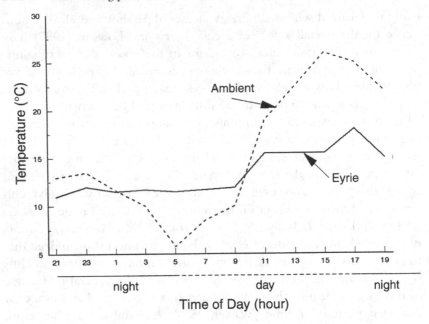

*Figure 6.18* Diurnal courses of ambient temperature and eyrie temperature on an east-facing cliff. This graph illustrates the degree to which eyrie temperature is buffered from high late-afternoon ambient temperature. Modified after Williams (1984).

1990). Interestingly, some raptors in tropical regions, such as Taita falcons (*Falco fasciinucha*) in Zimbabwe (Hartley *et al.*, 1993) and black eagles (*Aquila verreauxii*) in Namibia (Brown & Cooper, 1987), show no cliff aspect preference for nesting.

Williams (1984) believes that the easterly aspect buffers prairie falcon nestlings from ambient temperature extremes, especially in the early morning and late afternoon (Fig. 6.18). The cliff rapidly absorbs heat from solar radiation during the day, then acts as a heat source at night, losing this heat to the cooler night air (Williams, 1984). Snow petrels nest on north-facing inland cliffs in Antarctica, in part to take advantage of the sun during clear weather (Somme, 1977). Aspect may also affect the choice of nesting sites in the talus. For example, Kittlitz's murrelet in the Russian far east primarily selects south-facing talus slopes for nest sites because their greater exposure to the sun leads to more rapid snow melt in the spring (Day, 1995).

In other cases, bird species select cliffs of certain aspects because of the prevailing wind direction. Cape vultures in Namibia, for example, nest on northwest-facing cliffs to avoid high-intensity winds (Brown & Cooper, 1987). Other apparent aspect preferences may in reality be due to other factors only indirectly related to aspect. Watson (1997) concludes that the tendency for golden eagles to nest on northwest-facing to east-facing cliffs in the Scottish Highlands reflects the greater number of ledges on them that is due to a greater incidence of frost-shattering. The preference of peregrine falcons to nest on north-facing and east-facing cliffs in the British Isles may simply reflect the greater frequency of high cliffs with these aspects (Ratcliffe, 1993).

Compared with the large number of bird studies, few data are available on the aspect preferences of cliff mammals. One of the few investigations was on brush-tailed rock-wallabies in Australia (Short, 1982). Aspect was the second most important variable affecting habitat selection in this species, with north-facing aspects selected exclusively to avoid the shaded south-facing cliff faces.

## 6.8 Gravity

Gravity is rarely mentioned by ecologists as an important force organizing communities of organisms, but it may be one of the most important determinants of community structure on cliffs. Gravity will immediately remove the majority of the organic debris that is created on the cliff face, but retain and even concentrate it locally on ledges, in cracks and in crevices. Individual plants and animals persist in these microsites where nutrients accumulate. However, rockfall may instantly eliminate all evidence of soil, nutrients and the biota, all of which are then lost from the system. This result is quite different from that for level-ground habitats of low productivity such as arctic tundras, where debris, carcasses and excrement can accumulate (Booth, 1977; McKendrick et al., 1980).

Soil accumulations on ledges, in cracks and in crevices provide a repository for seed banks of cliff plants. Table 6.1 lists the species observed in the seed rain and Table 6.2 those observed in the seed bank for cliffs of the Niagara Escarpment. The seed bank is composed of many of the same species represented in the seed rain (Booth & Larson, 1999b), suggesting that a continuous resupply of the seed bank by the effects of gravity is important to the maintenance of the cliff vegetation. A far greater variety of species is available as recruits than are present as adult plants, however,

Table 6.1. *Partial species list of trees, shrubs and herbs found in the seed rain at two sites along the Niagara Escarpment, in 1993 and 1994.*

| Site | Plateau | | Cliff face | | Talus | |
|---|---|---|---|---|---|---|
| | Milton | Dufferin | Milton | Dufferin | Milton | Dufferin |
| **Trees** | | | | | | |
| *Acer saccharum* | + | + | + | + | + | + |
| *Acer spicatum* | + | | + | + | + | |
| *Betula alleghaniensis* | | + | + | + | + | + |
| *Betula papyrifera* | + | + | + | + | + | + |
| *Ostrya virginiana* | + | + | + | + | + | + |
| *Quercus rubra* | + | | + | + | | |
| *Thuja occidentalis* | + | + | + | + | + | + |
| **Shrubs** | | | | | | |
| *Cornus alternifolia* | | | | | | + |
| *Cornus rugosa* | | + | + | + | | |
| *Cornus sericea* | | | + | | | |
| *Diervilla lonicera* | | | + | | | |
| *Euonymus obovatus* | | | | | | |
| *Sambucus canadensis* | | | + | | + | |
| *Sambucus pubens* | | | + | + | + | + |
| *Vitis riparia* | | | + | | | |
| **Herbaceous** | | | | | | |
| *Alliaria officinalis* | | + | | + | | + |
| *Anthemis tinctoria* | | | | | | + |
| *Arabis glabra* | + | | | + | | |
| *Arabis hirsuta* | | + | + | + | | |
| *Aralia racemosa* | | | + | | | |
| *Asclepias syriaca* | + | | | + | | + |
| *Aster cordifolius* | + | | | + | + | |
| *Aster macrophyllus* | + | | | | + | |
| *Aster novae-angliae* | | | | + | + | |
| *Cirsium vulgare* | | | + | + | | |
| *Clematis virginiana* | | | + | | | |
| *Conopholis americana* | | + | | | | + |
| *Conyza canadensis* | | | + | | | |
| *Eupatorium rugosum* | + | + | + | + | + | + |
| *Fragaria virginiana* | | | + | | | |
| *Geranium robertianum* | | + | + | + | + | + |
| *Geum canadense* | | | + | | | + |
| *Hackelia virginiana* | | | + | + | | |
| *Impatiens pallida* | | | | | | + |
| *Lactuca serriola* | | | + | + | + | + |

Table 6.1 (*cont.*)

| Site | Plateau | | Cliff face | | Talus | |
|------|---------|---------|--------|----------|--------|----------|
| | Milton | Dufferin | Milton | Dufferin | Milton | Dufferin |
| *Leonurus cardiaca* | | | | + | + | |
| *Nepeta cataria* | | | | + | | |
| *Poa compressa* | | | + | + | | |
| *Ranunculus abortivus* | + | | | | | |
| *Rubus strigosus* | | + | + | + | | + |
| *Rumex crispus* | | | | + | | |
| *Satureja vulgaris* | | | + | | | |
| *Solanum dulcamara* | | + | + | + | + | |
| *Solidago canadensis* | + | + | + | + | + | + |
| *Solidago flexicaulis* | + | + | + | + | | + |
| *Sonchus asper* | | | + | | | |
| *Taraxacum officinale* | | | + | + | | |

*Source:* Data taken from Booth and Larson (1998).

suggesting that there is a progressive elimination of unsuccessful taxa, rather than a complete barrier to migration at one particular stage of development. Much of the seed bank of a cliff may eventually become food for invertebrate herbivores present on cliffs (Davis, 1951, Houle & Phillips, 1988), but experimental work has not been carried out to explore this.

Cliff plants must allocate resources to deal with the effects of gravity. This is often evident in the architecture of individuals growing on cliffs, which may be distinctly different from the architecture of the same species found in more benign environments. Aborted leaders, shoot dieback, extreme stunting and asymmetric growth (Fig. 6.19) are all commonly reported in cliff plants (Henderson, 1939; Davis, 1951; Briand *et al.*, 1991; 1992a, 1992b; Kelly *et al.*, 1992; Del Tredici *et al.*, 1992; Larson, Matthes-Sears & Kelly, 1993). In addition, there appears to be an inverse association between growth rate and longevity of trees, as described above (Davis, 1951; Schulman, 1954; Kelly *et al.*, 1992). Old specimens of bristlecone pine (*Pinus longaeva*) and eastern white cedar are typically as large as much younger individuals in the same forest. Older members of both species are deformed axially and radially (Fig. 6.20) and show significant amounts of cambial mortality. Stem dieback is also common in other woody genera on cliffs, such as *Betula, Sambucus, Cornus*

Table 6.2. *Partial species list for seeds found in the plateau, cliff face and talus seed banks at two sites along the Niagara Escarpment in 1993, 1994 and 1995*

| Site | Plateau Milton | Plateau Dufferin | Cliff face Milton | Cliff face Dufferin | Talus Milton | Talus Dufferin |
|---|---|---|---|---|---|---|
| **Trees** | | | | | | |
| *Acer saccharum* | + | | | + | | |
| *Acer spicatum* | | | + | | + | + |
| *Betula alleghaniensis* | + | + | + | + | + | + |
| *Betula papyrifera* | + | + | + | + | + | + |
| *Fagus grandifolia* | + | | + | + | | |
| *Fraxinus americana* | | + | + | + | | + |
| *Ostrya virginiana* | + | + | | + | + | |
| *Quercus rubra* | + | | | + | | |
| *Taxus canadensis* | | | | + | | |
| *Thuja occidentalis* | + | + | + | + | + | + |
| **Shrubs** | | | | | | |
| *Celastrus scandens* | | | + | | | |
| *Cornus alternifolia* | + | | | + | | |
| *Cornus rugosa* | + | + | + | + | + | |
| *Cornus sericea* | | | + | + | | |
| *Parthenocissus vitacea* | | + | + | + | + | |
| *Rhus typhina* | + | + | + | + | + | + |
| *Rubus strigosus* | + | + | + | + | + | + |
| *Sambucus canadensis* | | | + | + | + | |
| *Sambucus pubens* | + | + | + | + | + | + |
| *Vitis riparia* | | | + | + | + | + |
| **Herbaceous** | | | | | | |
| *Actaea alba* | | | | | + | |
| *Alliaria officinalis* | | | | + | | |
| *Amelanchier laevis* | | | + | + | | |
| *Aquilegia canadensis* | | | + | + | | |
| *Arabis hirsuta* | | | + | | | |
| *Arctium minus* | | | | + | | |
| *Aster novae-angliae* | | + | + | + | + | + |
| *Chenopodium album* | | | + | | | |
| *Cirsium arvense* | | | | + | | |
| *Conopholis americana* | | | + | | | |
| *Eupatorium rugosum* | + | + | + | + | + | + |
| *Fragaria virginiana* | | + | + | + | + | + |
| *Geranium robertianum* | | + | + | + | + | + |
| *Geum canadense* | + | + | | | | |
| *Hackelia virginiana* | | + | + | + | | |

Table 6.2 (*cont.*)

| Site | Plateau | | Cliff face | | Talus | |
|------|---------|---------|------------|---------|-------|---------|
| | Milton | Dufferin | Milton | Dufferin | Milton | Dufferin |
| *Leonurus cardiaca* | | + | + | + | + | + |
| *Nepeta cataria* | | | + | + | + | + |
| *Poa compressa* | | | + | + | | |
| *Polygonum convolvulus* | + | + | | | | |
| *Ranunculus abortivus* | | | + | + | | |
| *Rumex crispus* | | + | + | + | + | |
| *Solanum dulcamara* | + | + | + | + | + | + |
| *Solidago flexicaulis* | + | | | | | |
| *Solidago canadensis* | + | + | + | + | + | + |
| *Sonchus oleraceus* | | + | | + | | |
| *Taraxacum officinale* | | | + | + | | |
| *Verbascum thapsus* | | | | + | | |
| *Zea mays* | | | + | + | | |

*Source:* Data taken from Booth (1999).

and *Tilia*. For these organisms, extremely old shoot bases can give rise to shoots that themselves are quite short lived.

On the Niagara Escarpment, stem deformation of *Thuja occidentalis* can be initiated by rockfall-induced root mortality (Fig. 6.21) causing the death of discrete parts of the crown (Larson *et al.*, 1993; Fig. 6.22). Subsequent reorientation of growth to surviving roots suggests the existence of sectored hydraulic architecture. The experimental use of dyes in this species confirmed these suspicions (Larson *et al.*, 1994a). These results imply that other slow-growing conifers such as bristlecone pine may exhibit the same hydraulic pathways to cope with the effects of gravity on rocky slopes. Downslope erosion of the underlying substrate exposes bristlecone pine roots to desiccation and death (LaMarche, 1963, 1968), with a corresponding death of portions of the stem cambium. Rates of slope erosion have been calculated using these cambial death dates. Similar rates of erosion were calculated using the length of exposed root in stunted and deformed *Thuja occidentalis* on cliff faces near Troy, New York (Gilbert, 1871; Fig. 6.23, 6.24). This is the earliest known use of dendrochronology to date biogeophysical processes.

Gravity also influences the breeding success of sea bird colonies and the survival of other large and medium-sized vertebrates. On sites with minimal rockfall activity, for example, thick-billed murres which nest on

*Figure 6.19* A stunted, asymmetric and slow-growing specimen of *Thuja occidentalis* on a cliff along the Niagara Escarpment. All of these growth anomalies are imposed by gravity in this system. This tree was less than 1 m in length. Photo by P.E. Kelly.

level ledge sites have a higher breeding success than birds which nest on sloped surfaces (Birkhead *et al.*, 1985). Gravity, therefore, can interact with small-scale topographic features such as ledge width, depth and slope angle to control the success of individual families. For cliffs composed of highly fractured rock that experience a high rate of mass wasting, gravity can be a significant source of mortality to immature birds as well as small and large mammals (Geist, 1971; Allen, 1980). Mountain

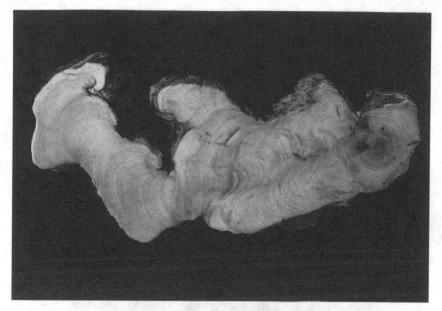

*Figure 6.20* Cross-section of a 935-year-old stem of *Thuja occidentalis* that had grown on a limestone cliff of the Niagara Escarpment. The total stem diameter was 25 cm, but all of the growth was asymmetric after the death of roots feeding one side of the axis. The photo clearly shows symmetric growth for the first 100 years or so, followed by the death of the cambium on one complete side of the axis. Photo by D.W. Larson.

goats (*Oreamnos americanus*), Stone's sheep (*Ovis dalli stonei*), and chamois (*Rupicapra rupicapra*) are all exposed to gravity-induced mortality, but the overall impact of these risks may be minimized by systematic re-use of the same network of secure trails on ledges and talus slopes.

## 6.9 Fire

While fire is not a factor controlling the organization of cliff-face communities, the absence of fire is. Fires on cliff faces are extremely rare because of the lack of fuel load. On most cliffs, there is no accumulation of litter and the patchy distribution of vegetation restricts the movement of fire. This allows ancient habitat on cliff faces to develop and persist through time. There is evidence that some plant and animal species persist in cliff habitats specifically because they represent a fire-free zone. For example, the conifer *Widdringtonia cedarbergensis* is particularly sensitive to fire, and its current distribution in South Africa reflects the

*Figure 6.21* Photograph of roots of *Thuja occidentalis* exposed by rockfall from cliffs of the Niagara Escarpment. Most of the roots exposed by rockfall do not penetrate deeply into the rock. Photo by D.W. Larson.

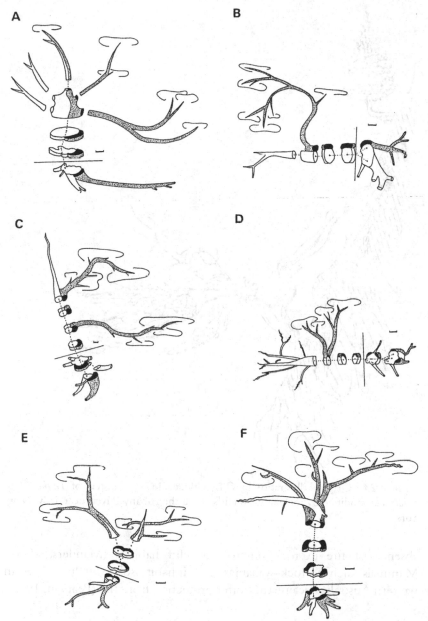

*Figure 6.22* Illustration of the impact of root mortality on crown mortality and architecture in cliff-face *Thuja occidentalis* growing on the Niagara Escarpment. Root death is followed by corresponding death of the crown to which it was hydraulically connected. The lag time is approximately 2.2 years. Stem F is the best example of the ecological impact of this sectored mortality as essentially two separate trees were produced from one initial stem. Taken from Larson *et al.* (1993), and used with permission of the University of Chicago Press.

*Figure 6.23* Original illustration by G.K. Gilbert (1871) of a stem of *Thuja occidentalis* modified by rockfall from cliffs along the Albany River, near Troy, New York.

absence of fire in rock outcrop or cliff habitats (Manders, 1986). Mammals such as rock-wallabies also inhabit isolated talus slopes in western Australia because they offer protection from fire (Pearson, 1992).

## 6.10 Maritime influence

Sea cliffs and inland cliffs are structured by very different processes. Some authors have argued that plant communities on sea cliffs are always poorly developed compared with those on inland cliffs (Davis, 1951; Bolton,

*Table of Cedar-root Chronology.*

| 1. Circum. of base of trunk, in inches | 2. Estimated age, in years | 3. Length of exposed root, in inches | 4. Recession of cliff per century, in inches |
|---|---|---|---|
| 11 | 210 | 72 | 34.2 |
| 19 | 363 | 117 | 32.2 |
| 30 | 573 | 150 | 26.2 |
| 16 | 306 | 72 | 23.5 |
| 19 | 363 | 72 | 19.8 |
| 29 | 554 | 108 | 19.5 |
| 6.5 | 124 | 24 | 19.3 |
| 7 | 134 | 24 | 17.9 |
| 9.5 | 183 | 30 | 16.3 |
| 19 | 363 | 56 | 15.4 |
| 37.5 | 716 | 84 | 11.5 |
| 19 | 363 | 38 | 10.5 |
| 6 | 115 | 12 | 10.4 |
| 8 | 153 | 12 | 7.8 |
| 9 | 172 | 12 | 7 |
| 28 | 535 | 36 | 6.7 |
| 9.5 | 183 | 12 | 6.5 |
| 15 | 286 | 9 | 3.1 |
| 7 | 134 | 2 | 1.5 |
| | | | 19) 289.3 |
| Average | | | 15.2 |

*Figure 6.24* Original table by G.K. Gilbert (1871) showing the application of dendrochronology to the problem of rock recession rates along the Albany River. Of particular ecological interest in the table are the age estimates, which reached 716 years for these cliff-face trees.

1981). While rapid erosion of sea cliffs may be in part responsible for this (see Section 6.3), it has generally been attributed to salt spray significantly limiting the growth of many plants, in particular woody species (Fig. 6.25). Unless fresh water is supplied at an exceptionally high rate by rain or mist, salt spray exposes plants on sea cliffs to severe desiccation stress. For this reason, halophytes make up a large proportion of species typical for maritime cliffs. Only where there is both a stable rocky substrate and frequent precipitation of fresh water, non-halophytic trees can form dense forests immediately adjacent to the sea, as along the west coast of North America. Experiments have shown that many of the herbaceous species adapted to inland cliffs cannot tolerate the effects of desiccation from high salt loads on maritime cliffs, while sea-cliff plants are more tolerant (Goldsmith, 1973b; Wilson & Cullen, 1986). Such differences are present not just in adults, but also at the seed stage: when comparing the germination behaviour of seeds from maritime cliff plants, from inland species that do not occur on cliffs, and from a single saltmarsh species, the results show that maritime cliff species are more tolerant of high levels of sea salt than the seeds of inland species (Fig.6.26; Goldsmith, 1973b).

The absence of sea–cliff species from inland cliffs in the same region has been attributed to the greater cold sensitivity of many maritime plants. This sensitivity, especially at the seedling stage, prevents these species from expanding their range to inland cliffs where temperatures are more extreme (Okusanya, 1979c). A second reason may be that plants adapted to sea cliffs have a lower competitive ability than plants adapted to inland cliffs (Goldsmith, 1973b; Wilson & Cullen, 1986).

## 6.11   Biotic interactions

For most ecosystems in the world, competition and predation are among the most important processes that control community structure. For cliffs the same is not necessarily true, or at least not for all groups of organisms. No reports have been published to show that plants on cliffs have conspicuous defence mechanisms of any type that work against the pressures of competition or herbivory. Their defence is primarily provided to them by the selection of cliffs as their habitat.

Cliffs are largely inaccessible to macroscopic herbivores, with the exception of species such as mountain goats or chamois. On cliffs in southern Germany, chamois preferably graze on cliff edges and accessible ledges, and 50–90 per cent of their food consists of grasses and sedges. Deer, which take a much larger proportion of herbaceous plants, do not graze near cliff edges at all (Herter, 1996). The recent introduction of chamois to this area (in the 1960s) has changed the composition of cliff-edge plant communities and represents an ongoing threat to certain rare species. *Daphne cneorum* and *Sesleria albicans* are among the species selectively taken by chamois and have disappeared from cliff edges where animal densities are high (Herter, 1993, 1996). However, the ability of even chamois to reach sheer cliff faces is limited and therefore most tall and vertical cliffs are free from herbivory by large grazing animals.

*Figure 6.25* A sea cliff in New Zealand illustrating the relative lack of woody vegetation on the cliff face. The herbaceous plants that occur there lack competitive ability when growing in productive level-ground sites. Figure courtesy of J.B. Wilson, taken from Wilson and Cullen (1986), and used with permission of S.I.R. Publishing.

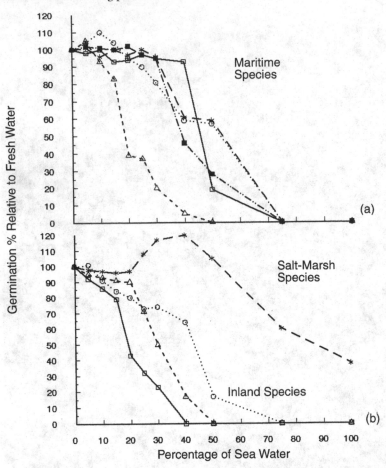

*Figure 6.26* Cumulative germination percentage of seeds of different species when exposed to sea water (100 per cent), fresh water (0 per cent) or intermediate solutions. The five maritime cliff species examined (a) were *Crithmum maritimum* (solid squares), *Daucus carota* ssp. *gummifer* (asterisks), *Inula crithmoides* (open circles), *Ligusticum scoticum* (open triangles), *Spergularia rupicola* (open squares). Panel (b) includes data for three inland species: *Daucus carota* ssp. *carota* (open circles), *Inula conyza* (open squares), and *Spergularia rubra* (open triangles). A strictly salt-marsh species, *Suaeda maritima* (asterisks) is also included. Redrawn from Okusanya (1979a), and used with permission of Blackwell Science Ltd, Publishers.

Less information has been published on the impacts of small and microscopic herbivores on cliff vegetation. Snails are known to feed on endolithic lichens on cliffs (Lüth, 1993) and slugs consume crustose lichens in the talus (Lawrey, 1980). A variety of insects, mites and other arthropods also feed on lichens, mosses and vascular plants that grow on cliffs (Lawrey, 1987; Lüth, 1993; Ficht et al., 1995).

One aspect that deserves special attention is the possible role being played by herbivores as vectors for the dispersal of cliff vegetation. Lüth (1993) notes that ants are important components of cliff ecosystems because they disperse the seeds of many cliff plants. Many of the woody species described in Chapter 4 have seeds or fruits that are well known to attract seed-eating birds and small mammals. Among them are species of *Juniperus, Taxus, Pinus, Cotoneaster, Ribes, Rubus, Prunus, Solanum, Sambucus, Cornus* and *Sorbus*. However, there is a conspicuous lack of published evidence showing that these interactions actually take place. An exception is Vogler's (1904) account of *Taxus baccata* in Switzerland. The restriction of this species to steep slopes and (usually) calcareous cliffs is attributed to two factors. First, yew trees in level-ground forests have been heavily harvested by people and browsed by wildlife, and the species persists on cliffs because both of these pressures are absent. Second, this species can germinate and grow on fully exposed bare rock as long as water is not limiting, but is dependent on herbivore dispersal since the seeds and the fleshy aril covering them do not provide for significant wind or water dispersal. Vogler states that the seeds are spread by birds that can travel large distances, and that this can explain the existence of individual cliff-face *Taxus* isolated from other populations by large distances. He also observes that this species occurs on ledges that are bird roosts, suggesting that the trees are first dispersed by the birds and then provided with nutrients from them for long periods of time. We suspect that Vogler is correct in these observations, and that all of the species listed above may be dependent on this mechanism to achieve and maintain their distribution on cliffs. Experimental work is necessary to test the validity of this idea.

Most available evidence indicates that both interspecific and intraspecific competition among plants is lower on cliffs than in surrounding habitats. This is because safe sites are widely spaced, root systems may be separated by rock layers, and a canopy cover is largely absent (Matthes-Sears & Larson, 1995). Most plant species that occur on cliffs cannot successfully compete with the vegetation of adjacent level-ground ecosystems (Davis, 1951; Snogerup, 1971; Karlsson, 1973). This lack of competitive ability appears to be further intensified by the extreme conditions, lack of space, and other hostile features of cliffs, leading Snogerup (1971) to propose that cliffs are 'evolutionary traps' (see below). This hypothesis is supported by data showing that cliff plants fail to thrive in more productive adjacent environments where competition from more aggressive plants is intense. For example, *Thuja occidentalis* and *Acer saccharum* have mutually exclusive distributions on the Niagara Escarpment, with *T. occidentalis* occurring on cliffs and cliff edges and *A. saccharum* occurring away from cliffs in level-ground forest (Bartlett & Larson, 1990; Bartlett *et al.*, 1991a, 1991b). *T. occidentalis* seedlings cannot become established in the leaf litter of *A. saccharum* and for this reason this species is absent from level-ground forest adjacent to cliffs (Bartlett *et al.*, 1991b; Fig. 6.27). *A. saccharum* seedlings are more sensitive to environmental stress and predators at cliff edges than *T. occidentalis*, resulting in 100 per cent mortality after three years (Bartlett *et al.*, 1991a). During establishment, *T. occidentalis* seedlings are more dependent on immediately fixed carbon, but their greater capacity to use dim light allows them to exploit the unfavourable cliff-edge habitat more efficiently than the more aggressive and faster-growing *A. saccharum* seedlings. The non-aggressive but physiologically opportunistic behaviour of these seedlings was interpreted as evidence in support of the slow-seedling hypothesis of Bond (1989). In the context of cliff ecology, the results are consistent with those of Goldsmith (1973a, 1973b), Okusanya (1979a), Malloch, Bamidele and Scott (1985), Coates and Kirkpatrick (1992), and Ware (1991), all of whom have shown that a lack of competitive ability in cliff plants limits their intrusion into adjacent level-ground habitats. A lack of tolerance to the stresses that occur on cliffs and cliff edges in turn excludes level-ground plants from cliff habitats. The physiological responses of adult *T. occidentalis* and *A. saccharum* in cliff habitats are more similar: both species display the same daily and seasonal pattern of net photosynthesis, stomatal conductance, predawn xylem water potential and leaf temperature (Fig. 6.28), and a physiological difference between them becomes evident only during peak drought conditions (Matthes-Sears & Larson, 1990). This shows that the habitat segregation between *T. occidentalis* and *A. saccharum* is established at the seedling stage. For *A. saccharum*, longer-term measures of tree mortality showed that it died out quickly from cliff edges and more slowly as distance from the cliff increased (Fig. 6.29). This result shows that the factors that exclude aggressive plants from cliffs may be different from the factors that exclude cliff-face species from the surrounding habitat.

The conclusions reached for plants about low predation and competition on cliffs are only partially applicable to animals. Cliffs do provide sea

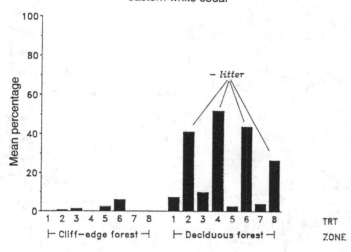

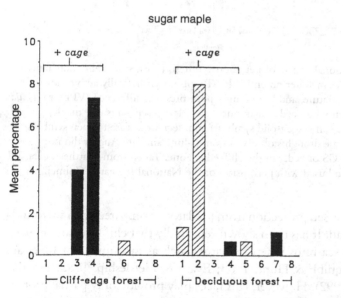

*Figure 6.27* Comparison of the emergence rates of eastern white cedar (*Thuja occidentalis*) and sugar maple (*Acer saccharum*) seeds under different experimental treatments (numbered 1–8), and in different cliff-edge or plateau zones on the Niagara Escarpment. In deciduous forest zones away from cliff edges, cedar seedlings only emerge in the absence of leaf litter. Conversely, for sugar maple, emergence was only high at cliff edges when seedlings were protected from predators. Taken from Bartlett *et al.* (1991b), and used with permission of Blackwell Science Ltd, Publishers.

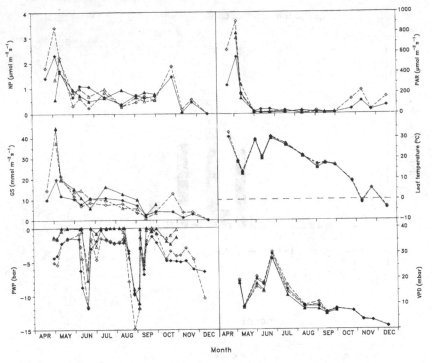

*Figure 6.28* Seasonal course of net photosynthesis (NP), stomatal conductance (GS), predawn xylem water potential (PWP), photosynthetically active radiation (PAR), leaf temperature, and leaf to air vapour pressure difference (VPD) in *Thuja occidentalis* (eastern white cedar, diamonds) and *Acer saccharum* (sugar maple, triangles) in both cliff-edge (solid symbols) and deciduous forest (open symbols) zones. The extreme droughts observed in mid-June and late August did not influence NP or GS of cedar in the cliff-edge zone. Taken from Matthes-Sears and Larson (1990), and used with permission of the National Research Council of Canada.

birds with increased protection from predators, compared with surrounding level ground. It has been shown repeatedly that cliff faces are increasingly used by sea birds when predators such as rats, minks or foxes are introduced (Squibb & Hunt, 1983; Evans & Nettleship, 1985; Freethy, 1987; Jones, 1992). However, cliff faces may provide only partial protection and some mammals such as foxes are adept at removing eggs from the cliff face itself by using crevices which run alongside breeding ledges (Birkhead & Nettleship, 1995). Predation from gulls and foxes may selectively remove thick-billed murres from cliffs which have a talus at their base. This is because fledgling failure is high where chicks have to cross talus to reach the sea (Gilchrist & Gaston, 1997a).

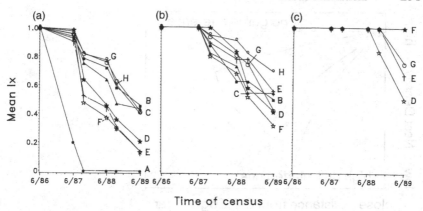

*Figure 6.29* Survivorship (lx) patterns for *Acer saccharum* (sugar maple) along cliff edges of the Niagara Escarpment. Panel (a) is for first-year seedlings, (b) for ages 6–9 years, and (c) for ages 10–13 years. Letter A is for plots 1–2 m from cliff edges; B, 3–5 m; C, 6–8m; D, 9–11 m; E, 12–14 m; F, 15–17 m; G, 18–20 m; H, 21–23 m. Taken from Bartlett *et al.* (1991a), and used with permission of the National Research Council of Canada.

While predation is lower on cliffs than in surrounding habitats, it is still an important force organizing bird communities within cliffs. Predation of thick-billed murres by ravens on East Digges Island, Northwest Territories, Canada, takes place in the top 30 per cent of the cliff face almost 70 per cent of the time because these areas are closest to the inland roosting, caching and foraging sites of ravens. This has led to the removal of low-density peripheral nesting sites near the cliff top and a shift to high-density nesting (Gaston & Elliott, 1996). High-density nesting provides greater protection for eggs and may lead to higher breeding performance (Coulson & Wooller, 1976; Gilchrist & Gaston, 1997a; Falk & Møller, 1997). The type of predator may also affect nest selection and spacing on cliffs. Spear and Anderson (1989) found different nesting patterns in yellow-footed gulls in Baja California, Mexico, depending on whether humans or other birds were the principal predators. In the absence of humans, gulls nest in dense colonies on beach berms where landward and seaward visibility is high, group defence is increased, and the rate of predation by ravens is low as a consequence. When humans are significant predators, however, the birds tend to nest in the talus in scattered non-colonial pairs because the difficulty in traversing talus reduces the foraging efficiency of egg-collecting humans (Spear & Anderson, 1989).

As with plants, animals may also select cliffs as habitats in order to

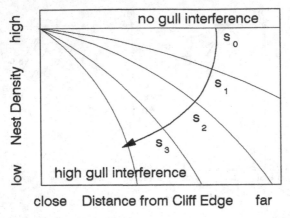

*Figure 6.30* The relationship between nest density of Atlantic puffins (*Fratercula arctica*) and distance from the edge of the cliffs in Newfoundland, Canada, when the puffins are exposed to varying amounts of interference from herring gulls (*Larus argentatus*). Modified from Nettleship (1972).

escape competition. In cliff-dwelling sea birds, competition for food resources is probably negligible and rarely constitutes a limiting factor (Furness & Barrett, 1985; Stempniewicz, 1995). Furness and Barrett (1985) found that food is 'superabundant' for sea bird island colonies in Norway comprised primarily of the black-legged kittiwake, herring gull, razorbill, shag, common murre, thick-billed murre, black guillemot, Atlantic puffin and great black-backed gull. The diets of most of the species differ in prey composition or fish size taken, suggesting that behavioural or anatomical differences rather than competition account for dietary choice. Similarly, Stempniewicz (1995) found that dovekies have no significant food competitors among birds. Fig. 6.30 illustrates the control of nest location and density in Atlantic puffins by the degree of competition from herring gulls in Newfoundland. As competition by gulls increases, the degree to which puffins depend on cliffs for nesting sites also increases (Nettleship, 1972; Birkhead & Nettleship, 1995). As with plants, some cliff-dwelling bird species may be non-competitive in level-ground habitats (Cullen, 1957; Guitián & Guitián, 1989).

While competition for resources such as food may be lower on cliffs than on level ground, the available nesting sites on a cliff face can be a valuable resource for which there is intense competition both among raptors and within sea bird communities. In Oregon, USA, red-tailed hawks (*Buteo jamaicensis*) migrate three to four weeks earlier in spring than Swainson's hawks (*Buteo swainsoni*) and establish nest sites earlier, but

Swainson's hawks then displace red-tailed hawks from their nesting territories. However, red-tailed hawks give up only marginal territory with fewer perches and maintain a core area of ideal habitat (Janes, 1994). In Alaska, ravens and gyrfalcons also compete for nesting sites on cliffs, sometimes alternating site use over several years. Gyrfalcons sometimes take over ravens' nests and even kill the adults (White & Cade, 1971). Late-migrating peregrine falcons try to dislodge gyrfalcons from their nests but are physically unable and are more successful against rough-legged hawks (White & Cade, 1971). Competition for nesting sites also occurs in sea birds. Birkhead and Nettleship (1987) found that common murres and thick-billed murres compete for cliff-ledge sites in eastern Canada, although the competition is not easily apparent because the two species nest in close proximity to each other. Breeding success is higher on wider ledges for both species; overall breeding success is consistently higher for common murres, which are more often successful in obtaining wider ledges and more aggressively defend their nesting sites. Their success in these confrontations with thick-billed murres has been attributed to their greater walking dexterity (Spring, 1971).

In sea bird colonies, increases in the population of one species can lead to a chain of reactions that can completely alter the composition of a colony (Buckley & Buckley, 1980). Belopol'skii (1961) observed that an increase in common murres forced thick-billed murres to peripheral nesting sites in a sea bird colony in the Barents Sea. The thick-billed murres subsequently pushed razorbills from nest sites under rocky ledges, which then forced the razorbills to lay eggs in Atlantic puffin burrows. The Atlantic puffins were forced to move to new, suboptimal sites with an increased risk of predation. In the Farallon Islands, California, Cassin's auklets nesting in crevices and hollows on talus slopes were observed evicting storm-petrels from their nesting cavities. The storm-petrels then in turn evicted pigeon guillemots and puffins from their nesting sites (Ainley et al., 1990b).

Raptors on inland cliffs may also rely on interspecific or intraspecific associations to establish nesting habitat on cliffs (Ellis & Groat, 1982). Gyrfalcons in Iceland (Woodin, 1980) and prairie falcons in the Mojave Desert (Boyce, 1987) nest primarily in abandoned ravens' nests and are dependent on ravens to select suitable cliff nest sites. Clearly, the habitat parameters required by ravens for breeding success on cliff faces are very similar to the parameters required by gyrfalcons and prairie falcons in these two areas. Griffon vultures (*Gyps fulvus*) are attracted to cliffs which are already colonized by other griffon vultures and are attracted to nesting sites painted to mimic vulture faeces (Sarrazin *et al.*, 1996). It is rare for griffon vultures to occupy previously unoccupied cliff faces if there are cliffs available with other roosting vultures, even if the suitable nesting habitat availability is similar between the two sites. Once a cliff is colonized it is never deserted. Sarrazin *et al.* (1996) believe that vultures are attracted to previously colonized cliffs because colonization is an indication of the quality of the food supply nearby.

## 6.12 Biological origins

There is remarkable agreement among the variety of people who have addressed the issue of the origins of the biological communities on cliffs. Almost every writer, regardless of country, language or taxonomic group studied, has concluded that cliff habitats are refuges from competition, predation, fire, human activities or climatic change in the surrounding landscape. Not that cliffs themselves have unchanging climates, but (as described in Chapters 2 and 4) cliffs have properties that buffer against climatic fluctuations generally. As a direct consequence of these features, cliffs have acted as reservoirs of relict biodiversity. Even where cliffs were exposed to significant glaciation, their biotas reflect their role as refuge from the environmental change that occurred during glacial retreat (Cooper, 1997) and from present-day harsh environmental conditions. To us this is an astounding fact, for most other ecologists fail to acknowledge the existence of cliffs as habitat at all, much less as habitat that supports interesting biotas. Ellenberg (1988) represents a good example of a work that emphasizes the stable and relict nature of cliff vegetation while simultaneously failing to recognize cliffs as separate habitat types. He writes:

> As long as the rock remains unchanged, its crevice dwellers form a stable community. Their roots soon form a mat running through the available fine soil, and scarcely any newcomers can find a foothold. It has already been stressed that steep rock faces cannot collect any humus except where there are small plant cushions. Under these circumstances, each further development endures for thousands of years. Rock crevice communities are thus permanent communities in the truest sense.

However, in the general account of the vegetation of central Europe from which this quote is taken, cliffs are not segregated out as an interesting or important habitat type.

Cliffs can take on the same ecological role as nunataks of glaciated regions (Davis, 1951; Bunce, 1968; Morisset, 1971; Brunton &

Lafontaine, 1974; Walker, 1987; Wiser, 1994; Young, 1996). The primary difference between cliffs in glaciated and non-glaciated parts of the world is that the former have had only 10 millennia to reach steady-state. The structural similarity of cliff communities in glaciated and non-glaciated regions reinforces the idea that cliffs return to their initial structure very rapidly after massive widespread disturbances. This idea is supported by data on the revegetation of abandoned limestone quarries along the Niagara Escarpment (Ursic et al., 1997). After only 65 years, most of the species that occur on vertical cliff faces in these quarries are also found on the naturally exposed cliffs of the Niagara Escarpment itself.

Similar conclusions about cliffs as refugia have been reached for other parts of the world as well, including regions that were never glaciated. For example, cliffs in the eastern Mediterranean are thought to have served as refugia during episodes of climate change. Present-day, widely spaced populations of plants such as Primula aucheri on these cliffs may be survivors of much more widespread populations that existed during wetter periods of the Holocene (Davis, 1951). Cliffs in Baden-Württemberg, Germany, as well are home to large numbers of glacial relict species which were widely distributed in the forest-free landscape of the immediate postglacial period, but were forced by the reinvading forest to retreat to naturally forest-free 'islands' in the landscape. Many of these species have a high light requirement, which has prevented them from expanding their range beyond cliffs until very recently, when secondary forest-free habitats have been increasingly created by humans (Lüth, 1993; Stärr et al., 1995). Cliffs as refugia for plants are also discussed by Wiser (1994) for the south-eastern USA and by Bunce (1968) and Polunin (1939) for the UK. Cliffs are refugia not only for plants, but also for animals. A classic example of a cliff relict animal is the pika, which is distributed on talus slopes throughout western North America (Udall, 1991). Some of the sites where pikas occur are separated by large areas of desert, yet the animals have a maximum dispersal distance of 3 km, even during cooler conditions. This has led to the conclusion that these disjunct populations are colonies of Pleistocene relicts (Hafner, 1993). Some individual talus pika populations possess unique dialects and fur colour (Udall, 1991). Fossil remains of these pikas have been used to reconstruct paleotemperatures in rocky habitats (Hafner, 1993).

While Davis (1951), Runemark (1969) and Türk (1994) have all discussed the relict nature of small and widely spaced cliff populations, Snogerup (1971) has presented the most thorough explanation of its causes. He begins his argument by observing that cliffs are conspicuous and common in most landscapes, making migration among them possible. He also notes that cliffs have a large endemic flora in most unglaciated parts of the world, and a large relict flora in places that were glaciated. He then argues that cliffs have a large degree of ecological constancy despite the vertical surface, the lack of soil, and extremes of some environmental conditions – extremes that vary over very small spatial scales. He then states that 'a comparatively unchanged type of vegetation has probably persisted in the cliffs through all the changes in the latter parts of vegetation history. This is further demonstrated by a comparison between present-day cliffs of different direction of exposure. In spite of the very different climate of southern and north exposed parts, the vegetation type is rather constant, though composed partly of different species.' The point is then made that species on cliffs often differ enormously from their closest relatives, and that narrow, strict and highly specialized conditions in the physical environment select for these differences in small, isolated populations. The more extreme the conditions, the more intense is the selection. It follows from this that those microsites on cliffs where soil accumulates and conditions for plant growth are locally favourable will probably be colonized by species that can normally grow well in situations involving intense competition. In other words, the 'good' microsites will select for taxa that can compete, while 'poor' microsites will select for species that cannot. Given that most cliffs consist mainly of 'poor' microsites, there is an open end to the evolutionary spectrum for organisms that can tolerate or avoid the stresses of such microsites. The existence of this 'open end' in the ecological gradient puts a strong selective advantage on any gene or gene combination that increases the ability of a taxon to occupy more extreme sites. As Snogerup explains:

> This makes cliffs function as an evolutionary trap because in any species, once it becomes specialized on the extreme side of the gradient, further specialization will be favoured by very strong selection pressure whereas any change in the opposite direction will mean that it has to face stronger competition. In the most extreme sites, which are yet empty, there will be no competition at all, but all other [growth] factors are at a minimum. Thus the evolutionary change will be guided in the direction towards and past the biolimit marked by the maximum adaptation of the present biota.

It follows that these maximum amounts of genetic adaptation to extremely small-scale habitat variability might occur very rapidly, and this has indeed been described by Abbott (1976) for *Senecio vulgaris* on a windswept north-facing sea cliff on Puffin Island, UK. It also follows that once the ability to survive and prosper on cliffs has evolved and the characters that yield high levels of competitive ability have been lost, these taxa may be incapable of reverting to forms that perform well in productive microsites in protected or sheltered areas. Such species are so tough physiologically and demographically that it is unlikely that even large environmental changes in the landscape will affect them. They will persist.

Evidence to support these ideas has not been easy to obtain. One example is present in the work of Walker (1987), Delcourt and Delcourt (1987) and Young (1996), who have argued that at the peak of the last glaciation, *Thuja occidentalis* and a broad array of other cliff species had a wide distribution in the south-eastern USA, but were restricted to marginal sites in rocky or wet habitat by competition from the more aggressive deciduous forest flora. As the glacial ice melted, this community, anchored by eastern white cedar, migrated north in hop-scotch fashion, following steep gorges and river valley cliffs. The populations were fragmented during the dry and warmer parts of the mid-Holocene period, and many died out completely. Small populations, some consisting of fewer than 100 adult trees, have persisted on north-facing cliffs and maintained high levels of genetic variability, perhaps attributable to the mechanisms described above. Genetic drift in these small populations imposed genetic bottlenecks (Young, 1996) that may have intensified the dependence of *Thuja occidentalis* on cliff environments during glacial maximum. Once the bottleneck was released in the post-glacial interval, the populations that succeeded to recolonize the large areas of eastern North America that now represent the main range of this species only represented a small subset of the genetic diversity that was present on these southern cliffs at glacial maximum. This interpretation is consistent with the high levels of genetic variability observed on these disjunct remnant populations of *Thuja occidentalis* (Walker, 1987; Young, 1996), in contrast with the lack of genetic variation that this species displays in the main part of its range (Matthes-Sears, Stewart & Larson, 1991).

Stöcker (1965), McVean and Ratcliffe (1962), and Allen (1971) all observe that cliffs are difficult to handle as large vegetation units because they are mosaics of many microsites which exist at scales much smaller than those used in vegetation surveys. The apparent contradiction between the view that cliffs are extremely heterogeneous and therefore cannot be handled by traditional community ecology, and the opposing view that cliffs have all the same vegetation even when some of the individual species components are different, is resolved by observing how difficult and locally intense has been the selection for plants and animals to live in these habitats. Once the microsites are seen as repeating components of the same kind of system, and as collectively defining the nature of cliff environments, then cliffs can be viewed as much more homogeneous ecologically, regardless of where the cliff occurs.

## 6.13  Biogeography

Traditional island biogeographic theory (MacArthur & Wilson, 1967) assumes that island biota is initially recruited from 'mainland' habitats that have been in some type of equilibrium for long periods of time. If the evidence of Delcourt and Delcourt (1987), Walker (1987) and Young (1996) is correct, then this assumption will only apply to a few cliff ecosystems that happen to be part of a very large complex of eroded topography. Most of the cliffs discussed so far do not fall into this category. Most cliffs have always existed as habitats fragmented to some extent. Whether the largest of these fragments can be sufficiently 'mainland' to function as a mainland source habitat cannot be determined without field studies of island biogeography on cliffs.

Only three studies have discussed the biota of cliffs in a biogeographical context. Ward and Anderson (1988) looked in part at the correlation between bird species richness, abundance and diversity and the length and height of 16 cliffs in south-central Wyoming, USA. They found a negative correlation between species richness or abundance and the length of exposed cliff face. Species richness and diversity were found to be positively correlated with cliff height. While these authors did not address the issue of distance between cliffs, their results for bird species do indicate that there is some relationship between the size of cliff faces and bird species abundance, richness and diversity.

The role of cliffs in determining the species richness of communities on the top of buttes in south-eastern Utah was also investigated by Johnson (1986). Five buttes with identical pools of organisms at their bases were chosen for study. Only half of the 47 mammal, ant and reptile species found at the base of the cliffs were found in the plateau at the cliff top. Cliff height was a significant factor in determining the number of ant and mammal species at the butte tops, while cliff microrelief was an important factor only for mammals. Buttes formed by taller, smoother cliffs support fewer species than shorter, broken-down cliffs. No species–area relationship was found for any of the taxa. Johnson theorizes that cliffs act as 'filter-bridges' that allow only some taxa to attain the butte summit. The physical characteristics of the cliff such as height and roughness are analogous to the water surrounding islands.

Haig (1997) has recently examined the extent to which the fragmentation of cliffs along the Niagara Escarpment influences community structure. She sampled cliffs that varied from 28 m in length to over 2000 m and measured species richness, relative abundance and some other components of community structure in the lichens, mosses and higher plants. The results showed that cliff size had no impact on any aspect of the structure of the community and she attributed these results to the idea that the cliff community is composed of microsites that do not increase in diversity as the size of the cliff increases. This means that a small cliff and a large cliff can support the same number and diversity of species, and these results support the ideas of Snogerup (1971) and Runemark (1969).

## 6.14  Succession

Cliffs represent highly stressful environments characterized mainly by biologically closed communities where the number of suitable microhabitats is limited, and where the organisms fit into the cliff habitats like 'corks in a bottle' (Davis, 1951). The number of possible colonizers is small and the successional trajectory predicted by Grime (1979) involved very few differences between initial and final states. This prediction is consistent with the evidence that rock ledges and inland cliffs have few alien exotic invaders (Crawley, 1986, 1987; Larson et al., 1989) and only a tiny portion of the total flora of cliffs is composed of annual plants. The flora and fauna of cliffs are generally predictable because the low base of productivity selects for that small fraction of the total biota that cannot handle lush conditions. These observations are consistent with those of Drake (1991), who showed that the species composition of artificial assemblages of algae and their predators became more predictable as their size (and hence the maximum total production) decreased. The results are also consistent with geographical-scale observations of Currie (1991, unpublished data; Figs. 6.31 and 6.32) that tree-species richness and the variance in tree-species richness both decline markedly

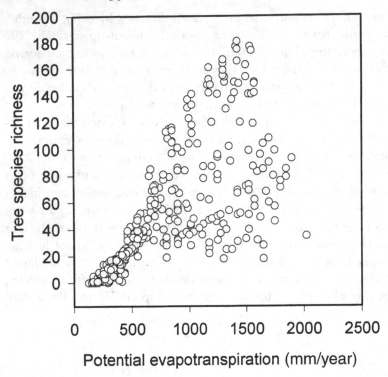

*Figure 6.31* The relationship between tree-species richness and potential evapotranspiration at the geographical scale covering many habitats between 25°N and 75°N. The results show that as available energy declines, so does the number of species that can exploit particular habitats. Taken from Currie (1991), and used with permission of the University of Chicago Press.

with declines in physical variables such as potential evapotranspiration. Only a small fraction of the available biota can exploit the exacting conditions on a newly exposed cliff. The length of time until newly created cliffs (such as quarries) are reinvaded is longer than in other habitats (such as old fields), but the final state is reached sooner: the flora of abandoned quarry walls may resemble that of natural cliffs after 70 years (Ursic *et al.*, 1996). Like similar 'marginal' plant communities in the Arctic, the cliff biota does not seem to progress beyond the initial phases of succession (Svoboda & Henry, 1987). On cliffs, succession is only a major influence on very young cliffs or cliff surfaces which have recently been exposed to disturbance. Kelly and Larson (1997a) recently showed that *Thuja occidentalis* forests of the Niagara Escarpment have approximately the same

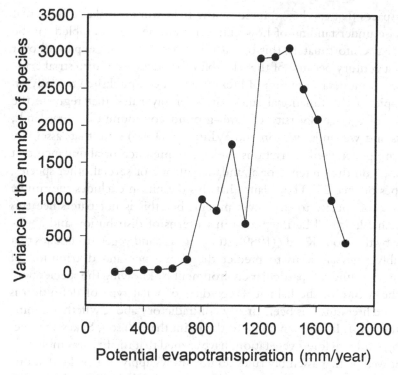

*Figure 6.32* The relationship between the variance in the number of species in various habitats between 25°N and 75°N and potential evapotranspiration. The diagram shows that at very low and very high levels of available energy, the variance declines. Original unpublished data from D. Currie.

age distribution of trees today as existed 220 years ago, and concluded that the cliffs of the Niagara Escarpment support a forest almost in steady-state.

## 6.15 Assembly of cliff communities

Unlike the situation that applies to level-ground vegetation where many of the species–species or species–environment contacts are hidden underground from view or are immersed within an enormously complex web of interactions involving scores of taxa, the factors that are important to the assembly of cliff communities ought to be more easily viewed and experimented with (Oettli, 1904; Booth & Larson, 1999a, 1999b). Low productivity, a sparse vegetation, and a tendency of species to root into small spaces that can be explored by the field worker make it likely that a precise understanding of how cliff communities are assembled can be obtained. Unfortunately, this opportunity has been little exploited over the past century because of the 'cliff-blind' tendencies of terrestrial ecologists. A moderate amount of literature has been published recently on the topic of the functional rules, or 'assembly rules' that regulate the genesis of a large diversity of level-ground communities – grasslands, forests and wetlands. Wilson and Whittaker (1995) state that 'assembly rules are generalised restrictions on species presence or abundance that are based on the presence or abundance of one or several other species, or types of species'. They claim that this definition excludes the simple response of species to the environment, but this is not true since this functional link will be integrated in patterns of distribution and abundance by the biota. Keddy (1989), studying wetland vegetation, states that assembly rules allow us to predict the abundance and distribution of organisms found in a particular environment by knowing the species pool and the nature of the habitat. Regardless of what type of definition is used, the literature has been largely contradictory about whether assembly rules exist. It has even come to the point that Wilson, Sykes and Peet (1995), studying forest vegetation, have argued that the first assembly rule might well be that assembly rules do not always apply! It may be that one of the reasons assembly rules have been so difficult to detect is that the productivity of most habitats is too great to avoid the historical effects described for experimental microcosms studied by Drake (1991).

The fact that productivity is constrained on cliffs has produced a much more predictable biota than has been found in other habitats. A review of British Isles flora shows that sea cliffs, embankments and walls have only 18 per cent of the total flora represented by aliens, while mountain summits and rock ledges on inland cliffs have no alien exotics at all (Crawley, 1986, 1987). This number is small compared with other habitats surveyed (waste ground 78 per cent; coniferous plantations 56 per cent; hedgerows 22 per cent) (Fig. 6.33). Crawley (1986, 1987) concludes that sites in a late stage of successional development, with little disturbance, low soil nutrient levels, high toxin levels, small seed banks, and an absence of predators and competitors, are difficult for aliens to invade. Similarly, Drake (1991) argued that the invasibility of experimental ecosystems is very limited when the size (and hence the maximum standing crop) of the system is small.

Therefore, in the context of the invasion and survival of propagules, the productivity of individuals and the factors that regulate longevity, the

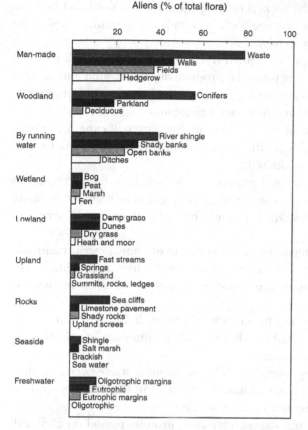

*Figure 6.33* Graph illustrating the percentage of the total flora made up by alien species in the UK for different habitat types. Note that upland rock summits and ledges, as well as upland scree slopes and shady rocks, have either zero values or very low values. Redrawn from Crawley (1987).

following rules appear to apply to the organisms that use cliffs as their principal habitat in their life cycle.

1. All species of a regional biota can be included in the propagule rain. Cliffs will be exposed to the same propagule rain as surrounding terrain. Constraints on species composition of the community will probably apply at later stages.
2. Not all species of the regional biota will be incorporated into the propagule bank. Propagules that require moderate winter temperatures or freedom from exposure to short-term heat will be excluded. Propagule size will be sorted by the physical dimensions of the cracks

and crevices on the rock. The propagule bank is not small because the heterogeneities on the rock accumulate propagules despite the loss of material by gravity.

3. The number of species that survive the juvenile phase will be small in areas of the lowest potential productivity (most solid and vertical microsites) and greatest in the areas with highest potential productivity (most fractured and horizontal microsites).

4. Species that occur in areas of low productivity will experience little intraspecific or interspecific competition. The opposite will be true for areas of high productivity.

5. Species with the longest amount of evolutionary time exposed to low-productivity microsites will be non-viable in high-productivity sites. These species will frequently be endemic or largely restricted to cliffs.

6. Species that are dependent on a high rate of annual recruitment will be selected against on cliffs. Species with low requirements for recruitment, or with the capacity for pulsed recruitment, will be selected for.

7. Populations of organisms on cliffs will be small, but will display high levels of genetic variability because of the intensity of selection by the abiotic environment.

8. Populations of organisms on cliffs will be more stable over time than populations in the surrounding terrain, but this stability will be directly tied to the rate of mass wasting.

9. Individual organisms that survive their juvenile period on cliffs will be slow growing, but long lived. Individuals living the longest will be in stable, low-productivity microsites.

10. Individual organisms that survive to adulthood on cliffs will not experience nutrient limitations, but will be extremely small for their age. Resource addition will not influence their performance.

11. Individuals at the juvenile and adult stages will not experience lethal amounts of predation or competition.

Although these general assembly rules may apply to the organization of the biotia on cliffs, it is usually rare that the species forming a community have co-evolved entirely within the confines of the observed habitat (Crawley, 1987). Many species use cliffs for only short periods of time in their life cycle, and for these species, only a subset of the above rules will probably apply. For example, many of the sea birds and raptors discussed above have limited competitive ability with other birds on level ground, and rely upon cliffs to provide protection from (or at least reduce the pressures of) competition and predation.

## 6.16 Summary

The factors that control the flora and fauna of cliffs appear to be surprisingly consistent from one geographical region to another. Cliffs support micro-organisms, algae, lichens, mosses and higher plants that are tolerant of certain environmental extremes such as lack of soil cover, lack of space for roots, presence of wind, absence of direct precipitation, and constant battles with the force of gravity. At the same time, many species appear to grow on cliffs because of relatively stable and predictable microclimates found there: radiation loads are muted compared with level ground, liquid water is expressed directly from cliff faces or concentrated at bedding planes, and temperatures are modest. Virtually all of the autotrophs on the cliffs lack competitive ability and, similarly, respond poorly to intense predation. Cliffs appear to be refuges from both of these normal ecological processes, at least at the macroscopic scales. No information is available yet to say whether endolithic detritus food chains are present. Macroscopic animals are probably very abundant on cliffs, but there is some evidence that the vertical substrate is a barrier to their movement when the cliffs are tall. Cliffs attract birds of prey and sea birds, but these birds almost always use the cliff as a nest site or food cache rather than as a food source. In all continents, birds occur on cliffs and depend on a lack of predation and competition for survival. It is very probable that certain parts of the flora and fauna of cliffs are functionally connected on the basis of seed dispersal. Experimental work needs to be done to demonstrate these links.

Cliff communities are assembled on the basis of several restrictions that apply to the biota as a whole. The assembly rules that apply to cliffs are extraordinarily different from the rules that apply to other types of communities that have been studied, again suggesting that cliffs are unique places in the terrestrial landscapes of the world.

# 7 · Interactions with humans

The attraction of humans to cliffs is not new. Cliffs have often formed the subject matter for visual arts and sculpture, particularly in China, Japan and Korea, where cliffs have always evoked strong emotional feelings (Fig. 7.1). The human population has always been fascinated with mountains, rock outcrops and cliffs. In many ways, this interest is a practical one, for cliffs and caves at their base have always offered protection and shelter from the elements and from rival populations of humans competing for similar resources. Hominid fossils from *Homo erectus* to *Homo sapiens neandertalensis* have been collected from caves at the base of cliff sites in Africa, Asia and Europe (Hoebel, 1966; Jurmain *et al.*, 1990) including Mt Carmel, Israel (Fig. 7.2). It is impossible to claim that cliffs and the caves within them were actively sought more than other habitats by palaeohumans because cave environments preserve artifacts that would rapidly degrade if exposed directly to weathering. Despite this, cave dwellings in Greece and other locations around the Mediterranean and into central Asia span the time period within which agriculture was started (Bogucki, 1996), and at the very least caves and the cliffs above them were sought out by humans as sites for occasional dwellings, tool manufacturing, and animal carcass reprocessing (Schepartz, personal communication). These sites include abundant remains of early tools and, later, evidence of fire. Similar sites in southern France (Ruspoli, 1987) are well known for abundant cave paintings made prior to the last glacial advance. The high degree of use of cliffs by palaeohumans gives weight to the 'urban cliff hypothesis' presented in the last two chapters, that the flora and fauna of cliffs may have partly joined with *Homo* species during the period of time in which our species was becoming less dependent on natural rockshelters and more capable of building our own of wood and/or stone. The 'urban cliff hypothesis' actually suggests that a large proportion of the human species still lives in concrete and glass versions

*Figure 7.1* Postage stamp from China showing a painting of a rock outcrop. Such images are very common in the history of oriental art.

of their ancestral cliffs, caves and talus slopes – even to the point of creating environments in which cliff species (both plants and animals, many of which we claim to despise) that were once rare and highly endemic now live as commensals with us. The idea that many native and alien field and crop plants originated in permanently open sites such as cliffs and river banks is not new. Marks (1983) has argued that North American field plants largely originated in the few permanently open sites (such as cliffs) that existed in the landscape at the time of European colonization, and that similar processes could have applied in Europe and Asia. Baker (1986) has noted that among the many species of plants currently used in temperate-zone agriculture, some such as *Brassica oleracea* have returned to their ancestral cliff habitats (Clapham *et al.*, 1952), even in unusual locations such as California.

Caves at the base of cliffs continued to be used by small populations of people in the Middle East throughout the early Holocene (Davis, 1951; Braidwood, 1967) and small cave dwellings continued to be used through to historical times in Malaysia and Tasmania (Henderson, 1939; Coates & Kirkpatrick, 1992). The Anasazi peoples of the American south-west built large villages into the base of cliffs, including Cliff Palace at Mesa Verde in south-western Colorado (Roberts, 1996), and Bandelier

*Figure 7.2* Mount Carmel in Israel is a limestone outcrop that supports a significant biotic community and was also a dwelling site for Neandertals for several hundred thousand years. Similar cave sites have been recorded throughout Asia, Europe, Africa and North America. Photo by S. Pfeiffer.

National Monument in New Mexico, USA (Fig. 7.3). Similar structures and dwellings carved into rock cliffs or built into natural cavities in cliffs have been recorded in many countries of the world (Fig. 7.4).

The development of agriculture in Europe, Asia and the Americas led to an increase in human population size and a corresponding reduction in the numbers of people leading a hunting-gathering existence (Bogucki, 1996). Accordingly, the protection provided by cliffs and caves was less needed and the number of people still dependent on cliffs for their survival or for the development of culture was significantly reduced. From approximately the tenth millennium BC to the beginning of the second millennium AD, the expansion of 'civilization' was fuelled by the conversion of productive forest, meadow, steppe and wetland into useful agricultural products or weapons (Kimmins, 1987). Throughout this period, monasteries and castles were built on the plateaus of mesas or rock outcrops, but the cliffs themselves were not exploited.

In North America the contact between Native Americans and Europeans in the late fifteenth century led to a rapid replacement of more-or-less sustainable ecological principles by Judeo-Christian values centred on consuming ecological resources. These values have

*Figure 7.3* Cliff dwellings at Bandelier National Monument, New Mexico, USA. Photo by U. Matthes.

maintained their momentum to the present and have resulted in the almost complete elimination of the original forest, prairie and wetland cover on the continent. Sites with low commercial value to Europeans tend to be the sites that now support presettlement ecosystems, and cliffs are among the most conspicuous of these (Larson & Kelly, 1991; Stahle & Chaney, 1994). Ironically, some of these sites with low commercial value have special significance to native populations. Along the Niagara Escarpment, Iroquois, Neutral, Huron, Petun and Odawa peoples used the limestone cliffs and talus slopes as human burial sites (Fox, 1990, 1991; Spence & Fox, 1992). Some high-value materials such as chert were also obtained from these cliffs, adding to the importance of the escarpment to the First Nations.

Historically, throughout eastern Asia, cliff faces also had strong religious significance to a number of cultures. The most obvious visible expression of these values is the 'hanging coffins' on cliff faces in China, Indonesia and the Philippines. Wooden coffins as old as 3600 years are suspended on wooden supports and on ledges hundreds of feet up cliff faces in the 'Three Gorges' area of China where the Yangtze, Shinning and Danning Rivers meet. It is thought that this practice brought the dead closer to heaven. On Sulawesi in Indonesia, a similar practice involved the placement of coffins on cliff faces and deep within caves on

*Figure 7.4* Egyptian stamp illustrating the construction of tombs and other ceremonial sites in cliffs along the Nile River.

sheer limestone cliffs (Volkman & Caldwell, 1990). Location on the cliff face was indicative of the wealth of the deceased person, with the wealthiest aristocrats being placed near the top. In other areas, graves have been hewn out of the cliff face, with wooden doors placed to seal the remains. In Luzon Province, Philippines, various tribal groups bury their dead above ground, and hanging coffins are found in caves on cliffs or on the cliff faces themselves. Cliff burials were thought to allow the spirits of the dead to roam free.

Many cliffs around the world are also sites where ancient cultures expressed themselves in the form of paintings and petroglyphs. Rock art is common to many cultures and was practised by aboriginal inhabitants in areas as geographically separated as South Africa (Lewis–Williams, 1981; Jolly, 1996; Solomon, 1996), Australia (Layton, 1992), India (Neumayer, 1983), Korea (Sasse, 1996), Polynesia (Ewins, 1995), and Europe (Ucko & Rosenfeld, 1967; Bower, 1996). Like human habitation sites on cliffs, it is unknown if there are many rock art sites on cliffs because the work is more easily preserved here or whether cliffs and the caves within them were actively sought out for artistic expression. If they were sought out, it is also difficult to determine whether cliffs and cave walls provided an excellent natural drawing board or whether cliffs held religious significance to the cultures involved. Some archaeologists believe rock art in most cultures depicted the trance-induced supernatural journeys of shamans (Bower, 1996; Solomon, 1996) and that it is possible that cliffs themselves were seen as gateways to other supernatural worlds. The frequent use of cliffs as burial sites does seem to support the statement that cliffs have been important spiritual guideposts in some cultures. The great difficulty in creating rock art deep in caves (Ruspoli, 1987) or high up cliff walls (Ewins, 1995) indicates that some of these images could only have been placed with deep and powerful intention and were not simply acts of luxurious enjoyment (Ucko & Rosenfeld, 1967).

One of the most important rock carvings in the history of Western science was constructed by Darius the Great and completed about 516 BC. On a 500 m tall limestone and sandstone cliff in central Persia (Iran) was carved a 30 by 40 m inscription of text and mathematical expressions translated into three languages, including cuneiform writing (Burton, 1997). It is believed that this site was selected by Darius the Great to ensure that it survived attacks by both the weather and human enemies. Creswicke Rawlinson deciphered the text by the middle of the nineteenth century and his translations played a central role in expanding modern concepts of mathematical and linguistic development in the Middle East.

Rock art, relics, middens and human remains are all physical signs of human exploitation of cliffs and associated caves for shelter, burial and artist expression, but cliffs have also provided a food supply source in some cultures. Until the 1900s, native Cup'ig on Nunivak Island in Alaska used walrus-skin ropes to descend vertical cliffs up to 150 m in height to gather nesting sea birds and their eggs for consumption (Hoffman, 1990; Pratt, 1990). The skins were also incorporated into clothing. Murres, horned puffins and tufted puffins were the main bird species harvested from these cliffs. The importance of this activity to the

Cup'ig Eskimo culture is suggested by the fact that three months of the Nunivak calendar are named for these cliff-dwelling birds (Pratt, 1990). Wild lettuce and other herbaceous plants were also collected from the cliff faces for consumption (Pratt, 1990). Haley (1984) describes the use by Inuit and Aleut hunters of long-handled, bowl-shaped nets crafted from braided sinew or snares of baleen strips to harvest cliff-nesting auklets, guillemots, murrelets, puffins, cormorants, gulls and particularly murres and kittiwakes directly from rookeries. These hunters were lowered down the cliff from the cliff edge.

Historically, in Europe, cliff faces in Scotland, the Faeroe Islands, Iceland and Greenland were farmed for sea birds (Evans & Nettleship, 1985; Freethy, 1987). Birds such as northern fulmars were captured by daring climbers known as 'fowlers' who gathered fulmars for food to avoid the effects of scurvy (Fisher, 1952). Murres and puffins have been found in archaeological sites across the northern hemisphere (Evans & Nettleship, 1985). Cliffs on islands such as St Kilda, Scotland, were traditionally used to harvest sea birds such as the now-extinct great auk, thick-billed murre and the common puffin until 1931. Dogs were used to root-out puffins from talus burrows, while the murres were captured by hunters with 5-m long fowling rods with a running noose at the end (Freethy, 1987). Some people on St Kilda became very adept climbers and apparently when the islanders were discovered by Victorian tourists, it was surmised that they possessed an extra prehensile toe for climbing. Hunters either ventured along narrow ledges or were lowered from the top at night with a white sheet over their heads. In the early morning light, murres apparently confused the sheet with a guano-stained rock until the cliff became too bright (Freethy, 1987). Perhaps the most dangerous occupation, however, may not have been that of the climber but of the gatherers in the boats below who were thrown 9–10-lb murres' carcasses from a height of 60–90 m! This practice of bird harvesting ended when alternative sources of oil, feathers and food were developed.

Gannets in particular were valued for their oil, which eased gout and muscular pain and was also used as cart grease (Nelson, 1978a). The birds had long been considered a valuable source of protein and the eggs were considered a delicacy. The feathers were also used to stuff feather beds (300 gannets for one bed!). In Scotland, Nelson (1978a) estimates that 1000 to 2000 gannets were killed annually on Bass Rock between 1511 and 1865, and on Ailsa Craig, cliff nets may have removed up to 1500 gannets a week. Gannets were captured with spring-loaded jaws on the end of bamboo poles. The gannets were then clubbed, passed up to the cliff edge, plucked, singed over a peat-fire, salted, stacked and then loaded onto boats (Nelson, 1978a).

On many western Pacific islands, millions of short-tailed albatross (*Diomedea albatrus*) that lived on cliffs were harvested for their feathers. By 1932, the only remaining breeding colony was on the Japanese island of Torishima (Haley, 1984). At that point the Japanese government declared the site a bird sanctuary, but before the edict was issued the local population attempted to kill all the birds. Ironically, the local village, its anchorage and population were destroyed by two volcanic eruptions during World War II. The bird was thought to be extinct, but ten pairs were observed on the island in 1954, which by 1982 had increased to 67 pairs with eggs (Haley, 1984). Common murres were similarly exploited in the Farallon Islands, California, USA, starting in the mid-nineteenth century. Almost 400 000 murre eggs were removed in 1854 using ropes, pulleys and baskets. This practice continued until 1959, when only 7000 eggs were removed (Haley, 1984; Ainley, 1990). Evans and Nettleship (1985) estimate that human exploitation has reduced populations of razorbills, common murres, thick-billed murres and Atlantic puffins by an order of magnitude since the nineteenth century.

Egg collecting was also a significant threat to bird populations, particularly to species such as the peregrine falcon whose nests are difficult to access and whose eggs are both beautiful and variable. This practice was popular between 1870 and 1920 amongst Victorian gentlemen and, at the time, to be an ornithologist was synonymous with being an egg collector (Ratcliffe, 1980). Eggs were kept for display in permanent collections.

## 7.1 Current use and exploitation of cliffs

### 7.1.1 Hiking

Back-packing and hiking are popular recreational activities around the world and they represent the primary methods by which cliffs and cliff edges are visited. Unfortunately, excessive numbers of hikers can negatively impact natural cliff areas where edges are used for lookouts (Stärr *et al.*, 1995) and caves and cliff bases under overhangs are used for shelter and fires (Herter, 1996). The relationship between trampling intensity and vegetation response has been studied in the USA for a variety of habitat types (Cole, 1995a, 1995b). One of the habitat types examined was subalpine heath dominated by *Phyllodoce empetriformis*. This species showed a linear decline in percentage cover as the intensity of trampling

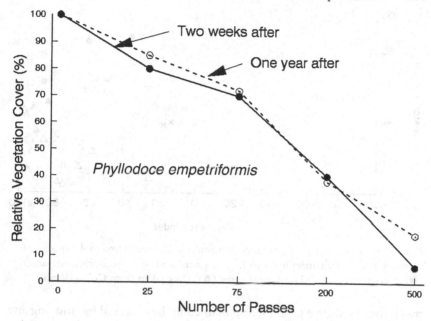

*Figure 7.5* Decline in percentage cover of *Phyllodoce empetriformis* as a function of trampling intensity in a subalpine heath community in Washington State, USA. The solid line represents the cover immediately after trampling; the dotted line represents cover after one year of recovery. Data from Cole (1995a).

increased (Fig. 7.5). It was almost completely eliminated after 500 passes in one year and did not recover the following year. For the 18 species examined, only one other slowly growing heath species, *Vaccinium scoparium*, showed a similar lack of recovery. All others showed at least some resiliency to trampling after one recovery year. The results also showed that chamaephytes were the plants least tolerant of trampling, and hemicryptophytes and geophytes were the most tolerant. These results suggested that cliff edges and faces could be particularly sensitive to hiking and climbing activity because they are dominated by chamaephytes of the Niagara Escarpment (Davis, 1951; Larson *et al.*, 1989; Fig. 7.6).

Studies of trampling on the herbaceous vegetation at cliff edges showed that the impact was scale dependent. Some species responded as described by Cole (1995a), but others actually increased in abundance after 500 passes (Taylor, Reader & Larson, 1993). It was concluded that indicator species could not be selected to monitor the effects of trampling on cliff-edge vegetation. An analysis of the entire vegetation community using multivariate methods showed that the community was

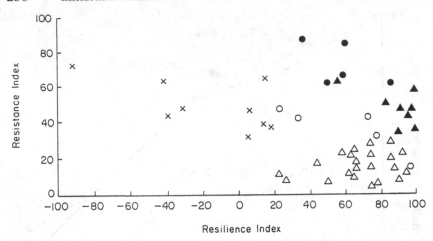

*Figure 7.6* Covariation of resistance and resilience to trampling in 49 species of plants classified as chamaephytes (×), erect graminoids (○), non-erect graminoids (●), erect forbs (△), and non-erect forbs (▲). Data taken from Cole (1995b).

much more sensitive to trampling than could be detected by studying any particular species (Taylor *et al.*, 1993; Fig. 7.7).

Univariate statistical tests and multivariate analyses were also used to explore the relationship between human hiking disturbance and vegetation structure along cliff edges of the Niagara Escarpment in Bruce Peninsula National Park, Ontario (Parikesit, Larson & Matthes-Sears, 1995). Earlier work on undisturbed cliffs had shown that the gradient from level-ground woodland to open cliff faces was the strongest detectable environmental gradient present (Larson *et al.*, 1989). In contrast to this result, Parikesit *et al.* (1995) found that the largest differences in vegetation structure along cliff edges in the National Park were associated with the intensities of hiking (Fig. 7.8). Sites with abandoned trails or no trails had a variable species composition, but sites exposed to heavy use had much less variability – almost as if the community was truncated in its patterns of distribution and abundance of species. No indicator species were found, however, that could be consistently and easily used to assess the impact of hiking disturbance. A more useful monitor of the impacts of hiking proved to be measure of soil depth, soil bulk density, litter cover, soil pH and soil nitrogen (Fig. 7.9). All of these variables were significantly changed by exposure to hiking. Soil depth and litter cover were reduced by hiking, while the other variables showed an increase in response to hiking.

In the upper Danube Valley in Germany, Herter (1993) also found that

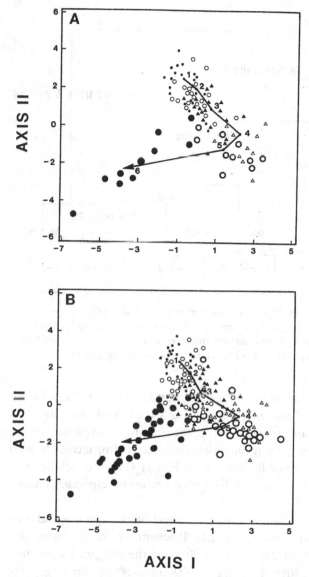

*Figure* 7.7   Multiple discriminant analysis of a cliff-edge understory community along the Niagara Escarpment exposed to no trampling (small dots), 50 passes in one season (small open circles ), 500 passes in one season (closed triangles ), 500 passes per year over 18 years (open triangles), 5000 passes per year over 18 years (large open circles ), and 25 000 passes per year for 18 years (large closed circles) on paths (A) and in the entire cliff-edge community (B). Data taken from Taylor *et al.* (1993).

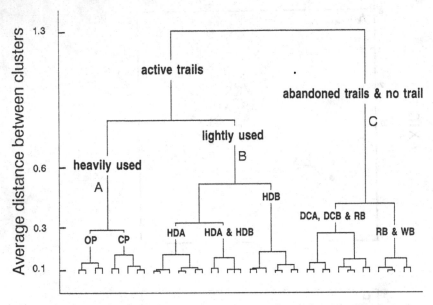

*Figure 7.8* Cluster analysis of cliff-edge understory forest along the Niagara Escarpment exposed to chronic trampling. Labels indicate the various levels of trampling intensity. The diagram illustrates that sites with similar disturbance levels have similar species composition. Data taken from Parikesit *et al.* (1995).

hiking at cliff edges affected the species composition of the adjacent grassland vegetation. Not surprisingly, areas with lots of trampling had more damage-resistant species. Areas of intense trampling, which included most lookouts, were denuded completely. Impacts were even observed away from major hiking routes. Herter (1993) concludes that hiking has made naturally rare cliff-edge communities especially endangered in this area.

Hiking has also been shown to have an effect on the age, size and productivity structure of cliff-edge forests. Recruitment of seedlings into cliff-edge populations of *Thuja occidentalis* along the Niagara Escarpment was stopped by intensities of hiking of 25 000 passes per year (Fig. 7.10; Larson, 1990). Annual productivity was higher in places experiencing high levels of disturbance (Fig. 7.11). Trees that persisted in disturbed habitats were larger and grew faster. This growth release was attributed not only to the reduction in intraspecific competition from saplings, but also to general thinning of the mid-canopy by intense ground disturbance at the cliff edge.

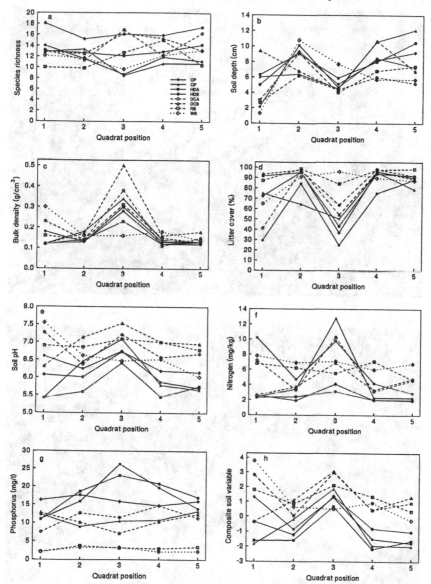

*Figure 7.9* Values of eight soil variables along cliff edges of the Niagara Escarpment exposed to trampling. Quadrat positions represent a gradient from the cliff edge (position 1) to the plateau forest (position 5); quadrat position 3 represents the centre of the trail. Different symbols and lines indicate different sites, which are arranged in order of decreasing trampling intensity in the legend in panel a. Data taken from Parikesit *et al.* (1995).

(a)

(b)

*Figure 7.10* Photographs comparing an (a) undisturbed and (b) trampled cliff-edge understory along the Niagara Escarpment. Photos by J. Cox and D.W. Larson, and used with permission of Springer-Verlag, from Taylor *et al.* (1993).

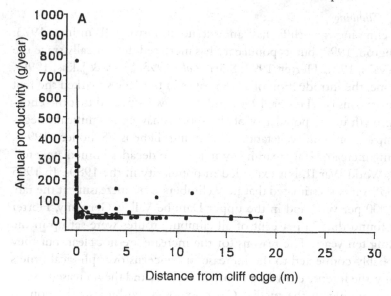

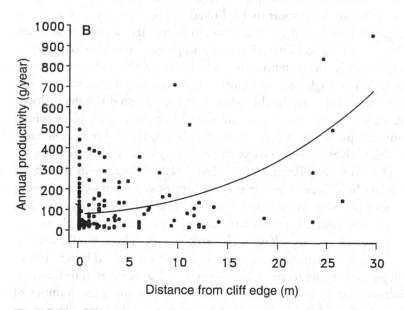

*Figure 7.11* Comparison of the annual productivity of cliff-edge *Thuja occidentalis* on the Niagara Escarpment in the absence (A) and presence (B) of trampling disturbance. Data taken from Larson (1990).

## 7.1.2 Climbing

Rock climbing on cliffs has an extended history (Bunting, 1973; Bonnington, 1992) but its popularity has increased dramatically since the 1950s (Valis, 1991; Herter, 1993; Stärr et al., 1995; Shaw & Jakus, 1996). In Europe, the introduction of nylon ropes in the 1950s marked the first phase of expansion (Herter, 1996) and it is now suspected that the enormous growth in the popularity of the sport is having a significant negative impact on cliff vegetation, including lichens (Schöller, 1994). Climbing increased only marginally in the two decades immediately following World War II, but exploded in popularity in the 1980s. In 1995 in the USA it was estimated that new climbers were increasing at the rate of 100 000 per year, and in the upper Danube Valley, Germany, Herter (1993) found that 75 per cent of all climbing routes were set up in the preceding ten years. The reasons for the increase are not clear, but they are probably connected to the increase in concerns over physical fitness generally, the intense crowding of urban centres, and the increased exposure of climbing in the media. Once exposed, enthusiasts develop an emotional bond to the sport that is linked to the excitement of risk, the challenge of discovery, the competition to be the first, a sense of wonder, physical well-being and a need for sensuous pleasure (Bonnington, 1992).

Unfortunately, the potential for disturbance to the cliff environment from climbing is high because climbers not only access the face, but also the cliff edge and base. In the talus, paths are created on steep slopes which are especially prone to erosion, and trampling takes place where equipment is put down (Stärr et al., 1995; Herter, 1996). On the cliff face, handholds such as crevices, cracks and even tiny holes and depressions are cleared by climbers (Herter, 1996). Herter (1996) documents the efforts of climbers he observed setting up new routes and purposely removing the vegetation. Vegetation that has not been cleared is then exposed to rope abrasion, which is particularly damaging to shrubs and plants that have little resistance to sideways shearing forces. These would include cushion plants which are weakly anchored in crevices (Herter, 1996). The degree of impact to an entire cliff face will depend on four interacting factors: cliff accessibility, the properties of the rock, the number of climbing lines, and the frequency and duration of climbing (Stärr et al., 1995).

Despite the increase in popularity, however, disproportionately few studies have addressed the potential impact of rock climbing on cliff flora or fauna. Apart from Herter (1993, 1996) and Stärr et al. (1995), Nuzzo (1995) studied the impact of rock climbing on the rare goldenrod Solidago *sciaphila* growing on vertical cliffs in Mississippi Palisades State Park in Illinois, USA. It was found that rock climbing strongly skewed the size and age distribution of the population (Table 7.1) in a similar way to that described by Larson (1990) for *Thuja occidentalis* along cliff edges. Total plant density was also reduced by climbing. Damage to *Solidago sciaphila* also occurred during the initial stages of climbing, while continued use prevented re-establishment (Nuzzo, 1995).

Further study in Mississippi Palisades State Park examined the response of plant communities on cliffs to the same intensities of climbing reported earlier (Nuzzo, 1996). The community study showed that only lichens declined significantly when exposed to rock climbing. Several problems apparent with this study, however, limit its usefulness. Unlike the work of Nuzzo (1995), in which cliff goldenrod density was sampled in unclimbed areas and along existing climbing lines, Nuzzo (1996) spread sampling out across wide bands of cliff face that included both climbed and unclimbed areas. Unfortunately, the methods used influenced the results in an unknown way: sections of cliff face used as 'unclimbed controls' were located under overhangs. Overhanging cliffs affect the light and temperature regimes of cliff faces and also have a different density of plants than vertical faces. This may have limited their usefulness as comparison sites.

The impact of rock climbing was also studied on sparsely vegetated granitic rock at Bon Echo Provincial Park, Ontario, Canada (Dougan & Associates, 1995). At this site, climbing lines could not be differentiated from non-climbing lines in terms of species richness or vegetation cover, although lichen abundance was reduced in climbed areas. Significant impacts from climbing were found in the talus at the bottom of climbing routes where vegetation degradation and soil loss were the primary impacts. While trail widening, erosion and soil loss were the primary impacts at the cliff top, only a portion of this damage could be attributed to rock-climbing activity.

The impact of rock climbing on the density and age structure of old-growth *Thuja occidentalis* was studied on cliffs of the Niagara Escarpment (Kelly & Larson, 1997a). Climbing had a significant negative impact on the density of *Thuja occidentalis* on cliffs. This was marked by the elimination of both young and extremely old individuals from the population. A high percentage of trees (20.2 per cent) on climbed cliff faces also showed obvious damage from human activity (i.e. cutting, chopping etc.) (Fig. 7.12). These signs of human activity were almost absent in unclimbed areas. Cliff-edge populations of *Thuja occidentalis* above climb-

Table 7.1 Solidago sciaphila mean values per m² in upper and lower cliff zones for cliffs never climbed, cliffs previously climbed, and cliffs currently climbed in north-eastern Iowa.

| | Upper zone of cliff | | | Lower zone of cliff | | |
|---|---|---|---|---|---|---|
| | Never climbed $n=18$ | Previously climbed $n=9$ | Currently climbed $n=6$ | Never climbed $n=69$ | Previously climbed $n=47$ | Currently climbed $n=36$ |
| Genet density (number/m²) | | | | | | |
| Total | 14.2 (a) | 12.0 (a) | 3.2 (b) | 0.8 (c) | 2.1 (b) | 0.2 (c) |
| Sterile genets | 12.1 (a) | 9.9 (a) | 2.2 (b) | 0.7 (c) | 1.5 (b) | 0.1 (c) |
| Fertile genets | 2.1 (a) | 2.1 (a) | 1.0 (b) | 0.1 (c) | 0.5 (b) | 0.1 (c) |
| Basal area (cm²/m²) | | | | | | |
| Total | 793.6 (a) | 678.0 (a) | 328.6 (b) | 53.2 (c) | 192.9 (b) | 31.2 (c) |
| Sterile genets | 485.0 (a) | 399.3 (a) | 107.3 (b) | 43.3 (c) | 121.4 (b) | 3.5 (c) |
| Fertile genets | 308.6 (a) | 278.9 (a) | 221.3 (a) | 9.5 (c) | 71.5 (b) | 27.7 (c) |
| Flowering ramet density (number/m²) | | | | | | |
| Total | 6.3 (a) | 8.8 (a) | 2.0 (b) | 0.2 (c) | 1.0 (b) | 0.3 (c) |
| Unbroken ramets | 5.2 (a) | 7.2 (a) | 1.2 (b) | 0.2 (c) | 1.0 (b) | 0.1 (c) |
| Broken ramets | 1.1 (a) | 1.6 (a) | 0.8 (a) | 0.0 (b) | 0.1 (b) | 0.2 (b) |
| Inflorescence length (cm/m²) | | | | | | |
| Total | 45.1 (a) | 72.2 (a) | 34.2 (a) | 2.3 (c) | 9.3 (b) | 2.1 (c) |
| Unbroken ramets | 42.5 (a) | 70.6 (a) | 31.2 (b) | 2.3 (c) | 9.2 (b) | 0.7 (c) |
| Broken ramets | 2.6 (a) | 1.7 (a) | 3.0 (a) | 0.0 (b) | 0.1 (b) | 1.4 (a) |

Notes:

$n=$ number of quadrats sampled. Different letters in parentheses indicate statistically different means ($p<0.05$).

Source: Taken and modified from Nuzzo (1995).

*Figure 7.12*  A living 781-year-old *Thuja occidentalis* located next to a popular rock-climbing route on the Niagara Escarpment. Fourteen axes on this tree have been sawn off, including the main leader. Photo by P.E. Kelly.

ing routes showed more damage than undisturbed cliff-edge sites, but these results are confounded by the presence of hiking and past stand disturbances from logging.

Two other studies attribute observed vegetation patterns in cliff environments of the Shawangunk Mountains of New York to natural processes even though this cliff is the most popular climbing destination in eastern North America (Roberts, 1996). The decline in foliose lichen populations on cliff faces of the Shawangunk Mountains was described by Smiley and George (1974), but this decline was attributed to air pollutants or loss of shading from cliff-top *Pinus strobus*. Similarly, Abrams and Orwig (1995) did not mention possible disturbance from intense rock-climbing activity in their discussion on the age structure of *Pinus rigida* on cliffs and rock outcrops in the same area.

In Germany, rock climbing in the upper Danube Valley has caused a number of qualitative and quantitative changes to the cliff vegetation. Damage to vegetation occurred within a few years of route establishment (Herter, 1993). This damage included a reduction in vegetation cover, the disappearance of non-competitive species sensitive to disturbance and specifically adapted to these extreme habitats, and an increase in ruderal species. Erosion of the cliff edge, cliff face and talus was also severe, including the clearing of soil and humus from crevices on the cliff face (Herter, 1993). A higher percentage of plants was damaged in plots used by climbers than in those used by hikers (Herter, 1996). Herter (1996) believes continued use of cliffs such as Eichfelsen will eventually lead to the elimination of all rare and protected species. In the Schwäbische Alb, Baden-Württemberg, rock climbing resulted in trampling on cliff ledges, the cliff base and at the cliff edge resulting in the destruction of lichen and moss communities (Lüth, 1993).

A study on the impact of rock climbing on abandoned quarry walls along the Neckar River near Heidelberg, Germany, revealed little damage. No cliff-edge damage was found due to the installation of climbing bolts at the cliff top which allows climbers to descend to the talus upon completion of the climb rather than disturb the cliff edge (Viemann, 1997). While some of the quarry walls supported no vegetation (usually under overhangs), others possessed a vertical zonation with respect to species composition and richness that was a function of the amount of shade from the forest canopy below. The most difficult climbs, which are rarely climbed, actually showed an increase in lichen cover over five years that was attributed to a decrease in $SO_2$ emissions (Viemann, 1997). Climbing at this quarry is permitted only between August and January.

Rock climbing has been shown to have a negative effect on cliff-dwelling raptors, especially on sea cliffs where climbing has become very popular (Crick & Ratcliffe, 1995). Peregrine falcon nests have been abandoned in areas of the UK such as the Peak District, Cumbria, North Wales and the Cheviots because of recreational climbing use. The impact of climbing varies depending on factors such as the point in the breeding cycle at which the disturbance occurs, the frequency of climbing, the closeness of the routes to the eyrie, and the size of the cliff (Ratcliffe, 1980). The impact is more pronounced on small cliffs because peregrines are more likely to find alternative nest sites on higher cliffs (Crick &

Ratcliffe, 1995). Peregrine falcons often abandon their nests in response to climbing, thus losing their eggs or young to chilling (Ratcliffe, 1980). In the Peak District, heavy climbing use has led Crick and Ratcliffe (1995) to conclude that presently abandoned nest sites are probably permanently abandoned. Nesting birds were also disturbed by rock climbing in the Schwäbische Alb, Germany (Lüth, 1993).

Another recreational pursuit on cliff faces involving ropework is rappelling or abseiling. Rappelling is essentially the opposite of rock climbing in that it is a technique which allows one to descend from the cliff top to the talus. It is likely that studies that describe the impact of rock climbing are essentially describing the impact of both climbing and rappelling as both techniques are practised at most cliffs. It is therefore impossible to describe impacts from rappelling alone. It is surmised that these impacts would involve similar amounts of disturbance to the cliff-edge community, except where there are bolts which allow rock climbers to avoid contact with the cliff edge. Disturbance from rappelling is probably less on the face as rappellers have less contact with the cliff than rock climbers. Disturbance at cliff edges may also occur as a result of hang gliding, which is a popular recreational pursuit, particularly in Europe (Lüth, 1993). While the actual action of hang gliding is not damaging to cliffs (although cliffs can be damaging to hang gliders!), hiking along cliff edges is required to access suitable launching sites.

### 7.1.3 Bonsai collecting

The process of miniaturizing trees for ornamental purposes originated in China around 200 AD and spread to Japan approximately 600–1000 years later (Del Tredici, 1989b). The term bonsai did not gain wide acceptance until the late nineteenth century. Previously, the decorative term *hachi-no-ki* was used to describe these trees, but bonsai became accepted because it alludes to an art form with aesthetic aspirations (Del Tredici, 1989b). In China and Japan, most of the prized ancient bonsai began life as naturally stunted trees rooted in marginal habitats such as cliffs, rocky slopes, mountain crevices and bogs (Koreshoff, 1984). At the end of the nineteenth century when bonsai became popular, individuals known as professional dwarf tree collectors wandered the countryside of Japan looking for old, attractive, natural trees on cliff faces to collect for bonsai. When a suitable tree was spotted, a ritual was performed in the rocks at the base of the cliff and offerings were made to deities to ensure safe passage. Ropes with grappling hooks were used to access the trees from the cliff bottom. This practice virtually disappeared once all the trees

*Figure 7.13* A specimen of *Thuja occidentalis* collected from the field and planted as a bonsai. Photo by P.E. Kelly.

were removed from these cliffs (Koreshoff, 1984). Following World War II, bonsai became more popular and nursery stock was used to meet the growing demand.

Bonsai has now expanded around the world. While the emphasis today is on pruning and shaping trees, collecting from the wild has become popular again because a good-sized trunk with many of the desirable characteristics of old age can be obtained immediately (Koreshoff, 1984; Fig. 7.13). Within the hobby, collecting from the wild is supported and encouraged (Koreshoff, 1984; Ota & Gun, 1988; Tanner, 1989; Hall, 1992; Biel, 1995). Cliffs, limestone pavements and rocky habitats of North America are popular locations for collection (Ota & Gun, 1988; Tanner, 1989; Biel, 1995). Along the Niagara Escarpment, stunted old-growth *Thuja occidentalis* have been excavated from both private and public land including National Park property. Particular trees on cliffs and limestone pavements have been known to 'disappear', leaving a hole and loose rock. Individual collectors with extracted stunted *Thuja occidentalis* have been stopped several times, including one occasion when the collectors were dressed in ceremonial garb. Tree removal is practised in many other areas (Tanner, 1989) and is difficult to prevent, even on public lands. The long-term effects of this practice on the populations of cliff-face trees are unknown but undoubtedly deleterious. It is interesting to note that in the early part of the twentieth century, Japan had removed virtually all of its accessible cliff trees for bonsai collecting (Koreshoff, 1984), thus eliminating old-growth forest from its cliff faces.

In addition to the collection of small trees from cliffs for use in bonsai, there was a tradition in Western horticulture from the late 1600s to the early 1900s in which plant collectors would engage in long and expensive expeditions to collect plants and shrubs from wilderness areas in Asia. Cox (1945) and others have described such plant-hunting expeditions and have made the point that many of the world's most familiar and commercially exploited types of house plants, garden plants and shrubs were originally endemic to areas of extreme topography in China, Japan and other countries in temperate or tropical Asia. Large numbers of species of *Rhododendron* were first collected from limestone cliffs in central China.

### 7.1.4 Poaching and hunting

Evidence for the removal of falcons from cliffs goes back to 1248–1250 AD, when Frederick II describes how to access peregrine falcon nests with a rope for the purposes of stealing young (Ratcliffe, 1980). Today, the popularity of falconry as a sport for aristocrats, especially amongst the wealthy in Arab countries (Flint, 1995), still places populations of cliff-dwelling falcons at risk to poachers. Falcon thieves and egg collectors are perceived as serious threats to populations of falcons. This threat is particularly pronounced in the countries of the former Soviet Union where lax border control and the disintegration of the state conservation system have provided a good environment for poaching, especially for the Saker falcon (*Falco cherrug*), a traditional bird for Arab falconry. The Kazakhstan government, perhaps realizing the difficulty in preventing poaching, accepted $2 000 000 from a Saudi prince for the establishment of a foundation for Saker falcon protection in exchange for the prince's exclusive right to catch 25 falcons a year (Flint, 1995). Reductions in peregrine falcon numbers in the UK (Mearns & Newton, 1988; Crick & Ratcliffe, 1995) and Queensland, Australia (Czechura, 1984), have also been attributed in part to falconry thieves and illegal egg collectors. The difference in success rates between inland and coastal clutches of peregrine falcons in Scotland has been attributed to higher nest robbery of inland sites (Mearns & Newton, 1988). Also, taller cliffs were thought to offer greater protection from collectors than shorter cliffs.

Game-keepers, hunters and farmers have also been recognized as threats to successful establishment and breeding of raptors in some regions (Czechura, 1984; Brown, 1991; Crick & Ratcliffe, 1995). The killing of mammalian predators with poisoned bait by farmers led to the elimination of bearded vultures (*Gypaetus barbatus*) from most of South Africa, even though they do not prey on stock (Brown, 1991). Current populations breed only in the mountains in and around Lesotho. Golden eagles (*Aquila chrysaetos*) are also persecuted because of the mistaken belief that they significantly reduce game and stock populations. In Montana, USA, bounties were placed on golden eagles because it was believed that they killed antelope, even though stomach analysis did not support this contention. Almost 300 golden eagles were shot in one month in 1948 (Watson, 1997). In the south-west USA, golden eagles were shot from airplanes because it was thought they were killing sheep. Although they received federal protection in 1963, Watson (1997) believes this has not stopped the killing. In Scotland, golden eagles were heavily persecuted because it was thought they reduced red grouse (*Lagopus lagopus scoticus*) numbers. There is little evidence to indicate eagles caused economic hardship in Scotland (Watson, 1997).

In Queensland, Australia, human persecution was considered to be the most significant cause of peregrine falcon mortality, largely due to pigeon fanciers who killed adults or young or interfered with peregrine eyries (Czechura, 1984). Pigeons are significant components of peregrine falcon diets (White *et al.*, 1988; Cade & Bird, 1990). Recent increases in the population of peregrine falcons in some areas since the early 1980s have been attributed not only to stricter conservation measures but also to increased food supply at the expense of homing pigeons. In the UK, homing pigeon numbers increased by approximately 150 per cent between 1977 and 1991, and pigeon fanciers have resorted to illegal control of peregrine falcon populations (Crick & Ratcliffe, 1995).

Some cultures also continue to harvest sea birds off cliffs, although improvements to the standard of living and better preserving techniques have reduced the frequency of this practice. Deep freezers, the advent of plastics as a replacement for feathers, and the high prices of cartridges and fuel have also contributed to this decline (Evans & Nettleship, 1985). Fowling for Atlantic puffins continues in the Faeroe Islands and Iceland, although some protection laws were introduced in the Faeroes in 1982 (Evans & Nettleship, 1985). On talus slopes, eggs and birds are pulled from nests using sticks attached with iron hooks. Herring nets are sometimes laid over the rocks to catch the birds as they emerge in the morning or are used on cliffs to catch flying birds (Harris, 1984). In Mexico, yellow-footed gull eggs are collected for eating from nest sites on talus slopes and beach-berms. Although talus slopes are suboptimal sites, their occupation by yellow-footed gulls is seen as an adaptation to human predation (Spear & Anderson, 1989). In certain regions of Canada, native populations may legally hunt murres, and in winter in Newfoundland and Labrador, the general populace may do the same. In the late 1970s, the annual kill was estimated to be 450 000 birds (Wendt & Cooch, 1984). Similar restrictions are in effect in Greenland but it is difficult to know if they are being enforced in remote areas (Evans & Nettleship, 1985).

Fishing nets also inadvertently kill many sea birds every year. Gill nets, long-line fishing and drift-nets are the biggest culprits (Evans & Nettleship, 1985; Freethy, 1987). Drift-nets are virtually indestructible and nets lost at sea may continue to be a hazard for many years. In the early 1970s, annual losses may have been as high as 350 000 birds in south-west Greenland (Evans & Nettleship, 1985) . Significant losses also occur in eastern Canada, northern Norway and western Ireland. The thick-billed murre is the principal species lost in western Greenland, while common murres are more commonly drowned off Newfoundland and Norway. Puffins, dovekies and black guillemots are also commonly drowned off Newfoundland, although it is likely that numbers decreased in the late 1990s with the moratorium on cod fishing in Canada. Razorbills are primarily caught along the west coast of Ireland. Numbers and species caught are dependent on both the density of birds and the depths at which the nets are set (Evans & Nettleship, 1985).

### 7.1.5 Quarries and development

Bare rock outcrops have been used as sources of building materials for thousands of years. The lack of overburden minimizes the time required to find and assess the quality of the resource. As human population size has increased, the requirement for building stone and aggregate has increased in proportion. Thus, many sites that are well known for their cliffs also have rock quarries nearby (Fig. 7.14). Unfortunately, quarry operations can destroy important natural cliff-face features (Churcher & Dods, 1979; Lüth, 1993). For example, a post-Wisconsin cave (circa 10 000 years BP) on the Niagara Escarpment that contained numerous skeletal remains including an extinct large pika (*Ochotona* sp.) skeleton (Churcher & Dods, 1979) was destroyed by quarry operations in the 1920s. In some cases, talus blocks are quarried directly for road building (Lüth, 1993). Suitable policies to accommodate the interests of quarry

(a)

(b)

*Figure 7.14* An abandoned (a) and an active (b) limestone quarry in southern Ontario. The quarrying operations ceased in the abandoned quarry in 1905. Photos by K. Ursic and D.W. Larson.

operators and conservationists still have not been developed. Of additional concern is that no studies have been conducted on the question of whether quarries have a negative influence on the biota or even the physical environment of the cliff communities near them. When one considers that most open pit mines disturb or redirect the flow of groundwater (see Chapter 2), it would be surprising to find that quarry construction was completely benign biologically.

Different cliff fauna seem to react differently to quarries and mining development. For example, prairie falcons (*Falco mexicanus*) in New Mexico are susceptible to gold-mining operations in cliff and talus habitats. Bednarz (1984) found no prairie falcon nests in the heavily mined Caballo Mountains compared with multiple nests in similar adjacent undisturbed mountain ranges. Bearded vultures (Donázar *et al.*, 1993) are also susceptible to human disturbance. The number of roads and inhabitants was found to be negatively correlated with breeding success in bearded vultures in Spain (Donázar *et al.*, 1993). Conversely, peregrine falcons (Ratcliffe, 1980; Norriss, 1995; Crick & Ratcliffe, 1995), kestrels (Shrubb, 1993) and golden eagles (Fala *et al.*, 1985) will nest on active quarry walls. In Victoria, Australia, White *et al.* (1988) report one peregrine falcon nest located 50 m from active quarrying including heavy machinery and a rock crusher. Another was located on a quarry wall in a shooting range. Cliff swallows and peregrine falcons are frequent users of buildings and bridges in urban settings. Cliff swallows seem to prefer bridges to natural cliffs in the American Midwest (Brown & Brown, 1990).

Quarry walls represent extreme environments in terms of nutrients, water supplies and exposure. While such sites experience episodic flooding and drought, some successful quarry restoration projects have been carried out in the UK (Usher, 1979). Few studies, however, have looked at the natural succession of plant or animal communities which colonize quarry walls upon abandonment. One such study on open cliff faces in abandoned limestone quarries near the Niagara Escarpment showed that quarry walls supported a predictable cliff flora within about 70 years (Fig. 7.15; Ursic *et al.*, 1996). In addition, it showed that the community of plants that was recruited in this period was similar to the naturally occurring community of plants that exists along the natural Niagara Escarpment. Lüth (1993) suggests that some abandoned quarries are as worthy of protection as natural cliff faces because they represent some of the few places where vegetation can develop naturally without human influence.

(a)

(b)

*Figure 7.15* Comparison of (a) a limestone quarry at the time of its abandonment in 1900, and (b) in 1993 after 90 years of natural revegetation. Photos by K. Ursic.

*Figure 7.16* The conversion of a section of cliff ecosystem into a commercial ski-hill operation. Such conversion of natural habitat into recreation areas is a significant threat to the maintenance of cliff communities. Photo by D.W. Larson.

The scenic nature of cliffs and the vistas which they provide also make them desirable to real-estate developers. Unfortunately, the corresponding increase in the value of commercial cliff-top properties makes them difficult to acquire for conservation purposes. Other cliff sites have actually been altered by intense commercial development. Figure 7.16 shows a downhill ski facility constructed by blasting portions of a cliff along the Niagara Escarpment. Cliffs surrounding large cities such as Hamilton, Ontario, Canada, are also bisected by systems of roads and municipal services that can significantly disrupt and fragment the flora and fauna of the cliffs..

Other cliffs have been or are being threatened with flooding. Large-scale hydroelectric developments in areas such as the south-western USA, the Nile delta and Norway have flooded large stretches of cliff and unknown numbers of floral and faunal species, as well as hundreds of archaeological sites on cliffs. Cliffs in the 'Three Gorges' area of China will also be inundated upon completion of a major hydro-electric dam. These include cliffs with hanging coffins and other cliff burial sites.

## 7.2 Conservation and management of cliffs

As commercial and agricultural exploitation of level-ground ecosystems has continued in the twentieth century, there has been a growing appreciation for the need to conserve and protect the small patches of natural habitat that remain. This growth accelerated in the years following World War II and was given even more impetus by the publication of the book *Silent Spring* by Rachel Carson (1962). Interest in conservation had preceded this book, but the goals in the early part of the twentieth century were more related to the conservation of game stocks for hunters rather than to the conservation of the landscape for its own sake.

Following the publication of *Silent Spring*, however, public awareness of the degree to which humans were altering the biosphere became acute. Institutions and groups such as the Environmental Protection Agency and the Audubon Society in the USA and the Nature Conservancy, were either born during this period or experienced unprecedented growth in membership. Such agencies and groups were involved with the identification and protection of valuable natural resources for the enjoyment of future generations. It became a high priority in the Western world to identify places with minimum amounts of human disturbance and a high degree of ecological 'integrity'. Cliffs and other areas of extreme topography are quickly becoming well known as areas that retain high levels of biodiversity (Maser *et al.*, 1979; Larson, Kelly & Matthes-Sears, 1994b; Schöller, 1994). Human impacts on their surfaces and at their edges are therefore important to understand.

Many natural areas, reserves and parks around the world have cliffs on their property. People are attracted to areas of extreme topography and enjoy a variety of recreational activities that exploit cliffs at different spatial scales. Photography and painting often feature cliffs, but usually from afar. Hiking, bicycling, skiing, all-terrain powered vehicle use (including snowmobiles) and horseback riding are recreational activities that are often enjoyed in the vicinity of cliffs. Rock climbing is the only recreational activity which involves direct contact with cliff faces.

As human population size increases, it is likely that the popularity of these activities will increase and cliff environments will be exposed to greater pressures, especially those in areas of high population densities. Although such disturbances are not necessarily worse than those observed in woodland, wetland or aquatic habitats, the fact that most cliffs have escaped disturbance until recently suggests that cliffs might deserve special consideration from a conservation and management perspective.

Management strategies have already been implemented for some cliffs. In the Shawangunk Mountains, a land trust known as the Mohonk Preserve has been established to foster land purchase, resource inventory and protection, research and education (Kiviak, 1991). The Mohonk Preserve is the most popular rock-climbing area in eastern North America (Roberts, 1996) and cliffs are a prominent part of the landscape. Recreation that does not place resources at risk is encouraged, although problems with high volumes of people hiking, climbing and backpacking are encountered. Despite these pressures, the Mohonk Preserve has managed to maintain presettlement forest on and near the cliffs, inclusive of populations of *Pinus rigida* and a wide variety of arctic disjunct species including *Asplenium montanum*. The active involvement of the local community and The Nature Conservancy has helped to maintain the interest in conservation management in this area.

A similar situation applies to the Niagara Escarpment. In Ontario, the Niagara Escarpment Commission (NEC), an agency of the provincial government, was formed in 1974 to co-ordinate land planning and economic development along a 700 km linear corridor of land stretching across southern Ontario. One hundred and fifteen conservation authorities, nature preserves, parks and public areas have been established within the zone and are governed by the regulations of the act. Development within a zone known as the Niagara Escarpment Planning and Development Area is carefully regulated, with the number and extent of permissible land uses increasing with distance from the cliff face. In 1990 the corridor was declared a UNESCO World Biosphere Reserve, which emphasizes research, monitoring, education and recreation within an area of biological importance.

In the western USA and in other countries of the world, national parks and nature preserves often include areas of extreme topography and these areas often have specific restrictions on agricultural and industrial development. The goal of most of these regulations is to ensure the maintenance of ecological structures and functions while also allowing a moderate amount of human visitation. In the western USA, where cliffs and rock outcrops are extremely abundant, a very large number of national parks and monuments exists (Fig. 7.17). In all of these, regulations exist to ensure the conservation and long-term management of the endemic species and other species that would be at risk from agricultural, industrial or recreational expansion.

In most areas, however, the protection of cliff faces has not been perceived as a priority amongst most resource managers although the poten-

*Figure 7.17* Rock pinnacles and cliffs in Chiracahua National Monument, Arizona, USA. Photo by D.W. Larson.

tial impact of rock climbing to cliff environments is becoming an issue and some management plans have been developed which specifically address rock-climbing issues. Devils Tower National Monument, Wyoming, USA, is one location where a climbing management plan has been implemented, largely in response to concerns expressed by Native Americans, who view Devils Tower as a sacred religious site, and biologists concerned over disturbance to prairie falcons nesting on its cliff faces. In Germany, a model centred on a rating scale for ecological and use factors was developed for protecting cliff environments in the Upper Danube Valley as the existing amount of protection was not considered sufficient (Herter, 1993). In core protected areas, recommendations include the reduction of mountain goat populations, the closing of hiking trails, the prohibition of climbing and the establishment of permanent plots to monitor the success of conservation measures (Herter, 1993, 1996). In reality, the most effective management of cliff resources should come from within the climbing community and some climbing organizations are becoming increasingly aware that decisions made by resource managers could limit their access to cliff resources. Organizations such as the Alpine Club of Canada and the American Mountain Guides Association have already been emphasizing the unique ecological properties of cliffs to their members.

Cliffs are popular destinations even where rock climbing is absent. Unfortunately, the ecological structure and function of the landscape matrix surrounding cliffs are often impacted. Trampling by humans along cliff edges removes the youngest age classes of trees that form ancient forests (Larson, 1990). Once the age structure of the forest is modified at the cliff edge, the alteration of other components of the biota is inevitable. Where hiking along cliff edges is prominent, it may become necessary to redirect hiking traffic to certain cliffs and block off access to others (Stärr et al., 1995). Unfortunately, even if disturbance can be stopped in an area, the impact will continue for some time. Impact in most cliff environments has been recent (i.e. climbing) and thus an equilibrium between impact and damage has not been reached. Degradation is likely to progress even when steps are taken to prevent it (Herter, 1993).

The persistence of some cliff species may actually be dependent on the nature of the vegetation growing at the cliff top. In Germany, some relict plant species may not be able to survive on cliffs which are forested at the top (Lüth, 1993). Plants that require a lot of light may disappear from cliffs where forests are planted in the plateau but are able to survive where cliff tops are used for pasture. Forests cast too much shade on the face, while intense agriculture contaminates runoff with fertilizers and pesticides, which may also affect the species composition on the cliff face (Lüth, 1993). In these instances, the management of cliffs is intrinsically a part of the management of the surrounding ecosystem.

## 7.3 Cliffs currently used as tools for education

Natural features with broad public appeal, such as cliffs, have not been adequately exploited as vehicles to demonstrate the societal value of scientific research and monitoring. The public's willingness to support more research and monitoring of key processes that are important to conservation and management will increase with increased awareness of the value of these natural features. Our experience within the parks and natural areas along the Niagara Escarpment is that public knowledge of a presettlement forest with trees exceeding 1000 years in age encourages the public to address issues of environmental degradation. Even topics that would normally appear to be strictly 'scientific' have continued to attract attention. Thus, popular articles about the ancient trees have appeared in periodicals such as *Discover*, *Audubon* and *International Wildlife*, but there has also been attendant coverage of cryptoendolithic organisms and other less mainstream topics such as dendrochronology (see Chapter 8).

Ecological monitoring of important natural areas has become vogue and while at present there are few reported terrestrial species of plants or animals that can be used as indicator species of environmental stress or disturbance (Cole, 1995a), land managers are eager to try to identify such indicators in the hope that they can be used to assess the ecological conditions that apply to the habitat as a whole. However, indicator species may not exist in some locations. Parikesit *et al.* (1995) showed that sampling an entire cliff-edge habitat is neither difficult nor time consuming, and that the results obtained apply immediately to the entire plant community without having to deal with awkward assumptions about how well the indicator taxa represent the biota as a whole. Community-scale sampling, however, is often viewed as 'research' and at the current time land managers have not been shown how such community-based results can represent the parameters to follow in the search for long-term trends in biological resource conservation.

The fact that the cliffs of the Niagara Escarpment support an ancient forest ecosystem provides the public and private sectors with an incentive to ask questions about local biodiversity in terms of both structure and function. Such questions would attract little attention in the absence of the undisturbed presettlement status of the cliffs. It really comes down to this: cliffs are extreme environments that instantaneously attract the interest of people. When this innate attraction is combined with the concept that cliffs are refuges from human and natural disturbance, people see the need to explore, study and monitor the integrity of such systems over time.

## 7.4 Summary

Despite the rugged terrain, cliff environments have served as focal points for a variety of past cultures. Even today, people retain an attraction to cliffs that may reflect our earlier dependence on them as a place of shelter. Early hominid use of caves in cliffs for habitation is well documented. Many cultures had a traditional reverence for cliffs, and cliffs and their immediate surroundings have been used as burial sites or rock art sites. Today, cliffs are popular recreational resources and serve as focal points for some nature reserves and national parks. Cliff edges are common locations for hiking trails and cliff faces have become extremely popular sites for rock climbing. Unfortunately, the few impact studies conducted in cliff environments indicate that both hiking and climbing are detrimental to some species in the cliff community. The collection of old-growth trees by bonsai collectors and the poaching of falcons and their eggs from cliffs represent significant threats to these species on cliff faces. Quarrying, mining and real-estate development affect the entire cliff community. Fortunately, compared to other elements of the landscape, cliffs are relatively undisturbed. Consequently, cliffs worldwide may support old-growth forests and undisturbed floral and faunal communities. These may serve as important benchmarks for ecological monitoring and tools for education on the structure of undisturbed communities. The lack of disturbance that cliffs experience from people, including the lack of formal scientific study, may be viewed as ironic from the viewpoint of the 'urban cliff hypothesis'. That is, that a habitat that was extremely important as a dwelling and refuge during the Palaeolithic period, and that was essentially recreated when mud, stone and wooden structures were built in villages, towns and cities, remains largely ignored by our own species.

# 8 · Summary, opportunities and synthesis

To a very great extent, cliffs are places that are of interest to everyone and no-one at the same time. This paradox attracts us and we think it will attract others when it becomes better known. People on all continents see images of cliffs in a wide variety of mass media and are consequently drawn as though pulled by a magnet to cliffs or habitats with extreme topography. Cliffs are sites with enormous spiritual value and may even be habitats that have given rise to a wide variety of our food plants, garden weeds and commensal animals. Yet these same sites have zero area when photographed from space, have attracted little scrutiny from scientists, and have received almost no legal protection from various forms of commercial exploitation. Some may be inclined to protest the last two statements based on the content of the book so far, but when one compares the vast and easily accessed literature for other habitat types, our conclusions are justified.

The literature that we have reviewed and discussed in the preceding chapters almost always focuses on particular organisms, groups of organisms or specific aspects of cliff ecosystems without considering them as 'places' in the same way as lakes are considered as distinctive habitats by limnologists or forests by forest ecologists. A result of the particular organization that we have selected is that we may have reinforced rather than eliminated the idea of separate structures and functions on cliffs. The intention of this chapter, therefore, is to summarize the most important aspects of cliff ecology as presented in the preceding pages and to try to integrate ideas about how cliffs appear to function as whole 'places' within landscapes. The chapter ends with an attempt to pinpoint research problems or hypotheses that could be examined in the future.

## 8.1 Summary of significant findings

### 8.1.1 Cliffs as ancient habitat

The previous chapters have detailed the abiotic and biotic characteristics of cliffs and the ecological processes that operate on them. The most conspicuous and important finding has been that cliffs everywhere represent relict, undisturbed, or perhaps even ancient habitat. A geographical model devised by Stahle and Chaney (1994) has been used to predict the location of old-growth forests in Arkansas, USA, and has shown that old-growth habitat is largely restricted to 'unproductive rocky land with steep topography'. Such predictions are borne out at a larger scale by the results of Davis (1951), who observed that cliffs are habitats that form environmental and anthropogenic refuges. Similar conclusions were reached by Jones (1994) in Australia, Chin (1977) in Malaya, and Curtis (1959) in the USA. Additional confirmation comes from locations where populations of old trees occur. For example, Landers and Graf (1975) report that the oldest forest stands in the state of Iowa, USA, were found on limestone and sandstone cliffs of the Mississippi River, and Sealy (1949) reported relict *Arbutus unedo* populations on cliffs in Ireland. Larson and Kelly (1991) showed that cliffs of the Niagara Escarpment supported the oldest and least disturbed forest ecosystem in eastern North America, and subsequent work has shown that some of the oldest trees (>1800 years) in the continent occur on these cliffs (Kelly *et al.*, 1992, 1994; Larson *et al.*, 1994b; Kelly & Larson, 1997a, 1997b).

Cliffs may support ancient and undisturbed communities (Klötzli, 1991) even in parts of the world that have a long history of agricultural and industrial development such as western Europe, Asia or Africa. Larson *et al.* (1999) have recently surveyed 15 of the eastern states of the USA as well as parts of Germany, France, England, Wales and New Zealand in an attempt to answer the question, 'Do cliffs generally support ancient woodland', where the word 'ancient' is taken to mean that the forest community has been self-regenerating for many centuries. Increment core samples were taken from 224 trees in this survey, and almost without exception the results showed that cliffs in these countries (or states) support a stunted forest of slow-growing ancient trees in a variety of age categories extending to over 1000 years. A wide variety of endemic taxa occur on these cliffs as well, along with an array of species or genera that recur at all or many of the sites. This study supports the views of travel writers Donahue (1996) and Zerrahn (1996) that cliffs support ancient trees in many sites that are popular tourist destinations. Many of these sites occur despite being close to areas of high population density and agricultural activity. This means that the global density of ancient forest habitat is far greater than previously considered.

### 8.1.2 Cliffs for environmental monitoring

The utility of ancient ecosystems to society, however, is far greater than the heuristic values associated with old trees in a striking topographic setting. Undisturbed forests contain old trees which preserve records of palaeoenvironmental change in their tree rings or their pollen. Past fluctuations in precipitation, temperature, sunshine, air pressure, air pollution, tectonic activity, hydrology, tree-line flux, fire frequency, insect outbreaks and glacial movement have all been successfully reconstructed from tree rings or pollen profiles.

While pollen records from organic deposits on cliffs have not been studied, many studies of tree rings have been carried out using plants growing on cliffs or rocks with extreme slope. The science of dating and reconstructing past events using tree rings is called dendrochronology, and dendrochronological applications have been reviewed by Fritts (1976), Schweingruber (1989) and Guyette, Cutter and Henderson (1989). Dendrochronological reconstructions of past events allow us to make better estimates of current and future environmental change. Improvement in the resolution of these estimates can save money for both industry and government. Schweingruber (1993), Ahmed and Sarangezai (1991) and Bhattacharyya, Lamarche and Telewski (1988) have reviewed the location and growth characteristics of dendrochronologically useful species worldwide and many of these taxa grow in extreme settings such as steep slopes and rocky substrate.

On the basis of the research reviewed in this book, we conclude that cliff sites worldwide ought to support slow-growing, widely spaced populations of woody plants of extreme age. In some areas, these plants may preserve local and regional climatic variation in their tree rings. Tree-ring chronologies can be reconstructed from these trees to facilitate palaeoenvironmental reconstructions. Our own experience along the Niagara Escarpment has shown that the time and money invested in developing tree-ring chronologies from *Thuja occidentalis* have been relatively small. The extreme age of the trees and the excellent preservation of coarse woody debris have made this task easier. The first stand of ancient *Thuja occidentalis* of presettlement origin was located in 1988 and an illustration of the age structure of part of this stand is shown in Fig. 8.1. Fieldwork in 1989 showed that a stunted, slow-growing, old-growth forest was present along the entire escarpment (Larson, 1990; Larson & Kelly, 1991; Figs. 8.2 and 8.3). These trees were successfully cross-dated soon after

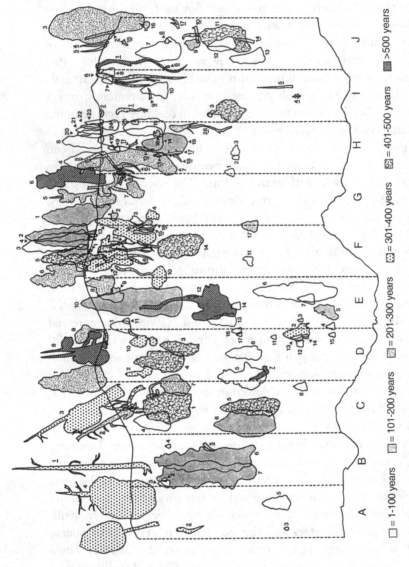

*Figure 8.1* Map of the distribution of trees in a typical uneven-aged population of *Thuja occidentalis* on a 50 m section of cliff along the Niagara Escarpment. Vertical dotted lines represent 5 m-wide sampling transects within which all the trees were sampled. Trees are grouped into 100-year age classes and included stems to 701 years. Taken from Kelly and Larson (1997b), and used with permission of Blackwell Science Ltd, Publishers.

☐ = 1-100 years   ▦ = 101-200 years   ▨ = 201-300 years   ▧ = 301-400 years   ▩ = 401-500 years   ▦ = >500 years

(a)

(b)

*Figure 8.2* Typical sections of the Niagara Escarpment, Ontario, Canada, as viewed from (a) an altitude of about 1000 m and (b) from the ground in winter. The sparse canopy of *Thuja occidentalis* represents the visible portion of an ancient undisturbed forest ecosystem. Photos by U. Matthes, and used with permission of the University of Chicago Press, from Kelly *et al.* (1992).

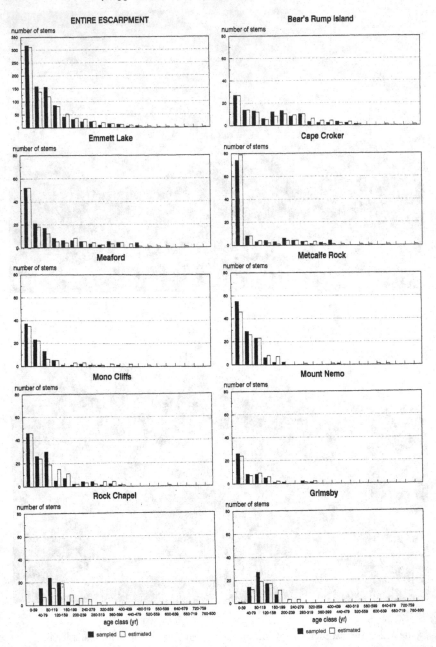

*Figure 8.3* Age-class frequency distributions for *Thuja occidentalis* trees at nine sites along the Niagara Escarpment. Trees were sampled in randomly positioned transects at each site, and absolute ring counts (sampled) as well as estimated ages for trees with partial cores (estimated) were plotted. Site names are organized

(Kelly *et al.*, 1992) and by 1994 a 1397-year temperature-sensitive chronology was published (Kelly *et al.*, 1994; Figs. 8.4, 8.5 and 8.6). Correlations with maximum summer temperature were also found which were similar to those found for *Thuja occidentalis* on rocky islands in Quebec (Archambault & Bergeron, 1992). The further addition of ring sequences from ancient woody debris in the talus has extended the Niagara Escarpment *Thuja occidentalis* chronology to 2787 years, the longest in eastern North America. Given the high level of debate over the frequency and duration of 'normal' versus 'anthropogenic' climatic fluctuations, we feel that many dendrochronologists and climatologists are missing opportunites to add significant and useful data by not sampling trees on local cliffs. We predict that if they look, they will find that cliffs frequently offer opportunities equal to the Niagara Escarpment.

### 8.1.3   Cliffs as refuges for biodiversity

In addition to their value as sources of proxy climate data, cliffs themselves are refugia for native flora and fauna (Ellenberg, 1988; Wardle, 1991). They are also core habitat for a wide variety of birds and other wildlife that are particularly sensitive to disturbance (Maser *et al.*, 1979). Thus, the retention of cliffs in regional landscapes can result in a maintenance of the ecological functions performed by these taxa. The preceding chapters include numerous reports of individual species that are either rare or endemic on cliffs. Davis (1951), Runemark (1969), Snogerup (1971) and Ratcliffe (1980) all make it perfectly clear that scientists and the public as well must recognize that many species that are highly rare or endemic naturally occur at low population densities in habitats that only occur sporadically in the landscape. The plants and animals that either exploit cliffs or are entirely restricted to them usually maintain quite small or, in the case of some sea birds, localized populations that are entirely dependent on the rocks maintaining their integrity. Attacks on the substrate, whether by chemical dissolution, quarrying, recreation or urban expansion, can create conditions wherein these small endemic or restricted populations simply vanish. Thus, species that are small, such as *Asplenium ruta-muraria* (wall rue) on cliffs of the Niagara

Caption for *Figure 8.3 (cont.)*
from north to south, with a plot for the entire data set shown in the upper left. Abundant seedling recruitment was found at most sites except on cliffs on unstable substrate or near large cities (Rock Chapel and Grimsby). Old trees were common everywhere. Illustration taken from Larson and Kelly (1991), and used with permission of the National Research Council of Canada.

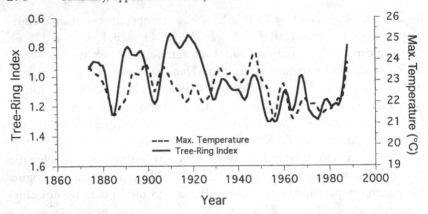

*Figure 8.4* Plot of standardized tree-ring indices for eastern white cedar (*Thuja occidentalis*) growing on cliffs of the Niagara Escarpment. Also plotted are mean maximum summer temperatures derived from several climate stations close to the tree-ring site. The indices are plotted on an inverse Y-axis to illustrate the inverse correlation between them and previous-year maximum summer temperatures. Illustration taken from Kelly *et al.* (1994), and used with permission of the National Research Council of Canada.

Escarpment, or *Woodsia ilvensis, Potentilla rupestris* and *Lychnis viscaria* (Jarvis & Pigott, 1973) in the UK, may be completely at risk unless the cliff habitats to which these species are restricted are viewed as special places that deserve special protection. Equally, birds and raptors that maintain very small populations, such as the bearded vulture (*Gypaetus barbatus*) in Namibia and South Africa or the golden eagle (*Aquila chrysaetos*) in North America, may be entirely dependent on large exposures of cliffs in order to maintain these populations. Many exceptionally rare organisms, such as the shrub *Microstrobos fitzgeraldii* (dwarf mountain pine) near Wentworth Falls in New South Wales, Australia, are located directly adjacent to dense human populations which threaten the existence of this species. Similarly, the conservation and continued existence of *Tilia cordata* and *T. platyphyllos* in the UK are entirely tied to their limestone cliffs being maintained as tightly protected habitat (Pigott, 1969). In the case of both of these last two species, every individual plant has been counted and monitored over the past several years, but even such monitoring of the individual organisms cannot protect the species against factors such as groundwater contamination. So, unless conservation efforts focus on the cliffs as entire habitats that nurture numerous species in a variety of trophic levels, these species may be lost entirely. Brewer-Carias (1986) reports that dozens of species are highly endemic to the

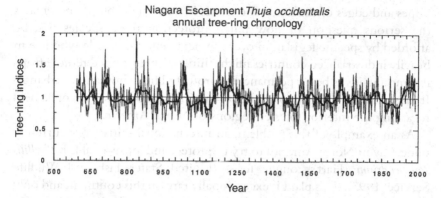

*Figure 8.5* A 1397-year tree-ring chronology for eastern white cedar (*Thuja occidentalis*) growing on cliffs of the Niagara Escarpment. Chronologies such as these allow for an examination of regional climatic change over an exceptionally long period of time. Illustration taken from Kelly *et al.* (1994), and used with permission of the National Research Council of Canada.

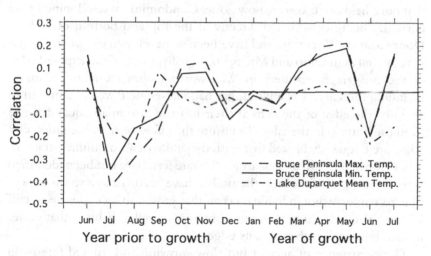

*Figure 8.6* Variation in the correlation between the tree-ring indices in cliff-face *Thuja occidentalis* and temperature. The solid line shows the variation in the correlation between the growth trend and minimum temperatures on the Bruce Peninsula; the dashed line shows the correlation with maximum summer temperatues at the same site. The dashed–dotted line shows the correlation between mean annual temperatures and cedar growth at non-cliff island sites on Lake Duparquet in western Quebec. All three lines show the same relationship and suggest a common physiological response of this species to the physical environment. Taken from Kelly *et al.* (1994), and used with permission of the National Research Council of Canada.

slopes and edges of the sandstone tepuis in Venezuela, but there remains the serious question of how much protection these species will be afforded by specific legislation over the next century. People who live in heavily industrialized countries might think of these eroded remote tropical mountains as being permanently protected from human disturbance. It must be immediately pointed out that similar thoughts were once held regarding the interior forested regions of North America.

As an example of this problem, an international initiative is currently underway in North America to try to protect and/or re-establish *Phyllitis scolopendrium* (Hart's tongue fern) (United States Fish and Wildlife Service, 1993). This plant is exceptionally rare on this continent and only occurs in Canada at the base of cliffs and on large-block talus slopes of the Niagara Escarpment around the Michigan Basin (Soper, 1954) and in a few other scattered locations in the USA. Meanwhile, prime sites with cliff edges, cliff faces and talus slopes of the Niagara Escarpment in the USA, and to a lesser extent in Canada, have been sold for real-estate development, mined for limestone and enveloped by cities. Most sites that once held this taxon are now gone. Condominium developments are currently being constructed directly at the top and bottom of cliffs in Door County, Wisconsin, and have been so developed elsewhere in that state as well as in Ohio and Michigan. Equally, most of the exposed edge of the Niagara Escarpment in Wentworth County, Ontario, Canada, including the City of Hamilton, has had real-estate development within 10 m of the edge of the cliffs at their top and within an equal distance from the bottom in the talus. To ensure the safety of such buildings, cliff edges are frequently blasted first with dynamite or ammonium nitrate. In other words, the cliffs as habitats for the rare fern have not been identified as protected, even if the ferns themselves have been. To us it seems impossible to conserve any individual taxon that grows on an exposure of cliff without also protecting the entire outcrop as well as habitats that buffer it from human disturbance at its edges.

The occurrence of ancient but slow-growing and stunted forests on exposed cliffs suggests that the traditional concept of old-growth forests as defined by Kimmins (1987) may have to be changed. Findley (1990) and Paullin (1932) present maps showing the location and density of old-growth forests in the USA from 1620 through to 1990. The map suggests that old-growth forests of the traditional type (tall lush canopy, abundant coarse woody debris, abundant lichens and mosses, non-competitive plants and animals) have all but disappeared due to timber harvesting and conversion of land to agriculture and industry. What the map does not present is an estimate of the fraction of the landscape that is rocky, exposed and harsh. It is very likely that such areas (including cliffs) were never considered to be useful as sources of harvestable timber or land suitable for agriculture, and were thus spared from 'development'. It is interesting to note in Paullin's maps that the area of land that held ancient woodland in the state of Iowa at the turn of the eighteenth century only included the steep valleys of the tributaries of the Mississippi River. We returned to some of these sites in the summer of 1997 and found that the cliffs along these river courses still support the same virgin woodlands that they held 200 years ago.

The information contained in this book should permit a new 'search image' to be adopted for the identification and analysis of old-growth forests. Instead of looking for sites with lush canopies and dense understory, we suggest that the bulk of the remaining unidentified old-growth forests of the world are in locations with rock outcrops or cliffs. The initial findings from field work carried out by us in 1997 suggest that these ideas are correct (Figs. 8.7 and 8.8). Cliffs in Europe have long been viewed as bare habitat without forest cover. This recent work shows not only that many cliffs support a rich forest community, but that these forests represent the most ancient intact woodland present in the entire landscape.

### 8.1.4  Cliffs as predictable habitats

Another conspicuous finding in this book is that cliffs attract a flora and fauna that have a considerable degree of global ecological, and sometimes taxonomic, similarity worldwide. True, there are fewer trees on sea cliffs than on inland cliffs, but for the most part the biota and their ecological controls appear to be similar regardless of the location or geological history of the cliff. These biological similarities appear to be controlled by six factors.

1. Gravity affects seedling or juvenile survival, propagule retention, litter retention and pedogenesis.
2. Most cliffs are not disturbed by humans, thus most cliffs are not systems in which direct anthropogenic effects have been significant.
3. Productivity is low for immobile organisms. This selects for organisms that are tolerant of low resource supply, or at least tolerant of having their growth restrained. It appears that cliffs provide an array of costs and benefits for immobile organisms but only an array of benefits for highly mobile ones, and for them productivity can be exceptionally

*Figure 8.7* A cliff face in the Verdon Gorge, France. Small, deformed, ancient trees of *Juniperus phoenicea* occur abundantly on these cliffs, along with a diverse array of lichens, mosses and other higher plants. Photo by D.W. Larson.

high! Low productivity is present in perennial plants, animals with limited dispersal ability, and micro-organisms on cliffs, and coarse woody debris is abundant and persistent.

4. Most of the organisms on cliffs would show higher productivity, survival, recruitment etc. on level ground, except that they apparently lack competitive ability and may be ecologically and evolutionarily confined to sites with low productivity. The absence of competition seems to select for non-aggressive taxa even in the case of top-carnivores such as raptors, whose nest-site selection reflects the vulnerability of eggs and young to high levels of predation found on level ground.

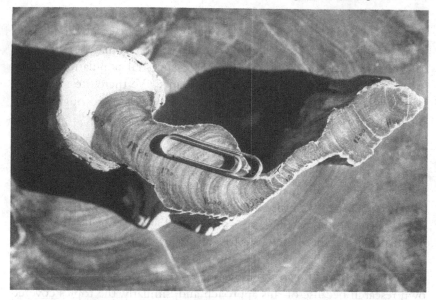

*Figure 8.8* Cross-section of a 1025-year-old *Juniperus phoenicea* collected from the cliff in Fig. 8.7. Photo by D.W. Larson.

5. Cliffs are not as hostile as they look, even if resource supply is low and/or very patchy at a small scale. Most environmental extremes are muted by the composition, weathering and aspect of the rock. Cliffs are (within a biological time scale) very stable, predictable and secure environments as the German quotation at the frontispiece of this book suggests. The rate of weathering and erosion of cliffs is the main factor limiting the longevity of the individual organisms and populations on cliffs.

6. Cliff biota show no evidence of successional or other developmental change with time. Thus, in a world where ecological organization is in a continuous state of flux, cliffs appear to be relatively constant.

At the beginning of this chapter we suggested that most cliffs are ancient or virgin ecosystems. Regardless of whether this view is accepted by the reader, it should be clear that, unlike level ground, cliffs retain a very distinctive flora and fauna that exploit the habitat as an environmental and anthropogenic refugium. Consequently, cliffs are important locations for the retention of local biodiversity and should be recognized more widely for this property. This recognition takes on an additional urgency if the 'urban cliff hypothesis' is true, for it means that there may be a tight evolutionary and cultural connection between the biota that humans surround themselves with as food and commensal organisms, and the spontaneous biota of cliff environments. If humans value biodiversity in their food systems (as well as in natural systems), it may be important to intensively explore cliffs for the stable ancestral populations of the species that have given rise to the crop plants. We therefore think that critical tests of all aspects of the 'urban cliff hypothesis' ought to be carried out soon.

### 8.1.5   Opportunities to exploit cliffs as educational tools

Cliffs in our view clearly offer the opportunity to explore the structure, function, genesis and maintenance of whole systems. The fact that the literature is sparse is explained mainly by the lack of specialist interest in whole-system ecology generally, and cliff ecology specifically. We argued in Chapter 1 that there are merits to adding the 'ecology of place' to existing research paths that emphasize 'processes' or 'taxa'. We think that the understanding of cliff ecosystems has been significantly expanded in our own research because of this approach and, similarly, the topics covered in this book illustrate the principles of adopting a whole-system view to any ecosystem. When one considers the commonness of cliffs worldwide, it becomes clear that educators can exploit these dramatic structures within an educational context.

## 8.2   Areas of uncertainty or deficiency

This book has summarized a wide variety of studies on cliffs which explore a variety of topics on cliff features, processes or taxa. In the absence of previous review papers summarizing this material, readers of this book have had to tolerate this first attempt to review and summarize the literature. It should now be clear that few of these studies have been carried out with the intent of understanding how cliffs as 'places' work. Thus, the configuration and scale of questions have been varied, and an enormous lack of consistency is present in the literature regarding the collection of qualitative versus quantitative data. We believe that we can now make some reasonable suggestions on research priorities if the long-term goal is to understand how cliffs work as separate landscape features.

### 8.2.1   Taxonomy and geography

The basis for ecological comparisons on cliffs is limited to macroscopic higher plants and raptors, because the coverage of other groups has been so patchy in the current literature. This is surprising in view of the high rate of recurrence of lichens, mosses, algae and invertebrates. Qualitative studies will therefore be useful in the near future for comparing species lists, and would be more useful if there was more effective sampling of the life-forms that occur on cliffs.

Certain geographical regions also stand out as being grossly under-reported, including China, Russia, west and east Africa, South America, and the western half of North America. Some of this under-reporting is doubtless due to our lack of inclusion of papers, reports, books or mono-graphs not published in the international scientific literature. We are certain that there is a much larger literature available than we have been able to secure, and we welcome the receipt of new or old work that we have not included in this volume. We also welcome the publication of other books or monographs that highlight the errors of fact or interpre-tation that we have made in this volume. When qualitative studies are considered, it would be most useful if investigators considered abiotic factors in the context of their work, even if full quantitative analyses were not carried out. This would allow for a better assessment of the controls of the structure and function of cliff communities. We should hasten to point out, however, that it is not particularly helpful for future workers to try to subjectively assess the environmental conditions on cliffs. There is already too much of this kind of information and, as we have shown above, such subjective assessments of factors such as light or moisture can be wrong.

Our research group could continue indefinitely to do research on the cliffs of the Niagara Escarpment and still we could encounter readers who think that the work has significance only at the local scale. Quantitative studies of cliff ecology in other geographical regions would provide a stronger basis upon which to claim similarities or differences in the organization of cliff communities. Large-scale controls of cliff biota by rock chemistry, geology, geomorphology, weathering and hydrology could be much easier to see at a larger scale.

### 8.2.2  Connections between cliffs and adjacent landscapes

More studies of the cliff edge and talus environments are necessary if an understanding of the entire cliff community is to be gained. Cliff edges are unusual in that they have the physical characteristics of deserts. Unfortunately, there are few formal comparisons of cliff-edge to adjacent cliff-face or level-ground habitats. These comparisons, even in the absence of detailed inventories of the flora and fauna of the cliffs them-selves, would offer insights into the small-scale habitat differences that control the physiology and demography of plant and animal species. Such work would be important from a conservation and management perspective because cliff edges are usually the locations where human and animal disturbance is the greatest. Talus slopes are exposed to many of the same environmental conditions as cliffs, but differ enormously in terms of physical stability. Thus, talus communities are organized mainly by disturbance, as opposed to cliff faces that are organized mainly by factors that slow growth such as environmental stress. These habitats provide a unique opportunity to study how disturbance organizes talus communities as most other disturbance studies focus on the impact of anthropogenic effects on plant and animal populations (Sousa, 1984).

Talus slopes also seem to harbour and conceal biodiversity, especially when one considers that most talus accumulates as blocks of rock separated by large chambers of air. Talus slopes may represent a terrestrial world roughly equivalent to the open oceans where a habitat with modest primary productivity at the surface has a stready rain of organic matter down the water (rock/air) column to the ocean floor (rocky pediment). Completely unknown organisms may occur down through the vertical profile of talus slopes as they do in the oceans. It is possible that the high density of top carnivores in the talus, including spiders and snakes, is attributable to this abundant but mainly invisible life extending from just below the rocky surface down to the pediment. Fibreoptic exploration of talus could confirm these speculations.

### 8.2.3 Archaeology

We have concluded in this book that human use of cliffs is ancient and widespread, but very few archaeologists have specifically noted the connection between cliffs and human biological or cultural evolution. For example, Solomon (1996) presents clear information about the exploitation of cliffs and caves for rock art in southern Africa, but never tries to interpret site selection by palaeohumans within an ecological context. The same is true for most other sites where human settlements in caves at the base of cliffs have been observed. What is the association between humans and cliffs? We have tried to argue in Chapter 7 that the connections are more frequent than one might expect by chance, but we did not set out to test this idea formally. Those skilled in the study of human origins, however, are in a much better position to evaluate formally the degree and form of the dependence of palaeohumans on cliffs. Even if the connection between humans and cliffs is an artifact caused by the exceptional preservation of relics in caves, it is still extremely useful to know the extent to which humans have exploited caves.

### 8.2.4 Food chains

There are numerous gaps in our understanding of the ecology of cliffs. From a taxonomic point of view, there is a need to survey the micro-organismic flora and fauna to understand the range of organisms that participate in the generation of the highly predictable communities of macroscopic plants and animals. Without this knowledge, it is difficult to know whether the predictable assemblages include plants and animals that exploit cliffs more or less independently from each other (that is, not a strictly interacting community) or whether cliffs support tight clusters of species that interact very strongly through the detritus food chain. If one examines Bormann and Likens (1979), one will see that a great deal of the knowledge about hardwood forest ecology emerged from dynamical observations of energy flow and nutrient cycling in the detritus food chain. Borman and Likens (1979) and Odum (1957) before them were able to illustrate some aspects of the trophic structure of small forested watersheds or streams by comparing the biota at different levels of organization. Given the wide variety of stable isotopic tracers that are available for use in ecology, however, such work is easily within reach and would add enormously to the value of the future work.

The results of Davis (1951) and the arguments presented by Larson *et al.* (1989) predict that obligate cliff food chains exist, but the authors of both papers are clearly biased in advocating these predictions. It is equally likely that cliffs recruit organisms that show minimal interaction with each other. In this case, cliffs might support biota abandoned by all other habitats. The assemblage of taxa on cliffs might therefore represent nothing more than slurries of taxa instead of tightly interdependent communities.

At first glance, the above question might appear to be restricted in its application to cliffs. Closer scrutiny, however, reveals an absence of studies that clearly show the difference between communities that are 'slurries' and those that are tightly and internally 'integrated'. There is a great deal of discussion about the requirement for biotic interaction in communities, but there is little evidence that the long-term stability of real communities is dependent on there being tight internal integration, even in non-glaciated tropical communities where there is a long and complex evolutionary history. Mooney (1991) points out that individual species rather than ecosystem processes are impacted in systems stressed by environmental change. If this is true, then following the response of individual taxa to large-scale stimuli such as global warming may be better than trying to follow the restructuring of the entire community.

### 8.2.5 Vertebrate communities

At the other end of the size scale, there are huge gaps in our understanding of the vertebrate community organization on cliffs. The preceding chapters have clearly shown that raptors exploit cliffs in similar ways in all landscapes, but the corresponding information about songbird and small mammal exploitation of cliffs is largely absent. This is true even though these vertebrates are more numerous. On the basis of biomass, these organisms probably play a larger role than raptors in the ecology of cliffs in areas such as seed dispersal and nutrient cycling. Ideally, a series of experiments ought to be carried out in which vertebrates that exploit cliffs are manipulated in such a way as to determine if their fecundity or survivorship is negatively or positively influenced by being moved to level ground, both in the presence and absence of competitors and predators. Such work would determine whether these organisms regard cliffs as 'prime' or 'optimal' habitat, or whether they exploit them as refugia in the same way as many plants and invertebrates.

### 8.2.6 Landscape-scale community organization

Given that cliffs form linear and vertical landscape features it is unlikely that theories of vertebrate community organization (for example the equilibrium theory of island biogeography: MacArthur & Wilson, 1967) in large level-ground landscapes will apply. Equally, there is no information about the way in which plant and animal communities on cliffs (as linear landscape elements) respond to disturbance, productivity or contamination in the surrounding landscape. One could argue that the shape of cliffs or escarpments would make them extremely vulnerable to encroachment of alien biota, or to disturbance from altered ecological processes in the surrounding habitat. It could also be argued that cliff biota, and indeed the entire cliff ecosystem, has so much structural and functional integrity that it is insensitive to a wide variety of disruptions in the surrounding landscape. Work is currently being carried out to evaluate the degree to which cliffs are sensitive to habitat fragmentation effects. While no results are yet available on this topic, we anticipate that the work will provide a useful insight into the question of the structural and functional integrity of cliff ecosystems.

## 8.3 Summary

This book is the first to have been written on the ecology of cliffs and we have attempted to present the current literature available on this topic. There are obvious gaps in our understanding to date, but this is not surprising considering that cliffs have rarely been studied as biological entities. We believe that these gaps in knowledge, however, do not detract from our attempts to understand cliffs as unique places in the landscape. Instead, they open up hundreds of opportunities for scientific inquiry. While attempts have already been made to answer some of the many questions posed about the structure and function of other habitats, this is not the case with cliffs. The study of cliffs is still in its infancy and every question resolved produces an array of new questions that demand resolution. It is an exciting time to be exploring the ecological complexities of a previously ignored 'place'.

# References

Abbott, R.J. (1976). Variation within common groundsel, *Senecio vulgaris* L. *New Phytologist*, **76**, 165–72.

Abrams, M.D. & Orwig, D.A. (1995). Structure, radial growth dynamics and recent climatic variations of a 320-year-old *Pinus rigida* rock outcrop community. *Oecologia*, **101**, 353–60.

Adam, M.D., Lacki, M.J. & Barnes, T.G. (1994). Foraging areas and habitat use of the Virginia big-eared bat in Kentucky. *Journal of Wildlife Management*, **58**, 462–9.

Adam, P., Stricker, P., Wiecek, B.M. & Anderson, D.J. (1990). The vegetation of sea cliffs and headlands in New South Wales, Australia. *Australian Journal of Ecology*, **15**, 515–47.

Adamović, L. (1909). *Die Vegetationsverhältnisse der Balkanländer*. Vol. XI in the series Die Vegetation der Erde, ed. A. Engler & O. Drude. Leipzig: Verlag Wilhelm Engelmann.

Ahmed, M. & Sarangezai, A. (1991). Dendrochronological approach to estimate age and growth rate of various species from Himalayan region of Pakistan. *Pakistan Journal of Botany*, **23**, 78–89.

Ainley, D.G. (1990). Introduction. In *Seabirds of the Farallon Islands*, ed. D.G. Ainley & R.J. Boekelheide, pp. 1-22. Stanford, CA: Stanford University Press.

Ainley, D.G., Boekelheide, R.J., Morrell, S.H. & Strong, C.S. (1990a). Pigeon Guillemot. In *Seabirds of the Farallon Islands*, ed. D.G. Ainley & R.J. Boekelheide, pp. 276–305. Stanford, CA: Stanford University Press.

Ainley, D.G., Boekelheide, R.J., Morrell, S.H. & Strong, C.S. (1990b). Cassin's Auklet. In *Seabirds of the Farallon Islands*, ed. D.G. Ainley & R.J. Boekelheide, pp. 306–38. Stanford, CA: Stanford University Press.

Allen, R.W. (1980). Natural mortality and debility. In *The Desert Bighorn*, ed. G. Monson & L. Summer, pp. 172–85. Tucson, AZ: University of Arizona Press.

Allen, T.F.H. (1971). Multivariate approaches to the ecology of algae on terrestrial rock surfaces in North Wales. *Journal of Ecology*, **59**, 803–26.

Alves, R.J.V. & Kolbek, J. (1993). Penumbral rock communities in campo-rupestre sites in Brazil. *Journal of Vegetation Science*, **4**, 357–66.

Anagnostidis, K., Economou-Amilli, A. & Roussomoustakaki, M. (1983). Epilithic and chasmolithic microflora (Cyanophyta, Bacillariophyta) from marbles of the Parthenon (Acropolis–Athens, Greece). *Nova Hedwigia*, **38**, 227–87.

Andersen, D.E. (1996). Intra-year reuse of great horned owl nest sites by barn owls in east-central Colorado. *Journal of Raptor Research*, **30**, 90–2.

Anderson, R.C., Fralish, J.S. & Baskin, J.M. (eds.) (1999). *Savannas, Barrens, and Rock Outcrop Plant Communities of North America*. Cambridge: Cambridge University Press.

Aramaki, M. (1978). Erosion of coastal cliff at Iwaki Coast in Kukushima prefecture, Japan. *Bulletin of the Association of Natural Science, Senshu University*, **11**, 5–36.

Archambault, S. & Bergeron, Y. (1992). An 802-year tree-ring chronology from the Quebec boreal forest. *Canadian Journal of Forest Research*, **22**, 674–82.

Arya, S.P. (1988). *Introduction to Micrometeorology*. San Diego: Academic Press.

Ashby, W.C., Vogel, W.G., Kolar, C.A. & Philo, G.R. (1984). Productivity of stony soils on strip mines. In *Erosion and Productivity of Soils Containing Rock Fragments: Proceedings of a Symposium*, ed. J.D. Nichols, P.L. Brown & W.J. Grant, pp. 31–44. Madison, WI: Soil Science Society of America.

Ashton, D.H. & Webb, R.N. (1977). The ecology of granite outcrops at Wilson's Promontory, Victoria. *Australian Journal of Ecology*, **2**, 269–96.

Audubon, J.J. (1989). *Audubon's Quadrupeds of North America*. Wellfleet, NJ: Secaucus. Reprint of Audubon, J.J. (1854). *Viviparous Quadrupeds of North America*, 2nd edn. Philadelphia: Audubon.

Baier, B., Banghard, K., Baumgärtner, D. *et al.* (1993). *Hohlwege: Entstehung, Geschichte und Ökologie der Hohlwege im westlichen Kraichgau*. Beihefte zu den Veröffentlichungen der Landesanstalt für Naturschutz und Landschaftspflege Baden-Württemberg. Vol. 72, ed. R. Wolf & D. Hassler. Karlsruhe: Ubstadt-Weiher: Verlag Regionalkultur.

Baker, H.G. (1986). Patterns of plant invasion in North America. In *Ecology of Biological Invasions of North America and Hawaii*, ed. H.A. Mooney & J.A. Drake, pp. 44–57. New York: Springer-Verlag.

Barden, L.S. (1988). Drought and survival in a self-perpetuating *Pinus pungens* population: equilibrium or non-equilibrium? *American Midland Naturalist*, **119**, 253–7.

Bartlett, R.M. & Larson, D.W. (1990). The physiological basis for the contrasting distribution patterns of *Acer saccharum* and *Thuja occidentalis* at cliff edges. *Journal of Ecology*, **78**, 1063–78.

Bartlett, R.M., Matthes-Sears, U. & Larson, D.W. (1990). Organization of the Niagara Escarpment cliff community. II. Characterization of the physical environment. *Canadian Journal of Botany*, **68**, 1931–41.

Bartlett, R.M., Matthes-Sears, U. & Larson, D.W. (1991a). Microsite- and age-specific processes controlling natural populations of *Acer saccharum* at cliff edges. *Canadian Journal of Botany*, **69**, 552–9.

Bartlett, R.M., Reader, R.J. & Larson, D.W. (1991b). Multiple controls of cliff-edge distribution patterns of *Thuja occidentalis* and *Acer saccharum* at the stage of seedling recruitment. *Journal of Ecology*, **79**, 183–97.

Bateman, R.M. (1961). Mammal occurrences in escarpment caves. *Ontario Field Biologist*, **15**, 16–18.

Bates, J.W. (1975). A quantitative investigation of the saxicolous bryophyte and lichen vegetation of Cape Clear Island, County Cork. *Journal of Ecology*, **63**, 143–62.

Batzli, G.O. (1975). The role of small mammals in arctic ecosystems. In *Small Mammals: their Productivity and Population Dynamics*, ed. F.B. Golley, K. Petrusewicz & L. Ryszkowski, pp. 243–68. Cambridge: Cambridge University Press.

Baur, B. & Baur, A. (1990). Experimental evidence for intra-and interspecific competition in two species of rock-dwelling land snails. *Journal of Animal Ecology*, **59**, 301–15.

Beadle, N.C.W. (1981). *The Vegetation of Australia.* Cambridge: Cambridge University Press.

Bédard, J. (1969). The nesting of the Crested, Least, and Parakeet Auklets on Saint Lawrence Island, Alaska. *Condor*, **71**, 386–98.

Bédard, J, Therriault, J.C. & Bérubé, J. (1980). Assessment of the importance of nutrient recycling by seabirds in the St. Lawrence estuary. *Canadian Journal of Fisheries and Aquatic Sciences*, **37**, 583–8.

Bednarz, J.C. (1984). The effect of mining and blasting on breeding Prairie Falcon (*Falco mexicanus*) occupancy in the Caballo Mountains, New Mexico. *Raptor Research*, **18**, 16–19.

Begg, R.J. (1981a). The small mammals of Little Nourlangie Rock, N.T. II. Ecology of *Antechinus bilarni*, the Sandstone Antechinus (Marsupialia: Dasyuridae). *Australian Wildlife Research*, **8**, 57–72.

Begg, R.J. (1981b). The small mammals of Little Nourlangie Rock, N.T. III. Ecology of *Dasyurus hallucatus*, the Northern Quoll (Marsupialia: Dasyuridae). *Australian Wildlife Research*, **8**, 73–85.

Begg, R.J. (1981c). The small mammals of Little Nourlangie Rock, N.T. IV. Ecology of *Zyzomys woodwardi*, the Large Rock-rat, and *Z. argurus*, the Common Rock-rat (Rodentia: Muridae). *Australian Wildlife Research*, **8**, 307–20.

Bell, R.A. (1993). Cryptoendolithic algae of hot semiarid lands and deserts. *Journal of Phycology*, **29**, 133–9.

Belopol'skii, L.O. (1961). *Ecology of Sea Colony Birds of the Barents Sea.* English translation by Israel Program for Scientific Translations, Jerusalem.

Berner, T. & Evenari, M. (1978). The influence of temperature and light penetration on the abundance of the hypolithic algae in the Negev Desert of Israel. *Oecologia*, **33**, 255–60.

Bhattacharyya, A., LaMarche, V.C. Jr & Telewski, F.W. (1988). Dendrochronological reconnaissance of the conifers of northwest India. *Tree-ring Bulletin*, **48**, 21–30.

Biel, J. (1995). Collecting eastern white cedar – *Thuja occidentalis*. *International Bonsai*, **17**(3), 25–7.

Biondi, E. (1988). Aspetti di vegetazione alo-nitrofila sulle coste del Gargano e delle Isole Tremiti. *Archivio Botanico Italiano*, **64**(1/2), 19–33.

Bioret, F., Bouzille, J.B., de Foucault, B., Géhu, J. & Godeau, M. (1987). Le systeme thermo-atlantique pelouses-landes-fourrés des falaises des îles sud-armoricaines (Groix, Belle-Ile, Yeu). *Documents Phytosociologiques*, **11**, 513–31.

Birkhead, T.R. (1977). The effect of habitat and density on breeding success in the Common Guillemot (*Uria aalge*). *Journal of Animal Ecology*, **46**, 751–64.

Birkhead, T.R., Greene, E., Biggins, J.D. & Nettleship, D.N. (1985). Breeding site characteristics and breeding success in Thick-billed Murres. *Canadian Journal of Zoology*, **63**, 1880–4.

Birkhead, T.R. & Nettleship, D.N. (1987). Ecological relationships between Common Murres, *Uria aalge*, and Thick-billed Murres, *Uria lomvia*, at the Gannet

Islands, Labrador. II. Breeding success and site characteristics. *Canadian Journal of Zoology*, **65**, 1630–7.

Birkhead, T.R. & Nettleship, D.N. (1995). Arctic Fox influence on a seabird community in Labrador: a natural experiment. *Wilson Bulletin*, **107**, 397–412.

Birks, H.J.B. (1973). *The Past and Present Vegetation of the Isle of Skye: a Paleoecological Study*. Cambridge: Cambridge University Press.

Birks, H.J.B. (1988). Long-term ecological change in the British uplands. In *Ecological Change in the Uplands*, ed. M.B. Usher & D.B.A.Thompson, pp. 37–56. Oxford: Blackwell Scientific Publications.

Black, D. (1982). The vegetation of the Boyd Plateau, New South Wales, Australia. *Vegetatio*, **50**, 93–111.

Bock, E. & Sand, W. (1993). The microbiology of masonry biodeterioration. *Journal of Applied Bacteriology*, **74**, 503–14.

Boekelheide, R.J., Ainley, D.G., Huber, H.R. & Lewis, T.J. (1990a). Pelagic Cormorant and Double-Crested Cormorant. In *Seabirds of the Farallon Islands*, ed. D.G. Ainley & R.J. Boekelheide, pp. 195–217. Stanford, CA: Stanford University Press.

Boekelheide, R.J., Ainley, D.G., Morrell, S.H., Huber, H.R. & Lewis, T.J. (1990b). Common Murre. In *Seabirds of the Farallon Islands*, ed. D.G. Ainley & R.J. Boekelheide, pp. 245–75. Stanford, CA: Stanford University Press.

Bogart, J.P., Cook, W.J. & Rye, L.A. (1995). *The Distribution of Ambystomatid Salamanders along the Niagara Escarpment*. Toronto: Ontario Heritage Foundation – Final Report.

Bogucki, P. (1996). The spread of early farming in Europe. *American Scientist*, **84**, 242–53.

Bolton, J.J. (1981). Community analysis of vertical zonation patterns on a Newfoundland rocky shore. *Aquatic Botany*, **10**, 299–316.

Bonardi, D. (1966). Contribution à l'étude botanique des inselbergs de Côte d'Ivoire forestière. Diplome d'études Superieures de Sciences Biologiques, Université d' Abidjan.

Bond, J.W. (1989). The tortoise and the hare: ecology of angiosperm dominance and gymnosperm persistence. *Biological Journal of the Linnean Society of London*, **36**, 227–49.

Bonnington, C. (1992). *The Climbers: a History of Mountaineering*. London: BBC Books/Hodder and Stoughton.

Booth, B.D. & Larson, D.W. (1999). Impact of language, history, and choice of system on the study of assembly rules. In *Ecological Assembly Rules: Perspectives, Advances, Retreats*, ed. E.Weiher & P. Keddy, pp. 206–29. Cambridge: Cambridge University Press.

Booth, B.D. & Larson, D.W. (1998). The role of seed rain in determining the assembly of a cliff community. *Journal of Vegetation Science*, **9**, 657–68.

Booth, B.D. (1999). Biological processes controlling the assembly of plant communities on cliff faces. PhD Thesis, University of Guelph, Ontario, Canada.

Booth, T. (1977). Muskox dung; its turnover rate and possible role on Truelove Lowland. In *Truelove Lowland, Devon Island, Canada: a High Arctic Ecosystem*, ed. L.C. Bliss, pp. 531–45. Edmonton: University of Alberta Press.

Bormann, F.H. & Likens, G.E. (1979). *Pattern and Process in a Forested Ecosystem*. Berlin: Springer-Verlag.

Bostick, P.E. (1971). Vascular plants of Panola Mountain, Georgia. *Castanea*, **36**, 194–209.

Boulinier, T. & Danchin, E. (1996). Population trends in Kittiwake *Rissa tridactyla* colonies in relation to tick infestation. *Ibis*, **138**, 326–34.

Bower, B. (1996). Visions on the rocks. *Science News*, **150**, 216–17.

Boyce, D.A. Jr (1987). Nest site characteristics of prairie falcons in the Mojave Desert, California. *Journal of Raptor Research*, **21**, 35–8.

Boyle, A.P., McCarthy, P.M. & Stewart, D. (1987). Geochemical control of saxicolous lichen communities on the Creggaun Gabbro, Letterfrack, Co. Galway, Western Ireland. *Lichenologist*, **19**, 307–17.

Brady, N.C. (1974). *The Nature and Properties of Soils.* New York: Macmillan Publishers.

Braidwood, R.J. (1967). *Prehistoric Men.* Glenview, IL: Scott, Foresman and Co.

Brewer-Carias, C. (1986). *La vegetación del mundo perdido.* Caracas: Julian Steyermark.

Briand, C.H., Posluszny, U. & Larson, D.W. (1992a). Comparative seed morphology of *Thuja occidentalis* (eastern white cedar) from upland and lowland sites. *Canadian Journal of Botany*, **70**, 434–8.

Briand, C.H., Posluszny, U. & Larson, D.W. (1992b). Differential axis architecture in *Thuja occidentalis* (eastern white cedar). *Canadian Journal of Botany*, **70**, 340–8.

Briand, C.H., Posluszny, U. & Larson, D.W. (1993). Influence of age and growth rate on radial anatomy of annual rings of *Thuja occidentalis* L. (eastern white cedar). *International Journal of Plant Sciences*, **154**, 406–11.

Briand, C.H., Posluszny, U., Larson, D.W. & Matthes-Sears, U. (1991). Patterns of architectural variation in *Thuja occidentalis* L. (eastern white cedar) from upland and lowland sites. *Botanical Gazette*, **152**, 494–9.

Broady, P.A. (1981). The ecology of sublithic terrestrial algae at the Vestfold Hills, Antarctica. *British Phycological Journal*, **16**, 231–40.

Brodeur, S. Morneau, F., Décarie, R. Negro, J.J. & Bird, D.M. (1994). Breeding density and brood size of rough-legged hawks in northwestern Québec. *Journal of Raptor Research*, **28**, 259–62.

Brodo, I.M. (1973). Substrate ecology. In *The Lichens*, ed. V. Ahmadjian & M.E. Hale, pp. 401–41. New York: Academic Press.

Brown, C.J. (1991). An investigation into the decline of the bearded vulture *Gypaetus barbatus* in southern Africa. *Biological Conservation*, **57**, 315–37.

Brown, C.J. & Bruton, A.G. (1991). Plumage colour and feather structure of the bearded vulture (*Gypaetus barbatus*). *Journal of Zoology London*, **223**, 627–40.

Brown, C.J. & Cooper, T.G. (1987). The status of cliff-nesting raptors on the Waterberg, SWA/Namibia. *Madoqua*, **15**, 243–9.

Brown, C.R. & Brown, M.B. (1990). The great egg scramble. *Natural History*, **2/90**, 34–40.

Brunton, D.F. & Lafontaine, J.D. (1974). An unusual escarpment flora in western Quebec. *Canadian Field Naturalist*, **88**, 337–44.

Buckley, F.G. & Buckley, P.A. (1980). Habitat selection and marine birds. In *Behavior of Marine Mammals*, Vol. 4: *Marine Birds*, ed. J. Burger, B.L. Olla & H.E. Winn, pp. 69–112. New York and London: Plenum Press.

Buffo, J., Fritschen, L.J. & Murphy, J.L. (1972). *Direct Solar Radiation on Various Slopes from 0 to 60 Degrees North Latitude.* USDA Forest Service Research Paper PNW-142. Portland, Oregon: US Department of Agriculture.

Bunce, R.G.H. (1968). An ecological study of Ysgolion Duon, a mountain cliff in Snowdonia. *Journal of Ecology*, **56**, 59–75.

Bunting, J. (1973). *Climbing.* London: Macmillan.

Burbanck, M.P. & Platt, R.B. (1964). Granite outcrop communities of the Piedmont Plateau in Georgia. *Ecology*, **45**, 292–306.

Burger, J. & Gochfield, M. (1985). Nest site selection by Laughing Gulls: comparison of tropical colonies (Culebra, Puerto Rico) with temperate colonies (New Jersey). *Condor*, **87**, 364–73.

Burton, D.M. (1997). *The History of Mathematics*, 3rd edn. New York: McGraw-Hill.

Burtt, E.H. Jr (1993). Cliff-facing behaviour of the Swallow-tailed Gull *Creagrus furcatus*. *Ibis*, **135**, 459–62.

Bustamante, J. (1994). Behavior of colonial Common Kestrels (*Falco tinnunculus*) during the post-fledging dependence period in southwestern Spain. *Journal of Raptor Research*, **28**, 79–83.

Cade, T.J. & Bird, D.M. (1990). Peregrine Falcons, *Falco peregrinus*, nesting in an urban environment: a review. *Canadian Field Naturalist*, **104**, 209–18.

Cadiou, B., Monnat, J.Y. & Danchin, E. (1994). Prospecting in the kittiwake, *Rissa tridactyla*: different behavioural patterns and the role of squatting in recruitment. *Animal Behaviour*, **47**, 847–56.

Camp, R.J. & Knight, R.L. (1997). Cliff bird and plant communities in Joshua Tree National Park, California, USA. *Natural Areas Journal*, **17**, 110–17.

Caris, J.P.T., Thewessen, T.J.M. & Felix, R. (1989). Genesis of the cliff-face near Bergen op Zoom in the southwest of the Netherlands. *Geologie en Mijnbouw*, **68**, 277–84.

Carson, R. (1962). *Silent Spring.* New York: Fawcett.

Ceballos, O. & Donázar, J.A. (1989). Factors influencing the breeding density and nest-site selection of the Egyptian vulture (*Neophron percnopterus*). *Journal of Ornithology*, **130**, 353–9.

Challinor, D. & Wingate, D.B. (1971). The struggle for survival of the Bermuda Cedar. *Biological Conservation*, **3**, 220–2.

Chapdelaine, G. & Brousseau, P. (1989). Size and trends of Black-legged Kittiwake (*Rissa tridactyla*) populations in the Gulf of St. Lawrence (Quebec) 1974–1985. *American Birds*, **43**, 21–4.

Chapman, L.J. & Putnam, D.F. (1973). *The Physiography of Southern Ontario*, 2nd edn. Toronto: Ontario Research Foundation.

Cherret, J.M. (1989). *Ecological Concepts.* Oxford: Blackwell Scientific Publications.

Chin, S.C. (1977). The limestone hill flora of Malaya I. *Garden's Bulletin, Singapore*, **30**, 165–219.

Churcher, C.S. & Dods, R.R. (1979). *Ochotona* and other vertebrates of possible Illinoian age from Kelso Cave, Halton Co., Ontario. *Canadian Journal of Earth Science*, **16**, 1613–20.

Churcher, C.S. & Fenton, M.B. (1968). Vertebrate remains from the Dickson limestone quarry, Halton, Co., Ontario, Canada. *Bulletin of the National Speleological Society*, **30**, 11–16.

Clapham, A.R., Tutin, T.G. & Warburg, E.F. (1952). *Flora of the British Isles*. Cambridge: Cambridge University Press.

Coates, F. & Kirkpatrick, J.B. (1992). Environmental relations and ecological responses of some higher plant species on rock cliffs in Northern Tasmania. *Australian Journal of Ecology*, **17**, 441–9.

Cole, D.N. (1995a). Experimental trampling of vegetation. I. Relationship between trampling intensity and vegetation response. *Journal of Applied Ecology*, **32**, 203–14.

Cole, D.N. (1995b). Experimental trampling of vegetation. II. Predictors of resistance and resilience. *Journal of Applied Ecology*, **32**, 215–24.

Cooper, A. (1984). Application of multivariate methods to a study of community composition and structure in an escarpment woodland in northeast Ireland. *Vegetatio*, **55**, 93–104.

Cooper, A. (1997). Plant species coexistence in cliff habitats. *Journal of Biogeography*, **24**, 483–94.

Copley, P.B. (1983). Studies on the Yellow-footed Rock-Wallaby, *Petrogale xanthopus* Gray (Marsupialia: Macropodidae) I. Distribution in South Australia. *Australian Wildlife Research*, **10**, 47–61.

Copley, P.B. & Robinson, A.C. (1983). Studies on the Yellow-footed Rock-Wallaby, *Petrogale xanthopus* Gray (Marsupialia: Macropodidae) II. Diet. *Australian Wildlife Research*, **10**, 63–76.

Coulson, J.C. & Wooller, R.D. (1976). Differential survival rates among breeding Kittiwake Gulls *Rissa tridactyla* (L.). *Journal of Animal Ecology*, **45**, 205–13.

Cowles, H.C. (1901). The physiographic ecology of Chicago and vicinity; a study of the origin, development, and classification of plant societies. *Botanical Gazette*, **31**, 73–108.

Cox, E.H.M. (1945). *Plant Hunting in China*. London: William Collins and Sons.

Cox, J.E. & Larson, D.W. (1993a). Spatial heterogeneity of vegetation and environmental factors on talus slopes of the Niagara Escarpment. *Canadian Journal of Botany*, **71**, 323–32.

Cox, J.E. & Larson, D.W. (1993b). Environmental relations of the bryophytic and vascular components of a talus slope plant community. *Journal of Vegetation Science*, **4**, 553–60.

Craig, T.H. & Craig, E.H. (1984). Results of a helicopter survey of cliff nesting raptors in a deep canyon in southern Idaho. *Raptor Research*, **18**, 20–5.

Cramp, S., Bourne, W.R.P. & Saunders, D. (1974). *The Seabirds of Britain and Ireland*. London: William Collins and Sons.

Crawford, R.M.M. (1989). *Studies in Plant Survival*. Oxford: Blackwell Scientific Publications.

Crawley, M.J. (1986). The population biology of invaders. *Philosophical Transactions of the Royal Society of London*, **B314**, 711–31.

Crawley, M.J. (1987). What makes a community invasible? In *Colonization, Succession, and Stability*, ed. A.J. Gray, M.J. Crawley & P.J. Edwards, pp. 429–53. Oxford: Blackwell Scientific Publications.

Crick, H.Q.P. & Ratcliffe, D.A. (1995). The peregrine *Falco peregrinus* breeding population of the United Kingdom in 1991. *Bird Study*, **42**, 1–19.

Croft, D.B. (1987). Socio-ecology of the Antilopine Wallaroo, *Macropus antilopinus*, in the Northern Territory, with observations on sympatric *M. robustus woodwardii* and *M. agilis*. *Australian Wildlife Research*, **14**, 243–55.

Cullen, E. (1957). Adaptations in the kittiwake to cliff-nesting. *Ibis*, **99**, 275–302.

Cullen, W.R., Wheater, C.P. & Dunleavy, P.J. (1998). Establishment of species-rich vegetation in reclaimed limestone quarry faces in Derbyshire, U.K. *Biological Conservation*, **84**, 25–33.

Currie, D.J. (1991). Energy and large-scale patterns of animal- and plant-species richness. *American Naturalist*, **137**, 27–49.

Curtis, J.T. (1959). *The Vegetation of Wisconsin*. Madison: University of Wisconsin Press.

Czechura, G.V. (1984). The peregrine falcon (*Falco peregrinus macropus* Swainson) in southeastern Queensland. *Raptor Research*, **18**, 81–91.

Dahl, E. (1951). On the relation between summer temperatures and the distribution of alpine vascular plants in the lowlands of Fennoscandia. *Oikos*, **3**, 22–50.

Davis, P.H. (1951). Cliff vegetation in the eastern Mediterranean. *Journal of Ecology*, **39**, 63–93.

Davis, P.H. (1965). *Flora of Turkey*. Cambridge: Cambridge University Press.

Day, R.H. (1995). New information on Kittlitz's Murrelet nests. *Condor*, **97**, 271–3.

Day, R.H. & Byrd, G.V. (1989). Food habits of the Whiskered Auklet at Buldir Island, Alaska. *Condor*, **91**, 65–72.

Deacon, H.J., Jury, M.R. & Ellis, F. (1992). Selective regime and time. In *The Ecology of Fynbos*, ed. R. Cowling, pp. 125–79. Capetown: Oxford University Press.

Debrot, A.O. & de Freitas, J.A. (1993). A comparison of ungrazed and livestock-grazed rock vegetations in Curaçao. *Biotropica*, **25**, 270–80.

Decker, F.R. (1931). Prairie falcon's nest. *The Oologist*, **48**, 96.

Delcourt, P.A. & Delcourt, H.R. (1987). *Long-Term Forest Dynamics of the Temperate Zone*. Ecological Studies Series, Vol. 63. New York: Springer-Verlag.

Del Tredici, P. (1989a). Ginkgos and multituberculates: evolutionary interactions in the Tertiary. *Biosystems*, **22**, 327–39.

Del Tredici, P. (1989b). *Early American Bonsai*. Jamaica Plain, MA: Arnold Arboretum of Harvard University.

Del Tredici, P., Ling, H. & Yang, G. (1992). The *Ginkgos* of Tian Mu Shan. *Conservation Biology*, **6**, 202–9.

Dias, J.M.A. & Neal, W.J. (1992). Sea cliff retreat in southern Portugal: profiles, processes and problems. *Journal of Coastal Research*, **8**, 641–54.

Donahue, T. (1996). The big air and the big times of France's Verdon Gorge. *Climbing*, **162**, 110–20.

Donázar, J.A. & Ceballos, O. (1988). Red Fox predation on fledgling Egyptian Vultures. *Journal of Raptor Research*, **22**, 88.

Donázar, J.A. & Fernandez, C. (1990). Population trends of the Griffon Vulture *Gyps fulvus* in Northern Spain between 1969 and 1989 in relation to conservation measures. *Biological Conservation*, **53**, 83–91.

Donázar, J.A., Hiraldo, F. & Bustamante, J. (1993). Factors influencing nest site selection, breeding density and breeding success in the bearded vulture (*Gypaetus barbatus*). *Journal of Applied Ecology*, **30**, 504–14.

Dougan & Associates (1995). *Life Science Inventory of the Mazinaw Rock Cliff Face, Bon Echo Provincial Park*. Toronto: Ministry of Natural Resources – Final Report.

Drake, J.A. (1991). Community-assembly mechanics and the structure of an experimental species ensemble. *American Naturalist*, **137**, 1–26.

Drury, W.H. & Nisbet, I.C.T. (1973). Succession. *Journal of the Arnold Arboretum*, **54**, 331–68.

Egler, F. (1947). Arid southeast Oahu vegetation, Hawaii. *Ecological Monographs*, **17**, 383–438.

Ellenberg, H. (1988). *Vegetation Ecology of Central Europe*. Cambridge: Cambridge University Press.

Ellis, D.H. & Groat, D.L. (1982). A prairie falcon fledgling intrudes at a peregrine falcon eyrie and pirates prey. *Raptor Research*, **16**, 89–91.

Emery, K.O. & Kuhn, G.G. (1982). Sea cliffs: their processes, profiles and classification. *Geological Society of America Bulletin*, **93**, 644–54.

Enderson, J.H. (1964). A study of the prairie falcon in the central Rocky Mountain region. *Auk*, **81**, 332–52.

Escudero, A. (1996). Community patterns on exposed cliffs in a Mediterranean calcareous mountain. *Vegetatio*, **125**, 99–110.

Escudero, A. & Pajarón, S. (1994). Numerical syntaxonomy of the Asplenietalia petrarchae in the Iberian Peninsula. *Journal of Vegetation Science*, **5**, 205–14.

Escudero, A. & Pajarón, S. (1996). La vegetación rupícola del Moncayo silíceo. Una aproximación basada en un Análisis Canónico de Correspondencias. *Lazaroa*, **16**, 105–32.

Escudero, A., Pajarón, S. and Gavilán, R. (1994). Saxicolous communities in the Sierra del Moncayo (Spain): a classificatory study. *Coenoses*, **9**, 15–24.

Etherington, J.R. (1981). Limestone heaths in south-west Britain: their soils and the maintenance of their calcicole–calcifuge mixtures. *Journal of Ecology*, **69**, 277–94.

Evans, P.G.H. & Nettleship, D.N. (1985). Conservation of the Atlantic Alcidae. In *The Atlantic Alcidae*, ed. D.N. Nettleship & T.R. Birkhead, pp. 427–88. London: Academic Press.

Everett, R. & Robson, K. (1991). Rare cliff-dwelling plant species as biological monitors of climate change. *Northwest Environment Journal*, **7**, 352–3.

Ewins, R. (1995). Proto-polynesian art? The cliff paintings of Vatulele, Fiji. *The Journal of the Polynesian Society*, **104**, 23–74.

Faegri, K. (1960). *Maps of the Distribution of Norwegian Plants. 1. The Coast Plants*. Oslo: Oslo University Press.

Fahey, B.D. & Lefebvre, T.H. (1988). The freeze–thaw weathering regime at a section of the Niagara Escarpment on the Bruce Peninsula, southern Ontario, Canada. *Earth Science Processes and Landforms*, **13**, 393–404.

Fala, R.A., Anderson, A. & Ward, J.P. (1985). Highwall-to-pole golden eagle nest site relocations. *Raptor Research*, **19**, 1–7.

Falk, K. & Møller, S. (1997). Breeding ecology of the Fulmar *Fulmarus glacialis* and the Kittiwake *Rissa tridactyla* in high-arctic northeastern Greenland, 1993. *Ibis*, **139**, 270–81.

Feldhamer, G.A., Gates, J.E. & Chapman, J.A. (1984). Rare, threatened, endangered and extirpated mammals from Maryland. In *Threatened and Endangered Plants and*

*Animals of Maryland*, ed. A.W. Norden, D.C. Forester & G.H. Fenwick, pp. 395–438. Maryland Natural Heritage Program Special Publication, 84–I. Towson, MD: Maryland Department of Natural Resources.

Ferris, F.G. & Lowson, E.A. (1997). Ultrastructure and geochemistry of endolithic microorganisms in limestone of the Niagara Escarpment. *Canadian Journal of Microbiology*, **43**, 211–19.

Ficht, B., Hepp, K., Künkele, G., Schilling, F. & Schmid, F. (1995). Lebensraum Fels. *Beihefte zu den Veröffentlichungen der Landesanstalt für Naturschutz und Landschaftspflege in Baden-Württemberg*, **82**, 49–162.

Findley, R. (1990). Will we save our own? *National Geographic*, **178**, 106–36.

Fisher, F.J.F. (1952). Observations on the vegetation of screes in Canterbury, New Zealand. *New Zealand Journal of Ecology*, **40**,156–67.

Fitter, A.H. & Hay, R.K.M. (1987). *Environmental Physiology of Plants*, 2nd edn. London: Academic Press.

Flint, A.L. & Childs, S. (1984). Physical properties of rock fragments and their effect on available water in skeletal soils. In *Erosion and Productivity of Soils Containing Rock Fragments: Proceedings of a Symposium*, ed. J.D. Nichols, P.C. Brown & W.J. Grant, pp. 91–103. Madison, WI: Soil Science Society of America.

Flint, V.E. (1995). Recent threats to large falcons *Falco sp.* in Russia and neighbouring countries. *Acta Ornithologica*, **30**, 23–4.

Foote, K.G. (1966). The vegetation of lichens and bryophytes on limestone outcrops in the Driftless area of Wisconsin. *Bryologist*, **69**, 265–92.

Ford, D. & Williams, P. (1989). *Karst Geomorphology and Hydrology*. London: Unwin Hyman.

Foster, D.R. & Boose, E.R. (1992). Patterns of forest damage resulting from catastrophic wind in central New England, USA. *Journal of Ecology*, **80**, 79–98.

Fox, B. (1991). Ancient Onenditagui. *Cuesta*, 1991, 11–13.

Fox, W.A. (1990). The Odawa. In *The Archeology of Southern Ontario to A.D. 1650*, ed. C.J. Ellis & N. Ferris, pp. 457–74. London, Ontario: London Chapter, Ontario Archeological Society, Number 5.

Freeland, W.J., Winter, J.W. & Raskin, S. (1988). Australian rock-mammals: a phenomenon of the seasonally dry tropics. *Biotropica*, **20**, 70–9.

Freethy, R. (1987). *Auks (An Ornithologist's Guide)*. New York: Facts on File.

Fritts, H. C. (1976). *Tree Rings and Climate*. London: Academic Press.

Fuls, E.R., Bredenkamp, G.J. & van Rooyen, N. (1992). Plant communities of the rocky outcrops of the northern Orange Free State, South Africa. *Vegetatio*, **103**, 79–92.

Fuls, E.R., Bredenkamp, G.J. & van Rooyen, N. (1993). Low thicket communities of rocky outcrops in the northern Orange Free State. *South African Journal of Botany*, **59**, 360–9.

Furness, R.W. & Barrett, R.T. (1985). The food requirements and ecological relationships of a seabird community in North Norway. *Ornis Scandinavica*, **16**, 305–13.

Garnier, B.J. & Ohmura, A. (1968). A method of calculating the direct shortwave radiation income of slopes. *Journal of Applied Meteorology*, **7**, 796–800.

Gaston, A.J. (1992). *The Ancient Murrelet*. London: T. and A.D. Poyser.

Gaston, A.J. & Elliot, R.D. (1996). Predation by Ravens *Corvus corax* on Brunnich's Guillemot *Uria lomvia* eggs and chicks and its possible impact on breeding site selection. *Ibis*, **138**, 742–8.

Gaston, A.J. and Nettleship, D.N. (1981). *The Thick-Billed Murres of Prince Leopold Island*. Canadian Wildlife Service Monograph Series, Number 6. Ottawa: Canadian Wildlife Service.

Gates, D.M. & Schmerl, R.B. (1975). *Perspectives on Biophysical Ecology*. New York: Springer-Verlag.

Géhu, J.M. (1964). Sur la végétation phanérogamique halophile des falaises Bretonnes. *Revue Générale de Botanique*, **71**, 73–8.

Géhu, J.M., Costa, M. & Uslu, T. (1990). Analyse phytosociologique de la végétation littorale des côtes de la partie turque de l'île de chypre dans un souci conservatoire. *Documents Phytosociologiques*, **12**, 203–34.

Geist, V. (1971). *Mountain Sheep*. Chicago: University of Chicago Press.

Gerrath, J.F., Gerrath, J.A. & Larson, D.W. (1995). A preliminary account of endolithic algae of limestone cliffs of the Niagara Escarpment. *Canadian Journal of Botany*, **73**, 788–93.

Gilbert, B.A. & Raedeke, K.J. (1992). Winter habitat selection of mountain goats in the North Tolt and Mine Creek drainages of the north central Cascades. In *Biennial Symposium of the Northern Wild Sheep and Goat Council*, Vol. 8, ed. J. Emmerich & W.G. Hepworth, pp. 305–24. Cody, WY.

Gilbert, G.K. (1871). Notes of investigations at Cohoes with reference to the circumstances of the deposition of the skeleton of Mastodon. *State Cabinet of Natural History*, 21st Annual Report, pp. 129–48. Albany, NY.

Gilchrist, H.G. & Gaston, A.J. (1997a). Effects of murre nest site characteristics and wind conditions on predation by Glaucous Gulls. *Canadian Journal of Zoology*, **75**, 518–24.

Gilchrist, H.G. & Gaston, A.J. (1997b). Factors affecting the success of colony departure by Thick-billed Murre chicks. *Condor*, **99**, 345–52.

Gildner, B.S. & Larson, D.W. (1992a). Seasonal changes in photosynthesis in the desiccation-tolerant fern *Polypodium virginianum*. *Oecologia*, **89**, 383–9.

Gildner, B.S. & Larson, D.W. (1992b). Photosynthetic response to sunflecks in the desiccation-tolerant fern *Polypodium virginianum*. *Oecologia*, **89**, 390–6.

Godwin, H. (1956). *The History of the British Flora*. Cambridge: Cambridge University Press.

Goldsmith, F.B. (1973a). The vegetation of exposed sea cliffs at South Stack, Anglesey. I. The multivariate approach. *Journal of Ecology*, **61**, 787–818.

Goldsmith, F.B. (1973b). The vegetation of exposed sea cliffs at South Stack, Anglesey. II. Experimental studies. *Journal of Ecology*, **61**, 819–29.

Gotelli, N.J. (1993). Ant Lion zones: causes of high-density predator aggregations. *Ecology*, **74**, 226–37.

Gotelli, N.J. (1996). Ant community structure: effects of predatory Ant Lions. *Ecology*, **77**, 630–8.

Grace, J. (1977). *Plant Response to Wind*. London: Academic Press.

Graham, A. (1973). *Vegetation and Vegetation History of Northern Latin America*. Amsterdam: Elsevier Scientific Publishers.

Graham, H. (1980). Impact of modern man. In *The Desert Bighorn*, ed. G. Monson, & L. Summer, pp. 288–309. Tucson, AZ: University of Arizona Press.

Grebence, B.L. & White, C.M. (1989). Physiographic characteristics of peregrine falcon nesting habitat along the Colorado River system in Utah. *Great Basin Naturalist*, **49**, 408–18.

Grime, J.P. (1979). *Plant Strategies and Vegetation Processes*. Chichester: John Wiley and Sons.

Guariguata, M.R. (1990). Landslide disturbance and forest regeneration in the Upper Luquillo Mountains of Puerto Rico. *Journal of Ecology*, **78**, 814–32.

Guilday, J.E. (1971). The Pleistocene history of the Appalachian mammal fauna. In *The Distributional History of the Biota of the Southern Appalachians*, Part III, ed. P.C. Holt, R.A. Paterson & J.P. Huddard, pp. 233–61. Blacksburg, VA: Virginia Polytechnic Institute and State University.

Guitián, J. & Guitián, P. (1989). La influencia de las colonias de aves marinas en la vegetación de los acantilados del noroeste Ibérico. *Boletim Sociedade Broteriana, Séries 2*, **62**, 77–86.

Guyette, R.P., Cutter, B.E. & Henderson, G.S. (1989). Long-term relationships between molybdenum and sulfur concentrations in redcedar tree rings. *Journal of Environmental Quality*, **18**, 385–9.

Häyren, E. (1940). Die Algenvegetation der Sickerwasserstreifen auf den Felsen in Südfinnland. *Societas Scientiarum Fennica Commentationes Biologicae*, **7**, 1–19.

Hafner, D.J. (1993). North American Pika (*Ochotona princeps*) as a Late Quaternary biogeographic indicator species. *Quaternary Research*, **39**, 373–80.

Haig, A.R. (1997). *The Effect of Habitat Fragmentation on Lichen and Plant Species Richness of Niagara Escarpment Cliffs in Ontario, Canada*. MSc Thesis, University of Guelph, Canada.

Haigh, M.J. (1987). The Holon: hierarchy theory and landscape research. In *Geomorphological Models*, Catena supplement 10, ed. F. Ahnert, pp. 181–92. Cremlingen, Germany: Catena Verlag.

Haley, D. (1984). Introduction. In *Seabirds*, ed. D. Haley, pp. 12–19. Seattle: Pacific Search Press.

Hall, B. (1992). Collected bonsai symposium. *International Bonsai*, **14**(2), 34–42.

Hallberg, H.P. & Ivarsson, R. (1965). Vegetation of coastal Bohuslän. *Acta Phytogeographica Suecica*, **50**, 111–22.

Hambler, D.J. (1961). A poikilohydrous, poikilochlorophyllous angiosperm from Africa. *Nature*, **191**, 1415–16.

Hambler, D.J. (1964). The vegetation of granitic outcrops in western Nigeria. *Journal of Ecology*, **52**, 573–94.

Hanson, C.T. & Blevins, R.L. (1979). Soil water in coarse fragments. *Journal of the Soil Science Society of America*, **43**, 819–20.

Harris, M.P. (1984). *The Puffin*. Staffordshire, UK: T. & A.D. Poyser.

Harris, M.P., Wanless, S. & Barton, T.R. (1996). Site use and fidelity in the common guillemot *Uria aalge*. *Ibis*, **138**, 399–404.

Harris, M.P., Wanless, S., Barton, T.R. & Elston, D.A. (1997). Nest-site characteristics, duration of use and breeding success in the Guillemot *Uria aalge*. *Ibis*, **139**, 468–76.

Hartley, R.R., Bodington, G., Dunkley, A.S. & Groenewald, A. (1993). Notes on the breeding biology, hunting behavior, and ecology of the Taita falcon in Zimbabwe. *Journal of Raptor Research*, **27**, 133–42.

Hatch, S.A. (1989). Diurnal and seasonal patterns of colony attendance in the northern fulmar, *Fulmarus glacialis*, in Alaska. *Canadian Field Naturalist*, **103**, 248–60.

Haynes, L.A. (1992). Mountain Goat habitat of Wyoming's Beartooth Plateau: implications for management. In *Biennial Symposium of the Northern Wild Sheep and Goat Council*, Vol. 8, ed. J. Emmerich & W.G. Hepworth, pp. 325–39. Cody, WY.

Hays, L.L. (1987). Peregrine falcon nest defense against a golden eagle. *Journal of Raptor Research*, **21**, 67.

Healy, T. & Kirk, R.M. (1982). Coasts. In *Landforms of New Zealand*, ed. J.M. Soons & M.J. Selby, pp. 81–104. Auckland: Longman Paul.

Hedderson, T.A. & Brassard, G.R. (1990). Microhabitat relationships of five co-occurring saxicolous mosses on cliffs and scree slopes in eastern Newfoundland. *Holarctic Ecology*, **13**, 134–42.

Hellmers, H., Horton, J.S., Juhren, G. & O'Keefe, J. (1955). Root systems of some chaparral plants in southern California. *Ecology*, **36**, 667–79.

Henderson, M.R. (1939). The flora of the limestone hills of the Malay Peninsula. *Journal of the Malayan Branch, Royal Asiatic Society*, **17**, 13–87.

Hepburn, I. (1943). A study of the vegetation of sea-cliffs in north Cornwall. *Journal of Ecology*, **31**, 30–9.

Herrington, R.E. (1988). Talus use by amphibians and reptiles in the Pacific Northwest. In *Management of Amphibians, Reptiles and Small Mammals in North America*, USDA Forest Service General Technical Report RM-166, ed. R.C. Szaro, K.E. Severson & D.R. Patton, pp. 216–21. Flagstaff, AZ: USDA Forest Service.

Herter, W. (1993). *Gefährdung der Xerothermvegetation des Oberen Donautals – Ursachen und Konsequenzen*. Veröffentlichungen des Projekts 'Angewandte Ökologie', **7**, 163–76. Karlsruhe: Landesanstalt für Umweltschutz Baden-Württemberg.

Herter, W. (1996). *Die Xerothermvegetation des Oberen Donautals – Gefährdung der Vegetation durch Mensch und Wild sowie Schutz- und Erhaltungsvorschläge*. Veröffentlichungen des Projekts 'Angewandte Ökologie', **10**, 1–274. Karlsruhe: Landesanstalt für Umweltschutz Baden-Württemberg.

Hétu, B. (1992). Coarse cliff-top aeolian sedimentation in northern Gaspésie, Québec, Canada. *Earth Surface Processes and Landforms*, **17**, 95–108.

Hoebel, E.A. (1966). *Anthropology: the Study of Man*. New York: McGraw-Hill.

Hoffman, B.W. (1990). Bird netting, cliff-hanging, and egg gathering: traditional procurement strategies on Nunivak Island. *Arctic Anthropology*, **27**, 66–74.

Hofman, F., Nowak, R. & Winkler, S. (1974). Substrate dependence of calcareous and silicate rock inhabiting lichens of the island Čiovo, Yugoslavia. *Journal of the Hattori Botanical Laboratory*, **38**, 313–25.

Holmen, H. (1965). Subalpine tall herb vegetation, site and standing crop. *Acta Phytogeographica Suecica*, **50**, 240–8.

Hoogstraal, H. (1978). Biology of ticks. In *Tick-Borne Diseases and Their Vectors*, ed. J.K.H. Wilde, pp. 3–14. Edinburgh: Centre for Tropical Veterinary Medicine.

Hora, F.B. (1947). The pH range of some cliff plants on rocks of different geological origin in the Cader Idris area of north Wales. *Journal of Ecology*, **35**, 158–65.

Horsup, A. (1994). Home range of the allied rock-wallaby, *Petrogale assimilis*. *Wildlife Research*, **21**, 65–84.

Horsup, A. & Marsh, H. (1992). The diet of the allied rock-wallaby, *Petrogale assimilis*, in the wet–dry tropics. *Wildlife Research*, **19**, 17–33.

Hosking, E. & Lockley, R.M. (1984). *Seabirds of the World*. Beckenham, UK: Croom Helm.

Hotchkiss, A.T., Woodward, H.H., Muller, L.F. & Medley, M.E. (1986). *Thuja occidentalis* L. in Kentucky. *Transactions of the Kentucky Academy of Science*, **47**, 99–100.

Houle, G. & Delwaide, A. (1991). Population structure and growth–stress relationship of *Pinus taeda* in rock outcrop habitats. *Journal of Vegetation Science*, **2**, 47–58.

Houle, G. & Phillips, D.L. (1988). The soil seed bank of granite outcrop plant communities. *Oikos*, **52**, 87–93.

Hovis, J., Snowman, T.D., Cox, V.L., Fay, R. & Bildstein, K.L. (1985). Nesting behavior of Peregrine Falcons in west Greenland during the nestling period. *Raptor Research*, **19**, 15–19.

Huntley, B. (1979). The past and present vegetation of the Caenlochan National Nature Reserve, Scotland. I. Present vegetation. *New Phytologist*, **83**, 215–83.

Hurlbert, S.H. (1984). Pseudoreplication and the design of ecological field experiments. *Ecological Monographs*, **54**, 187–211.

Ingestad, T. & Lund, A-B. (1986). Theory and techniques for steady-state mineral nutrition and growth of plants. *Scandinavian Journal of Forest Research*, **1**, 439–53.

Ishizuka, K. (1974). Maritime vegetation. In *The Flora and Vegetation of Japan*, ed. M. Numata, pp. 239–310. Amsterdam: Elsevier.

Jaag, O. (1945). Untersuchungen über die Vegetation und Biologie der Algen des nackten Gesteins in den Alpen, im Jura und im schweizerischen Mittelland. *Beiträge zur Kryptogamenflora der Schweiz*, **9**, 1–559.

Jackson, G. & Sheldon, J. (1949). The vegetation of magnesian limestone cliffs at Markland Grips near Sheffield. *Journal of Ecology*, **37**, 38–50.

Jahns, H.M. & Fritzler, E. (1982). Flechtenstandorte auf einer Blockhalde. *Herzogia*, **6**, 243–70.

Janes, S.W. (1985). Habitat selection in raptorial birds. In *Habitat Selection in Birds*, ed. M.L. Cody, pp. 159–88, Orlando, FL: Academic Press.

Janes, S.W. (1994). Partial loss of red-tailed hawk territories to Swainson's hawks: relations to habitat. *Condor*, **96**, 52–7.

Jarvis, S.C. (1974). Soil factors affecting the distribution of plant communities on the cliffs of Craig Breidden, Montgomeryshire. *Journal of Ecology*, **62**, 721–33.

Jarvis, S.C. & Pigott, C.D. (1973). Mineral nutrition of *Lychnis viscaria*. *New Phytologist*, **72**, 1047–55.

Jeffries, D. (1985). Analysis of the vegetation and soils of glades on calico rock sandstone in northern Arkansas. *Bulletin of the Torrey Botanical Club*, **112**, 70–3.

Jeffries, M.J. & Lawton, J.H. (1984). Enemy free space and the structure of ecological communities. *Biological Journal of the Linnean Society of London*, **23**, 269–86.

Johnson, D. (1986). Desert buttes: natural experiments for testing theories of island biogeography. *National Geographic Research*, **2**, 152–66.

Jolly, P. (1996). Symbiotic interaction between black farmers and south-eastern San. *Current Anthropology*, **37**, 277–305.

Jones, D.P. & Graham, R.C. (1993). Water-holding characteristics of weathered granite rock in chaparral and forest ecosystems. *Journal of the Soil Science Society of America*, **57**, 256–61.

Jones, I.L. (1992). Factors affecting survival of adult least auklets (*Aethia pusilla*) at St. Paul Island, Alaska. *The Auk*, **109**, 576–84.

Jones, J.R., Cameron, B. & Fisher, J.J. (1993). Analysis of cliff retreat and shoreline erosion: Thompson Island, Massachusetts, USA. *Journal of Coastal Research*, **9**, 87–96.

Jones, W. (1994). *The Biology and Management of the Dwarf Mountain Pine (Microstrobos fitzgeraldii) in New South Wales*. New South Wales National Parks and Wildlife Service Species Management Report, 13. Huntsville, NSW: New South Wales National Parks and Wildlife Service.

Jones, W.G., Hill, K.D. & Allen, J.M. (1995). *Wollemia nobilis*, a new living Australian genus and species in the Araucariaceae. *Telopea*, **6**, 173–6.

Jung, E. (1961). Die Waldgesellschaften der hinteren Sächsischen Schweiz am Beispiel des Grossen Zschandes. *Arbeitsgemeinschaft Sächsischer Botaniker*, **3**, 75–112.

Jurko, A. & Peciar, V. (1963). Pflanzengesellschaften an schattigen Felsen in den Westkarpaten, *Vegetatio*, **11**, 199–209.

Jurmain, R., Nelson, H. & Turnbaugh, W.A. (1990). *Understanding Physical Anthropology and Archeology*, 4th edn. St Paul, MN: West Publishing Co.

Kallio, P., Laine, U. & Mäkinen, Y. (1969). Vascular flora of Inari Lapland. 1. Introduction and Lycopodiaceae–Polypodiaceae. *Reports from the Kevo Subarctic Research Station*, **5**, 1–108.

Kallio, P., Laine, U. & Mäkinen, Y. (1971). Vascular flora of Inari Lapland. 2. Pinaceae and Cupressaceae. *Reports from the Kevo Subarctic Research Station*, **8**, 73–100.

Kampp, K. (1990). The Thick-billed Murre population of the Thule district, Greenland. *Arctic*, **43**, 115–20.

Karlsson, L. (1973). Autecology of cliff and scree plants in Sarek National Park, northern Sweden. *Växtekologiska Studier*, **4**, 1–203.

Keddy, P. A. (1989). *Competition*. London: Chapman and Hall.

Keever, C., Oosting, H.J. & Anderson, L.E. (1951). Plant succession on exposed granite of Rocky Face Mountain, Alexander County, North Carolina. *Bulletin of the Torrey Botanical Club*, **78**, 401–21.

Kelly, P.E., Cook, E.R. & Larson, D.W. (1992). Constrained growth, cambial mortality, and dendrochronology of ancient *Thuja occidentalis* on cliffs of the Niagara Escarpment: an eastern version of bristlecone pine? *International Journal of Plant Sciences*, **153**, 117–27.

Kelly, P.E., Cook, E.R. & Larson, D.W. (1994). A 1397-yr tree-ring chronology of *Thuja occidentalis* from cliff faces of the Niagara Escarpment, southern Ontario, Canada. *Canadian Journal of Forest Research*, **24**, 1049–57.

Kelly, P.E. & Larson, D.W. (1997a). Effects of rock climbing on populations of presettlement eastern white cedar (*Thuja occidentalis*) on cliffs of the Niagara Escarpment, Canada. *Conservation Biology*, **11**, 1125–32.

Kelly, P.E. & Larson, D.W. (1997b). Dendroecological analysis of the population dynamics of an old-growth forest on cliff-faces of the Niagara Escarpment, Canada .*Journal of Ecology*, **85**, 467–78.

Kelly, W. E. (1980). Predator relationships. In *The Desert Bighorn*, ed. G. Monson & L. Summer, pp. 186–96. Tucson, AZ: University of Arizona Press.

Kimmins, H. (1987). *Forest Ecology*. New York: MacMillan Publishers.

Kinnear, J.E., Onus, M.L. & Bromilow, R.N. (1988). Fox control and rock-wallaby population dynamics. *Australian Wildlife Research*, **15**, 435–50.

Kiviat, E. (1991). *The Northern Shawangunks, an Ecological Survey*. New Paltz, NY: Mohonk Preserve, Inc.

Klötzli, F. (1991). Niches of longevity and stress. In *Modern Ecology*, ed. G. Esser & D. Overdieck, pp. 97–110. Amsterdam: Elsevier.

Kolasa, J. & Pickett, S.T.A. (1989). Ecological systems and the concept of biological organization. *Proceedings of the National Academy of Science*, **86**, 8837–41.

Komar, P.D. & Shih, S.-M. (1993). Cliff erosion along the Oregon Coast: a tectonic-sea level imprint plus local controls by beach processes. *Journal of Coastal Research*, **9**, 747–65.

Koponen, S. (1990). Spiders (Araneae) on the cliffs of the Forillon National Park, Québec. *Naturaliste Canadien*, **117**, 161–5.

Koreshoff, D. (1984). *Bonsai, its Art, Science, History and Philosophy*. Brisbane, Australia: Boolarong Publications.

Küppers, M. (1984). Carbon relations and competition between woody species in a central European hedgerow. I. Photosynthetic characteristics. *Oecologia*, **64**, 332–43.

Küppers, M. (1985). Carbon relations and competition between woody species in a central European hedgerow. IV. Growth form and partitioning. *Oecologia*, **66**, 343–52.

Lack, D. (1968). *Ecological Adaptations for Breeding Birds*. London: Methuen.

Lacki, M.J., Adam, M.D. & Shoemaker, L.G. (1993). Characteristics of feeding roosts of Virginia Big-eared Bats in Daniel Boone National Forest. *Journal of Wildlife Management*, **57**, 539–43.

LaMarche, V.C. Jr (1963). Origin and geologic significance of buttress roots of Bristlecone Pines, White Mountains, California. *U.S. Geological Survey Professional Paper*, **475–C**, C149–C150.

LaMarche, V. C. Jr (1968). Rates of slope degradation as determined from botanical evidence, White Mountains, California. *U.S. Geological Survey Professional Paper*, **352**, 341–77.

Lammers, T.G. (1980). The vascular flora of Starr's Cave State Preserve. *Proceedings of the Iowa Academy of Science*, **87**, 148–58.

Landers, R.Q. & Graf, D. (1975). The oldest Iowa tree, II. Eastern Red Cedar on Cedar River Bluffs. *Proceedings of the Iowa Academy of Science*, **82**, 123.

Lange, O.L. (1953). Hitze- und Trockenresistenz der Flechten in Beziehung zu ihrer Verbreitung. *Flora*, **140**, 39–97.

Larson, D.W. (1980). Patterns of species distribution in an *Umbilicaria* dominated community. *Canadian Journal of Botany*, **58**, 1269–79.

Larson, D.W. (1990). Effects of disturbance on old-growth *Thuja occidentalis* at cliff edges. *Canadian Journal of Botany*, **68**, 1147–55.

Larson, D.W., Doubt, J. & Matthes-Sears, U. (1994a). Radially sectored hydraulic pathways in the xylem of *Thuja occidentalis* as revealed by the use of dyes. *International Journal of Plant Sciences*, **155**, 569–82.

Larson, D.W. & Kelly, P.E. (1991). The extent of old-growth *Thuja occidentalis* on cliffs of the Niagara Escarpment. *Canadian Journal of Botany*, **69**, 1628–36.

Larson, D.W., Kelly, P.E. & Matthes-Sears, U. (1994b). Calling nature's bluff: the secret world of cliffs. In *Encyclopaedia Britannica 1995 Yearbook of Science and the Future*, ed. D. Calhoun, pp. 28–47. Chicago: Encyclopedia Britannica Inc.

Larson, D.W., Matthes, U., Gerrath, J.A. *et al.* (1999). Ancient stunted trees on cliffs. *Nature*, **398**, 382–3.

Larson, D.W., Matthes-Sears, U. & Kelly, P.E. (1993). Cambial dieback and partial shoot mortality in cliff-face *Thuja occidentalis*: evidence for sectored radial architecture. *International Journal of Plant Sciences*, **154**, 496–505.

Larson, D.W., Matthes-Sears, U. & Kelly, P.E. (1999). The cliff ecosystem of the Niagara Escarpment. In *Savannas, Barrens, and Rock Outcrop Plant Communities of North America*, ed. R.C. Anderson, J.S. Fralish & J.M. Baskin, pp. 362–74. Cambridge: Cambridge University Press.

Larson, D.W., Spring, S.H., Matthes-Sears, U. & Bartlett, R.M. (1989). Organization of the Niagara Escarpment cliff community. *Canadian Journal of Botany*, **67**, 2731–42.

Lauterer, P. (1991). Psyllids (Homoptera, Psylloidea) of the limestone cliff zone of the Pavlovske Vrchy hills (Czechoslovakia). *Casopsis Moravskeho Muzea Vedy Prirodni*, **76**, 241–63.

Lavranos, J.J. (1995). *Aloe whitcombei* and *A. collenetteae*, two new cliff-dwelling species from Oman, Arabia. *Cactus and Succulent Journal*, **67**, 30–3.

Lawrey, J.D. (1980). Correlations between lichen secondary chemistry and grazing activity by *Pallifera varia*. *Bryologist*, **83**, 328–34.

Lawrey, J.D. (1984). *Biology of Lichenized Fungi*. New York: Praeger.

Lawrey, J.D. (1987). Nutritional ecology of lichen/moss arthropods. In *Nutritional Ecology of Insects, Mites and Spiders*, ed. F. Slansky Jr & J.G. Rodriguez, pp. 209–33. New York: John Wiley & Sons.

Lawton, R.O. (1982). Wind stress and elfin stature in a montane rain forest tree: an adaptive explanation. *American Journal of Botany*, **69**, 1224–30.

Layton, R. (1992). *Australian Rock Art*. Cambridge: Cambridge University Press.

Leclerc, J.C., Couté, A. & Dupuy, P. (1983). Le climat annuel de deux grottes et d'une église du Poitou, ou vivent des colonies pures d'algues sciaphiles. *Cryptogamie Algologie*, **4**, 1–19.

Leedy, R.R. (1972). The status of the prairie falcon in western Montana: special emphasis on possible effects of chlorinated hydrocarbon insecticides. MS Thesis, University of Montana, Missoula.

Lewis, D.C. & Burgy, R.H. (1964). The relationship between oak roots and groundwater in fractured rock as determined by tritium tracing. *Journal of Geophysical Research*, **69**, 2579–88.

Lewis-Williams, J.D. (1981). *Believing and Seeing*. London: Academic Press.

Lindqvist, O.V. (1964). The spider fauna of the cliffs in eastern Finnish Lapland. *Reports from the Kevo Subarctic Research Station*, **1**, 288–91.

Lisci, M. (1997). Flora vascolare dei muri in aree urbane della Toscana centro-meridionale. *Webbia*, **52**, 43–66.

Lisci, M. & Pacini, E. (1993a). Plants growing on the walls of Italian towns. 1. Sites and distribution. *Phyton*, **33**, 15–26.

Lisci, M. & Pacini, E. (1993b). Plants growing on the walls of Italian towns. 2. Reproductive ecology. *Giornale Botanico Italiano*, **127**, 1053–78.

Lovelock, J.E. (1979). *Gaia*. Oxford: Oxford University Press.

Luckman, B.H. (1976). Rockfalls and rockfall inventory data: some observations from Surprise Valley, Jasper National Park, Canada. *Earth Surface Processes*, **1**, 287–98.

Lundqvist, J. (1968). Plant cover and environment of steep hillsides in Pite Lappmark. *Acta Phytogeographica Suecica*, **53**, 1–153.

Lüth, M. (1993). *Felsen und Blockhalden*. Biotope in Baden-Württemberg, 6. Karlsruhe: Umweltministerium Baden-Württemberg.

Lynn, D.H. & Olson, V. (1990). *Microbes of the Niagara Escarpment May–August 1990*. Toronto: Ontario Heritage Foundation – Final Report.

MacArthur, R.H. & Wilson, E.O. (1967). *The Theory of Island Biogeography*. Princeton, NJ: Princeton University Press.

Maccarone, A.D. & Montevecchi, W.A. (1981). Predation and caching of seabirds by red foxes (*Vulpes vulpes*) on Baccalieu Island, Newfoundland. *Canadian Field Naturalist*, **95**, 352–3.

Makirinta, U. (1985). Vegetation types and exposure on acid rocks in south Häme, south Finland. *Colloques Phytosociologiques, Végétation et Géomorphologie*, **13**, 469–84.

Malloch, A.J.C. (1971) Vegetation of the maritime cliff tops on the Lizard and Land's End Peninsulas, West Cornwall. *New Phytologist*, **70**, 1155–97.

Malloch, A.J.C., Bamidele, J.F. & Scott, A.M. (1985). The phytosociology of British sea-cliff vegetation with special reference to the ecophysiology of some maritime cliff plants. *Vegetatio*, **62**, 309–18.

Malloch, A.J.C. & Okusanya, O.T. (1979). An experimental investigation into the ecology of some maritime cliff species. I. Field observations. *Journal of Ecology*, **67**, 283–92.

Manders, P.T. (1986). An assessment of the current status of the Clanwilliam Cedar (*Widdringtonia cedarbergensis*) and the reasons for its decline. *South African Forestry Journal*, **139**, 48–53.

Manry, D. (1993). Cliff-hanger in Morocco. *International Wildlife*, **23**, 40–3.

Marks, P.L. (1983). On the origin of the field plants of the northeastern United States. *American Naturalist*, **122**, 210–28.

Marloth, R. (1913). *The Flora of South Africa*, Vol. 1. Capetown: Darter Bros. & Co.

Maser, C., Rodiek, J.E. & Thomas, J.W. (1979). Cliffs, talus, and caves. In *Wildlife Habitats in Managed Forests of the Blue Mountains of Oregon and Washington*, USDA Handbook 553, ed. J.W. Thomas, pp. 96–103. Washington, DC: United States Department of Agriculture.

Matheson, J. (1995). Organization of forest bird and small mammal communities of the Niagara Escarpment, Canada. MSc Thesis, University of Guelph, Canada.

Matheson, J.D. & Larson, D.W. (1998). Influence of cliffs on bird community diversity. *Canadian Journal of Zoology*, **76**, 278–87.

Matsukura, Y. (1990). Notch formation due to freeze–thaw action in the north-facing valley cliff of the Asama Volcano region, Japan. *Geographical Bulletin*, **32**, 118–24.

Matsukura, Y. & Yatsu, E. (1982). Wet–dry slaking of tertiary shale and tuff. *Transactions of the Japan Geomorphological Union*, **3**, 25–39.

Matthes-Sears, U., Gerrath, J.A., Gerrath, J.F. & Larson, D.W. (1999). Community structure of epilithic and endolithic algae and cyanobacteria on cliffs of the Niagara Escarpment, Ontario, Canada. *Journal of Vegetation Science*.

Matthes-Sears, U., Gerrath, J.A. & Larson, D.W. (1997). Abundance, biomass, and productivity of endolithic and epilithic lower plants on the temperate-zone cliffs of the Niagara Escarpment, Canada. *International Journal of Plant Sciences*, **158**, 451–60.

Matthes-Sears, U., Kelly, P.E. & Larson, D.W. (1993). Early-spring gas exchange and uptake of deuterium-labelled water in the poikilohydric fern *Polypodium virginianum*. *Oecologia*, **95**, 9–13.

Matthes-Sears, U. & Larson, D.W. (1990). Environmental controls of carbon uptake in two woody species with contrasting distributions at the edge of cliffs. *Canadian Journal of Botany*, **68**, 2371–80.

Matthes-Sears, U. & Larson, D.W. (1991). Growth and physiology of *Thuja occidentalis L.* from cliffs and swamps: is variation habitat or site specific? *Botanical Gazette*, **152**, 500–8.

Matthes-Sears, U. & Larson, D.W. (1995). Rooting characteristics of trees in rock: a study of *Thuja occidentalis* on cliff faces. *International Journal of Plant Sciences*, **156**, 679–86.

Matthes-Sears, U. & Larson, D.W. (1999). Limitations to seedling growth and survival by the quantity and quality of rooting space: implications for the establishment of *Thuja occidentalis* on cliff faces. *International Journal of Plant Sciences*, **160**, 122-8.

Matthes-Sears, U., Nash T.H. III & Larson, D.W. (1986). The ecology of *Ramalina menziesii*. IV. *In situ* photosynthetic patterns and water relations of reciprocal transplants between two sites on a coastal–inland gradient. *Canadian Journal of Botany*, **64**, 1183–7.

Matthes-Sears, U., Nash, T.H. III & Larson, D.W. (1987). Salt loading does not control $CO_2$ exchange in *Ramalina menziesii* Tayl. *New Phytologist*, **106**, 59–69.

Matthes-Sears, U., Nash, C.H. & Larson, D.W. (1995). Constrained growth of trees in a hostile environment: the role of water and nutrient availability for *Thuja occidentalis* on cliff faces. *International Journal of Plant Sciences*, **156**, 311–19.

Matthes-Sears, U., Neeser, C. & Larson, D.W. (1992). Mycorrhizal colonization and macronutrient status of cliff-edge *Thuja occidentalis* and *Acer saccharum*. *Ecography*, **15**, 262–6.

Matthes-Sears, U., Stewart, S.C. & Larson, D.W. (1991). Sources of allozymic variation in *Thuja occidentalis* in southern Ontario, Canada. *Silvae Genetica*, **40**, 100–5.

Maycock, P.F. & Fahselt, D. (1992). Vegetation of stressed calcareous screes and slopes in Sverdrup Pass, Ellesmere Island, Canada. *Canadian Journal of Botany*, **70**, 2359–77.

McKendrick, J.D., Batzli, G.O., Everett, K.R. & Swanson, J.C. (1980). Some effects of mammalian herbivores and fertilization on tundra soils and vegetation. *Arctic and Alpine Research*, **12**, 565–78.

McVean, D.N. & Ratcliffe, D.A. (1962). *Plant Communities of the Scottish Highlands*. Monographs of the Nature Conservancy, Number 1. London: Her Majesty's Stationery Office.

Mearns, R. & Newton, I. (1988). Factors affecting breeding success of Peregrines in south Scotland. *Journal of Animal Ecology*, **57**, 903–66.

Meese, R.J. & Fuller, M.R. (1989). Distribution and behaviour of passerines around Peregrine *Falco peregrinus* eyries in western Greenland. *Ibis*, **131**, 27–32.

Merola, M. (1995). Observations of the nesting and breeding behavior of the rock wren. *Condor*, **97**, 585–7.

Michalik, S. (1991). Distribution of plant communities as a function of the relative inolation of the Czyzówki rocky ridge in the Ojców National Park. *Acta Societatis Botanicorum Poloniae*, **60**, 327–38.

Montagne, C., Ruddell, J. & Ferguson, H. (1992). Water retention of soft siltstone fragments in a Ustic Torriorthent, central Montana. *Journal of the Soil Science Society of America*, **56**, 555–7.

Monteith, J.L. & Unsworth, M.K. (1990). *Principles of Environmental Physics*, 2nd edn. London: Edward Arnold.

Mooney, H.A. (1991). Emergence of the study of global ecology: is terrestrial ecology an impediment to progress? *Ecological Applications*, **1**, 2–5.

Moore, J. (1987). A survey of the breeding falcons of Eqalungmiut Nunaat, West Greenland in 1984. *Journal of Raptor Research*, **21**, 111–15.

Morisset, P. (1971). Endemism in the vascular plants of the Gulf of St. Lawrence region. *Naturaliste Canadien*, **98**, 167–77.

Morisset, P., Bedard, J. & Lefabvre, G. (1983). *The Rare Plants of Forillon National Park*. Québec: Parks Canada.

Morneau, F., Brodeur, S., Décarie, R., Carrière, S. & Bird, D.M. (1994). Abundance and distribution of nesting golden eagles in Hudson Bay, Québec. *Journal of Raptor Research*, **28**, 220–5.

Morton, J.K. & Venn, J.M. (1984). *The Flora of Manitoulin Island*. University of Waterloo Biology Series, 28. Waterloo, Ontario: University of Waterloo Press.

Morton, J.K. & Venn, J.M. (1987). *The Flora of the Tobermory Islands*. University of Waterloo Biology Series, 31. Waterloo, Ontario: University of Waterloo Press.

Moss, M.R. & Nickling, W.G. (1980). Geomorphological and vegetation interaction and its relationship to slope stability on the Niagara Escarpment, Bruce Peninsula, Ontario. *Géographie Physique et Quaternaire*, **24**, 95–106.

Murphy, E.C., Hoover-Miller, A.A., Day, R.H. & Oakley, K.L. (1992). Intracolony variability during periods of poor reproductive performance at a Glaucous-winged Gull colony. *Condor*, **94**, 598–607.

Nekola, J.C., Smith, T.A. & Frest, T.J. (1996). Land snails of Door Peninsula natural habitats. Final Report to the Wisconsin chapter of the Nature Conservancy, March 4, 1996.

Nelson, J.B. (1978a). *The Gannet*. Berkhamsted, UK: T. and A.D. Poyser.

Nelson, J.B. (1978b). *The Sulidae*. Oxford: Oxford University Press.

Nelson, R.W. (1990). Status of the Peregrine Falcon, *Falco peregrinus pealei*, on Langara Island, Queen Charlotte Islands, British Columbia, 1968–1989. *Canadian Field Naturalist*, **104**, 193–9.

Nettleship, D.N. (1972). Breeding success of the common Puffin (*Fratercula arctica* L.) on different habitats at Great Island, Newfoundland. *Ecological Monographs*, **42**, 239–68.

Neumayer, E. (1983). *Prehistoric Indian Rock Paintings*. Oxford: Oxford University Press.

Nordhagen, R. (1943). Sikilsdalen og Norges fjellbeiter. *Bergens Museums Skrifter*, **22**, 1–607.

Norriss, D.W. (1995). The 1991 survey and weather impacts on the Peregrine

*Falco peregrinus* breeding population in the Republic of Ireland. *Bird Study*, **42**, 20–30.

Nuzzo, V.A. (1995). Effects of rock climbing on cliff goldenrod (*Solidago sciaphila* Steele) in northwest Illinois. *American Midland Naturalist*, **133**, 229–41.

Nuzzo, V. (1996). Structure of cliff vegetation on exposed cliffs and the effect of rock climbing. *Canadian Journal of Botany*, **74**, 607–17.

Oberdorfer, E. (1977). *Süddeutsche Pflanzengesellschaften. Teil I*, 2nd edn. Stuttgart: Fischer.

Obst, J. (1994). Tree nesting by the gyrfalcon (*Falco rusticolus*) in the Western Canadian Arctic. *Journal of Raptor Research*, **28**, 4–8.

Odasz, A.M. (1994). Nitrate reductase activity in vegetation below an arctic bird cliff, Svalbard, Norway. *Journal of Vegetation Science*, **5**, 913–20

Odum, H.T. (1957). Trophic structure and productivity of Silver Springs, Florida. *Ecological Monographs*, **27**, 55–112.

Oettli, M. (1904). Beiträge zur Ökologie der Felsflora. *Jahrbuch der St. Gallischen Naturwissenschaftlichen Gesellschaft für das Vereinsjahr 1903*, pp. 182–352.

Ogden, V.T. & Hornocker, M.G. (1977). Nesting density and reproductive success of prairie falcons in south-western Idaho. *Journal of Wildlife Management*, **41**, 1–11.

Ohsawa, M. & Yamane, M. (1988). Pattern and population dynamics in patchy communities on a maritime rock outcrop. In *Diversity and Pattern in Plant Communities*, ed. H.J. During, M.J.A. Werger & J.H. Willems, pp. 209–20. The Hague: SPB Academic Publishing.

Oke, T.R. (1987). *Boundary Layer Climates*, 2nd edn. London: Methuen and Co.

Okusanya, O.T. (1979a). An experimental investigation into the ecology of some maritime cliff species. II. Germination studies. *Journal of Ecology*, **67**, 293–304.

Okusanya, O.T. (1979b). An experimental investigation into the ecology of some maritime cliff species. III. Effect of sea water on growth. *Journal of Ecology*, **67**, 579–90.

Okusanya, O.T. (1979c). An experimental investigation into the ecology of some maritime cliff species. IV. Cold sensitivity and competition studies. *Journal of Ecology*, **67**, 591–600.

O'Neill, R.V., DeAngelis, D.L., Waide, J.B. & Allen, T.F.H. (1986). *A Hierarchical Concept of Ecosystems*. Princeton, NJ: Princeton University Press.

Onions, C.T. (1968). *The Shorter Oxford English Dictionary*, 3rd edn. Oxford: Oxford University Press.

Oosting, H.J. & Anderson, L.E. (1937). The vegetation of a barefaced cliff in western North Carolina. *Ecology*, **18**, 280–92.

Oosting, H.J. & Anderson, L.E. (1939). Plant succession on granite rock in eastern North Carolina. *Botanical Gazette*, **100**, 750–68.

Oppenheimer, H.R. (1956). Pénétration active des racines de buissons Méditerranéens dans les roches calcaires. *Bulletin of the Research Council of Israel*, **5D**, 219–24.

Oppenheimer, H.R. (1957). Further observations on roots penetrating into rocks and their structure. *Bulletin of the Research Council of Israel*, **6D**, 18–31.

Orwin, J. (1972). The effect of environment on assemblages of lichens growing on rock surfaces. *New Zealand Journal of Botany*, **10**, 37–47.

Ota, S. & Gun, K. (1988). Developing collected juniper bonsai. *International Bonsai*, **10**(4), 5–15.

Parikesit, P., Larson, D.W. & Matthes-Sears, U. (1995). Impacts of trails on cliff-edge forest structure. *Canadian Journal of Botany*, **73**, 943–53.

Paullin, C.O. (1932). *Atlas of the Historical Geography of the United States*. Washington, DC: Carnegie Institution of Washington, and the American Geological Society of New York.

Pawlowski, B., Medwecka-Kornaś, A. & Kornaś, J. (1966). Review of terrestrial and freshwater plant communities. In *The Vegetation of Poland*, ed. W. Szafer, pp. 241–534. Oxford: Pergamon Press.

Pax, F. (1908). *Grundzüge der Pflanzenverbreitung in den Karpathen*. Leipzig: Verlag Wilhelm Engelmann.

Peacock, M.M. & Smith, A.T. (1997). The effect of habitat fragmentation on dispersal patterns, mating behavior, and genetic variation in a Pika (*Ochotona princeps*) metapopulation. *Oecologia*, **112**, 524–33.

Pearson, D.J. (1992). Past and present distribution and abundance of the Black-footed Rock-wallaby in the Warburton region of Western Australia. *Wildlife Research*, **19**, 605–22.

Peltier, L. (1950). The geographical cycle in periglacial regions as it is related to climatic geomorphology. *Annals of the Association of American Geographers*, **40**, 214–36.

Penhallurick, R.D. (1993). House Sparrows nesting on cliffs in Scilly. *British Birds*, **86**, 435–6.

Pentecost, A. (1980). The lichens and bryophytes of rhyolite and pumice–tuff rock outcrops in Snowdonia, and some factors affecting their distribution. *Journal of Ecology*, **68**, 251–67

Pérez, F.L. (1991). Soil moisture and the distribution of giant Andean rosettes on talus slopes of a desert paramo. *Climate Research*, **1**, 217–31.

Pérez, F.L. (1994). Geobotanical influence of talus movement on the distribution of caulescent Andean rosettes. *Flora*, **189**, 353–71.

Philips, J.R. & Dindal, D.L. (1977). Raptor nests as a habitat for invertebrates: a review. *Raptor Research*, **11**, 87–96.

Phillips, D.L. (1981). Succession in granite outcrop shrub–tree communities. *American Midland Naturalist*, **106**, 313–17.

Pigott, C.D. (1969). The status of *Tilia cordata* and *Tilia platyphyllos* on the Derbyshire limestone. *Journal of Ecology*, **57**, 491–504.

Pigott, C.D. (1989). Factors controlling the distribution of *Tilia cordata* Mill at the northern limits of its geographical range. IV. Estimated ages of the trees. *New Phytologist*, **112**, 117–21.

Pigott, C.D. & Huntley, J.P. (1978). Factors controlling the distribution of *Tilia cordata* at the northern limits of its geographical range I. Distribution in north-west England. *New Phytologist*, **81**, 429–41.

Pigott, C.D. & Huntley, J.P. (1980). Factors controlling the distribution of *Tilia cordata* at the northern limits of its geographical range. II. History in north-west England. *New Phytologist*, **84**, 145–64.

Pigott, C.D. & Pigott, S. (1993). Water as a determinant of the distribution of trees at the boundary of the Mediterranean zone. *Journal of Ecology*, **81**, 557–66.

Pigott, C.D. & Walters, S.M. (1954). On the interpretation of the discontinuous distributions shown by certain British species of open habitats. *Journal of Ecology*, **42**, 95–116.

Platt, S.W. (1974). Breeding status and distribution of the prairie falcon in northern New Mexico. MS Thesis, Oklahoma State University, Stillwater.

Polunin, N. (1939). Arctic plants in the British Isles. *Journal of Botany, London*, **77**, 371–413.

Porembski, S., Barthlott, W., Dörrstock, S. & Biedinger, N. (1994). Vegetation of rock outcrops in Guinea: granite inselbergs, sandstone table mountains and ferricretes – remarks on species numbers and endemism. *Flora*, **189**, 315–26.

Porembski, S., Brown, G. & Barthlott, W. (1996). A species-poor tropical sedge community: *Afrotrilepis pilosa* mats on inselbergs in West Africa. *Nordic Journal of Botany*, **16**, 239–45.

Porter, R.D. & White, C.M. (1973). The peregrine falcon in Utah, emphasizing ecology and competition with the prairie falcon. *Brigham Young University Science Bulletin*, **28**, 1–74.

Pratt, K.L. (1990). Economic and social aspects of Nunivak Eskimo 'cliff-hanging'. *Arctic Anthropology*, **27**, 75–86.

Press, A.J. (1989). The abundance and distribution of Black Wallaroos *Macropus bernardus* and Common Wallaroos *Macropus robustus* on the Arnhem Land Plateau Australia. In *Kangaroos, Wallabies and Rat-Kangaroos*, ed. G. Grigg, P. Jarman & I. Hume, pp. 783–6. New South Wales, Australia: Surrey Beattey and Sons.

Prior, K.A. (1990). Turkey Vulture food habits in southern Ontario. *Wilson Bulletin*, **102**, 706–10.

Proctor, M.C.F. (1980). Estimates from hemispherical photographs of the radiation climates of some bryophyte habitats in the British Isles. *Journal of Bryology*, **11**, 351–66.

Ranck, G.L. (1968). *The Rodents of Libya*. Smithsonian Institution Publication number 275. Washington, DC: Smithsonian Institute.

Ratcliffe, D.A. (1960). The mountain flora of Lakeland. *Proceedings of the Botanical Society of the British Isles*, **4**, 1–25.

Ratcliffe, D.A. (1980). *Bird Life of Mountains and Uplands*. Cambridge: Cambridge University Press.

Ratcliffe, D.A. (1993). *The Peregrine Falcon*, 2nd edn. London: Poyser.

Reitan, O. (1986). Cliff-related breeding of passerines in a woodland area. *Fauna Norvegica Series C, Cinclus*, **9**, 68–73.

Rejmánek, M. (1971). Ecological meaning of the thermal behaviour of rocks. *Flora*, **160**, 527–61.

Richard, J.-L. (1972). La végétation des crêtes rocheuses du Jura. *Berichte der Schweizerischen Botanischen Gesellschaft*, **82**, 68–112.

Riddiford, N. & Potts, P. (1993). Exceptional claw-wear of Great Reed Warbler. *British Birds*, **86**, 572.

Riebe, H. (1996). Nationalpark und Landschaftsschutzgebiet Sächsische Schweiz. In *1996 Jahrbuch, Verein zum Schutz der Bergwelt*, pp. 77–94. München: Verein zum Schutz der Bergwelt.

Riley, J. (1993). There are bones down there. *Cuesta*, 1993, 11–13.

Rinkevich, S.E. & Gutiérrez, R.J. (1996). Mexican spotted owl habitat characteristics in Zion National Park. *Journal of Raptor Research*, **30**, 74–8.

Rioux, J. & Quézel, P. (1949). Contribution à l'étude des groupements rupicoles endémiques des Alpes-Maritimes. *Vegetatio*, **2**, 1–13.

Rishbeth, J. (1948). The flora of Cambridge walls. *Journal of Ecology*, **36**, 136–48.

Ritter, D.F. (1978). *Process Geomorphology*. Dubuque, IA: Wm. C. Brown.

Roberts, D. (1996). High adventures in the Gunks. *Smithsonian*, **25**, 30–41.

Robinson, A.C., Lim, L., Canty, P.D., Jenkins, R.B. & Macdonald, C.A. (1994). Studies of the Yellow-footed Rock-wallaby, *Petrogale xanthopus* Gray (Marsupialia: Macropodidae). Population studies at Middle Gorge, South Australia. *Wildlife Research*, **21**, 473–81.

Rodriguez-Estrella, R. & Brown, B.T. (1990). Density and habitat use of raptors along the Rio Bavispe and Rio Yaqui, Sonora, Mexico. *Journal of Raptor Research*, **24**, 47–51.

Rodwell, J.S. (1992). *British Plant Communities*. Cambridge: Cambridge University Press.

Rohrer, J.R. (1982). Bryophytes of Hanging Rock, Avery and Watanga Counties, North Carolina. *Castanea*, **47**, 221–9.

Rohrer, J.R. (1983). Vegetation pattern and rock type in the flora of the Hanging Rock area, North Carolina. *Castanea*, **48**, 189–205.

Rumble, M.A. (1987). Avian use of scoria rock outcrops. *Great Basin Naturalist*, **47**, 625–30.

Runde, D.E. & Anderson, S.H. (1986). Characteristics of cliffs and nest sites used by breeding Prairie Falcons. *Raptor Research*, **20**, 21–8.

Rune, O. (1953). Plant life on serpentines and related rocks in the north of Sweden. *Acta Phytogeographica Suecica*, **31**, 1–39.

Runemark, H. (1969). Reproductive drift, a neglected principle in reproductive biology. *Botanisk Notiser*, **122**, 90–129.

Ruspoli, M. (1987). *The Cave of Lascaux: the Final Photographs*. New York: Abrams.

Rutherford, M.C. (1972). Notes on the flora and vegetation of the Omuverume Plateau-mountain, Waterberg, South West Africa. *Dinteria*, **8**, 3–55.

Růžička, V. (1990). The spiders of stony debris. *Acta Zoologica Fennica*, **190**, 333–7.

Růžička, V. (1996). Species composition and site distribution of spiders (Araneae) in a gneiss massif in the Dyje river valley. *Revue Suisse de Zoologie hors serie*, 561–9.

Růžička, V., Hajer, J. & Zacharda, M. (1995). Arachnid population patterns in underground cavities of a stony debris field (Araneae, Opiliones, Pseudoscorpionidea, Acari: Prostigmata, Rhagidiidae). *Pedobiologia*, **39**, 42–51.

Růžička, V. & Zacharda, M. (1994). Arthropods of stony debris in the Krkonose Mountains, Czech Republic. *Arctic and Alpine Research*, **26**, 332–8.

Sarrazin, F., Bagnolini, C., Pinna, J. L. & Danchin, E. (1996). Breeding biology during establishment of a reintroduced Griffon vulture *Gyps fulvus* population. *Ibis*, **138**, 315–25.

Sasse, W. (1996). Prehistoric rock art in Korea: Pan'gudae. *Korea Journal*, **36**, 75–91.

Schade, A (1923). Die kryptogamischen Pflanzengesellschaften an den Felswänden der Sächsischen Schweiz. *Berichte der Deutschen Botanischen Gesellschaft*, **41**, 49–59.

Schöller, H. (1994). Das Naturdenkmal 'Eschbacher Klippen' im östlichen Hintertaunus – ein aussergewöhnlicher Flechtenbiotop im Konflikt mit modernen Freizeitinteressen. *Botanik und Naturschutz in Hessen*, **7**, 5–21.

## 326 · References

Schulman, E.R. (1954). Longevity under adversity in conifers. *Science*, **119**, 396–9.

Schweingruber, F.H. (1989). *Tree Rings*. Dordrecht: Kluwer Academic Publishers.

Schweingruber, F.H. (1993). *Trees and Wood in Dendrochronology*. Berlin: Springer-Verlag.

Scolaro, J.A., Laurenti, S. & Gallelli, H. (1996). The nesting and breeding biology of the South American Tern in northwest Patagonia. *Journal of Field Ornithology*, **67**, 17–24.

Sealy, J.R. (1949). Arbutus unedo. *Journal of Ecology*, **37**, 365–88.

Sellers, W.D. (1965). *Physical Climatology*. Chicago: The University of Chicago Press.

Shaw, W.D. & Jakus, P. (1996). Travel cost models of the demand for rock climbing. *Agricultural Resource Economics Review*, **25**, 133–42.

Short, J. (1982). Habitat requirements of the Brush-tailed Rock-wallaby, *Petrogale penicillata*, in New South Wales. *Australian Wildlife Research*, **9**, 239–46.

Shrubb, M. (1993). Nest sites in the Kestrel *Falco tinnunculus*. *Bird Study*, **40**, 63–73.

Shure, D.J. & Ragsdale, H.L. (1977). Patterns of primary succession on granite outcrop surfaces. *Ecology*, **58**, 993–1006.

Simmons, N.M. (1980). Behaviour. In *The Desert Bighorn*, ed. G. Monson & L. Summer, pp. 124–44. Tucson, AZ: University of Arizona Press.

Sinclair, B.J. & Marshall, S.A. (1986). The madicolous fauna in southern Ontario. *Proceedings of the Entomological Society of Ontario*, **117**, 9–14.

Sinnemann, C. (1992). Cliff evaluation in the South Okanagan. Work Term Report, Department of Geography, University of Victoria, Victoria, BC.

Small, R.J. (1989). *Geomorphology and Hydrology*. Harlow, UK: Longman Group.

Smiley, D. & George, C.J. (1974). Photographic documentation of lichen decline in the Shawangunk Mountains of New York. *Bryologist*, **77**, 179–87.

Smith, C.A. (1986). Bi-level analysis of habitat selection by Mountain Goats in coastal Alaska. In *Proceedings of the Fifth Biennial Symposium of the Northern Wild Sheep and Goat Council*, ed. G. Joslin, pp. 366–80. Missoula, MT.

Snazell, R. & Bosmans, R. (1997). A little known *Zodarion* (Araneae : Zodariidae) new to Britain. *Journal of Zoology, London*, **241**, 285–9.

Snogerup, S. (1971). Evolutionary and plant geographical aspects of chasmophytic communities. In *Plant Life of South-West Asia*, ed. P.H. Davis, P.C. Harper & I.C.Hedge, pp. 157–70. Aberdeen: The Botanical Society of Edinburgh.

Solomon, A. (1996). Rock art in Southern Africa. *Scientific American*, **275**, 106–13.

Somme, L. (1977). Observations on the Petrel (*Pagodroma nivea*) in Vestfjella, Dronning Maud Land. *Norsk Polarinstitut Arbok*, **1976**, 285–92.

Soons, J.M. & Selby, M.J. (1982). *Landforms of New Zealand*. Auckland: Longman Paul.

Soper, J.H. (1954). The Hart's tongue fern in Ontario. *American Fern Journal*, **44**, 129–47.

Sousa, W.P. (1984). The role of disturbance in natural communities. *Annual Review of Ecology and Systematics*, **15**, 353–91.

Spear, L.B. & Anderson, D.W. (1989). Nest-site selection by Yellow-footed Gulls. *Condor*, **91**, 91–9.

Spence, M.W. & Fox, W.A. (1992). The Winona rockshelter burial. *Ontario Archaeology*, **53**, 27–44.

Spring, L. (1971). A comparison of functional and morphological adaptations in the Common Murre (*Uria aalge*) and Thick-billed Murre (*Uria lomvia*). *Condor*, **73**, 1–27.

Squibb, R.C. & Hunt, G.L. Jr (1983). A comparison of nesting-ledges used by seabirds on St. George Island. *Ecology*, **64**, 727–34.

Stahle, D.W. & Chaney, P.L. (1994). A predictive model for the location of ancient forests. *Natural Areas Journal*, **14**, 151–8

Stärr, A., Banzhaf, P., Gottschlich, G. *et al.* (1995). Neufassung der Gefährdungsgrade felsbesiedelnder Farn- und Blütenpflanzen der Schwäbischen Alb – Eine auf Felsbiotope bezogene Rote Liste. *Veröffentlichungen der Landesanstalt für Naturschutz und Landschaftspflege Baden-Württemberg*, **70**, 99–120.

Stempniewicz, L. (1995). Predator–prey interactions between Glaucous Gull *Larus hyperboreus* and Little Auk *Alle alle* in Spitsbergen. *Acta Ornithologica*, **29**, 155–68.

Stöcker, G. (1965). Vegetationskomplexe auf Felsstandorten, ihre Auflösung und Systematisierung der Komponenten. *Feddes Repertarium*, **142**, 222–36.

Stone, E.L. & Kalisz, P.J. (1991). On the maximum extent of tree roots. *Forest Ecology and Management*, **45**, 59–102.

Stoutjesdijk, P. (1974). The open shade, an interesting microclimate. *Acta Botanica Neerlandica*, **23**, 125–30.

Sunamura, T. (1992). *Geomorphology of Rocky Coasts*. New York: John Wiley and Sons.

Sutter, R. (1969). Ein Beitrag zur Kenntnis der soziologischen Bindung süd-südostalpiner Reliktendemismen. *Acta Botanica Croatica*, **28**, 349–65.

Svoboda, J. & Henry, G.H.R. (1987). Succession in marginal environments. *Arctic and Alpine Research*, **19**, 373–84.

Tanner, O. (1989). Bonsai: a way of looking at trees with different eyes. *Smithsonian*, **20**, 138–53.

Taylor, K.C., Reader, R.J. & Larson, D.W. (1993). Scale-dependent inconsistencies in the effects of trampling on a forest understory community. *Environmental Management*, **17**, 239–48.

Tinkler, K. (1986). Canadian landform examples. 2. Niagara Falls. *Canadian Geographer*, **30**, 367–71.

Tovell, W. (1992). *Guide to the Geology of the Niagara Escarpment*. Concord, Ontario: Niagara Escarpment Commission.

Tranquillini, W. (1979). *Physiological Ecology of the Alpine Timberline*. Berlin: Springer-Verlag.

Trenhaile, A.S. (1987). *The Geomorphology of Rock Coasts*. Oxford: Oxford University Press.

Trenhaile, A.S. (1990). *The Geomorphology of Canada*. Toronto: Oxford University Press.

Tschanz, B. & Hirsbrunner-Scharf, M. (1975). Adaptations to colony life on cliff ledges: a comparative study of Guillemot and Razorbill chicks. In *Function and Evolution in Behaviour*, ed. G. Baerends, C. Beer & A. Manning, pp. 358–80. Oxford: Clarendon Press.

Türk, W. (1994). Das 'Höllental' im Frankenwald – Flora und Vegetation eines floristisch bemerkenswerten Mittelgebirgstales. *Tuexenia*, **14**, 17–52.

Ucko, P.J. & Rosenfeld, A. (1967). *Palaeolithic Cave Art*. New York: McGraw-Hill.

## 328 · References

Udall, J.R. (1991). The pika hunter. *Audubon*, **93**, 60–70.

Underwood, B.A. (1991). Thermoregulation and energetic decision-making by the honeybees *Apis cerana, Apis dorsata* and *Apis laboriosa. Journal of Experimental Biology*, **157**, 19–34.

United States Fish and Wildlife Service (1993). *American Hart's-Tongue Recovery Plan*. Atlanta: US Fish and Wildlife Service.

Uno, G.E. & Collins, S.L. (1987). Primary succession on granite outcrops in southwestern Oklahoma. *Bulletin of the Torrey Botanical Club*, **114**, 387–92.

Ursic, K., Kenkel, N.C. & Larson, D.W. (1997). Revegetation dynamics of cliff faces in abandoned limestone quarries. *Journal of Applied Ecology*, **34**, 289–303.

Usher, M.B. (1979). Natural communities of plants and animals in disused quarries. *Journal of Environmental Management*, **8**, 223–36.

Valis, T. (1991). *Escarpment Climbing History*. Toronto: Borealis Press.

Vasina, W.G. & Straneck, R.J. (1984). Biological and ethological notes on *Falco peregrinus cassini* in central Argentina. *Raptor Research*, **18**, 123–30.

Vatev, I.T. (1987). Notes on the breeding biology of the long-legged buzzard (*Buteo rufinus*) in Bulgaria. *Journal of Raptor Research*, **21**, 8–13.

Viemann, G. (1997). Veränderungen der epilithischen Flechtenvegetation durch anthropogene Einflüsse im Naturschutzgebiet 'Ehemaliger Buntsandsteinbruch an der Neckarhalde Ziegelhausen'. Thesis, Department of Botany, Universität Heidelberg.

Vogel, C. (1976). *Ökologie, Lebensweise und Sozialverhalten der grauen Languren in verschiedenen Biotopen Indiens*. Advances in Ethology, 17. Berlin: Parey-Verlag.

Vogler, P. (1904). Die Eibe (*Taxus baccata* L.) in der Schweiz. *Jahrbuch der St. Gallischen Naturwissenschaftlichen Gesellschaft für das Vereinsjahr 1903*, pp. 436–91.

Volkman, T.A. & Caldwell, I. (1990). *The Celeber of Sulawesi*. Berkeley, CA: Periplus Edition.

Vyatkin, P.S. (1993). Nesting of the Northern Fulmar on the western Bering Sea coast. *Condor*, **95**, 226–7.

Wade, L.K. & McVean, D.N. (1969). *Mt. Wilhelm Studies 1. The Alpine and Subalpine Vegetation*. Research School of Pacific Studies Publication BG/1. Canberra: The Australian National University.

Wagner, W.L., Weller, S.G. & Sakai, A.K. (1994). Description of a rare new cliff-dwelling species from Kaua'i, *Schiedea attenuata* (Caryophyllaceae). *Novon*, **4**, 187–90.

Wakelyn, L.A. (1987). Changing habitat conditions on Bighorn Sheep ranges in Colorado. *Journal of Wildlife Management*, **51**, 904–12.

Walker, G.E. (1987). Ecology and Population Biology of *Thuja occidentalis* L. in its Southern Disjunct Range. PhD Thesis, University of Tennessee, Knoxville, TN.

Walters, T.W. & Wyatt, R. (1982). The vascular flora of granite outcrops in the central mineral region of Texas. *Bulletin of the Torrey Botanical Club*, **109**, 344–64.

Ward, J.P. & Anderson, S.H. (1988). Influences of cliffs on wildlife communities in southcentral Wyoming. *Journal of Wildlife Management*, **52**, 673–8.

Wardle, P. (1991) *Vegetation of New Zealand*. Cambridge: Cambridge University Press.

Ware, S. (1991). Influence of interspecific competition, light and moisture levels on growth of rock outcrop *Talinum* (Portulacaceae). *Bulletin of the Torrey Botanical Club*, **118**, 1–5.

Warham, J. (1990). *The Petrels*. London: Academic Press.

Washburn, A.L. (1973). *Periglacial Processes and Environments*. London: Edward Arnold.

Watson, A., Payne, S. & Rae, R. (1989). Golden Eagles *Aquila chrysaetos*: land use and food in northeast Scotland. *Ibis*, **131**, 336–48.

Watson, J. (1997). *The Golden Eagle*. London: T. and A.D. Poyser.

Weick, F. (1980). *Birds of Prey of the World*. Hamburg: Parey-Verlag..

Weidinger, K. (1996). Patterns of colony attendance in the Cape Petrel *Daption capense* at Nelson Island, South Shetland Islands, Antarctica. *Ibis*, **138**, 243–9.

Weimerskirch, H., Jouventin, P. & Stahl, J.C. (1986). Comparative ecology of the six albatross species breeding on the Crozet Islands. *Ibis*, **128**, 195–213.

Wendt, S. & Cooch, F.G. (1984). The kill of murres in Newfoundland in the 1977–78, 1978–79 and 1979–80 hunting seasons. *Canadian Wildlife Service Progress Note*, **146**, 1–10.

Wentworth, T.R. (1981). Vegetation on limestone and granite in the Mule Mountains, Arizona. *Ecology*, **62**, 469–82.

White, C.M. & Cade, T.J. (1971). Cliff-nesting raptors and ravens along the Colville River in arctic Alaska. *The Living Bird*, **10**, 107–50.

White, C.M., Emison, W.B. & Bren, W.M. (1988). Atypical nesting habitat of the Peregrine Falcon (*Falco peregrinus*) in Victoria, Australia. *Journal of Raptor Research*, **22**, 37–43.

White, W.B. (1988). *Geomorphology and Hydrology of Karst Terrains*. London: Oxford University Press.

Whitmore, T.C. (1984). *Tropical Rainforests of the Far East*, 2nd edn. London: Clarendon Press.

Whittaker, R.H. (1975). *Communities and Ecosystems*, 2nd edn. New York: Macmillan.

Williams, A.T., Davies, P. & Bomboe, P. (1993). Geometrical simulation studies of coastal cliff failures in Liassic strata, south Wales, U.K. *Earth Surface Processes and Landforms*, **18**, 703–20.

Williams, R.N. (1981). Breeding ecology of prairie falcons at high elevations in central Colorado. MS Thesis, Brigham Young University, Provo, Utah.

Williams, R.N. (1984). Eyrie aspect as a compensator for ambient temperature fluctuations: a preliminary investigation. *Raptor Research*, **18**, 153–5.

Wilmanns, O. & Rupp, S. (1966). Welche Faktoren bestimmen die Verbreitung alpiner Felsspaltenpflanzen auf der Schwäbischen Alb? *Veröffentlichungen der Landesstelle für Naturschutz und Landschaftspflege Baden-Württemberg*, **34**, 62–86.

Wilson, J.B. & Cullen, C. (1986). Coastal cliff vegetation of the Catlins region, Otago, South Island, New Zealand. *New Zealand Journal of Botany*, **24**, 567–74.

Wilson, J.B., Sykes, M.T. & Peet, R.K. (1995). Time and space in the community structure of a species-rich limestone grassland. *Journal of Vegetation Science*, **6**, 729–40.

Wilson, J.B. & Whittaker, R.J. (1995). Assembly rules demonstrated in a saltmarsh community. *Journal of Ecology*, **83**, 801–7.

Wilson, U.W. & Manuwal, D.A. (1986). Breeding biology of the Rhinoceros Auklet in Washington. *Condor*, **88**, 143–55.

Wingate, D.B. (1985). *The Restoration of Nonsuch Island as a Living Museum of Bermuda's Pre-colonial Terrestrial Biome*. I.C.B.P. Technical Publication No. 3, pp. 225–38. Bermuda: Department of Agriculture and Fisheries.

Winterringer, G.S. & Vestal, A.G. (1956). Rock-ledge vegetation in southern Illinois. *Ecological Monographs*, **26**, 105–30.

Wiser, S.K. (1994). High-elevation cliffs and outcrops of the Southern Appalachians: vascular plants and biogeography. *Castanea*, **59**, 85–116.

Woodhouse, R.M & Nobel, P.S. (1982). Stipe anatomy, water potentials and xylem conductances in seven species of ferns (Filicopsida). *American Journal of Botany*, **69**, 135–40.

Woodin, N. (1980). Observations on gyrfalcons (*Falco rusticolus*) breeding near Lake Myvatn, Iceland, 1967. *Raptor Research*, **14**, 97–124.

Wootton, J.T. (1991). Direct and indirect effects of nutrients on intertidal community structure: variable consequences of seabird guano. *Journal of Experimental Marine Biology and Ecology*, **151**, 139–53.

Worthington, E.B. (1975). *The Evolution of IBP*. Cambridge: Cambridge University Press.

Wunder, J. & Möseler, B.M. (1996). Kaltluftströme auf Basaltblockhalden und ihre Auswirkung auf Mikroklima und Vegetation. *Flora*, **191**, 335–44.

Yarranton, G.A. & Green, W.G.E. (1966). The distributional pattern of crustose lichens on limestone cliffs at Rattlesnake Point, Ontario. *Bryologist*, **69**, 450–61.

Yosef, R. (1991). Foraging habits, hunting and breeding success of Lanner Falcons (*Falco biarmicus*) in Israel. *Journal of Raptor Research*, **25**, 77–81.

Young, A. (1972). *Slopes*. Edinburgh: Oliver and Boyd.

Young, J. M. (1996). The Cliff Ecology and Genetic Structure of Northern White Cedar (*Thuja occidentalis* L.) in its Southern Disjunct Range. MSc Thesis, University of Tennessee Knoxville, TN.

Zehnder, A. (1953). Beitrag zur Kenntnis von Mikroklima und Algenvegetation des nackten Gesteins in den Tropen. *Berichte der Schweizerischen Botanischen Gesellschaft*, **63**, 5–26.

Zerrahn, S. (1996). Das Wunder vor der Haustür. *Zug*, August 1996, 36–8.

Zohary, M. (1973). *Geobotanical Foundations of the Middle East*, 2nd edn. Stuttgart: Gustav Fischer.

# Index